Machine Learning for Pattern Recognition

Machine Learning for Pattern Recognition

Editors

Chih-Lung Lin
Bor-Jiunn Hwang
Shaou-Gang Miaou
Yuan-Kai Wang

Basel • Beijing • Wuhan • Barcelona • Belgrade • Novi Sad • Cluj • Manchester

Editors

Chih-Lung Lin
Hwa Hsia University of Technology
New Taipei City
Taiwan

Bor-Jiunn Hwang
Ming Chuan University
Taoyuan
Taiwan

Shaou-Gang Miaou
Chung Yuan Christian University
Taoyuan
Taiwan

Yuan-Kai Wang
Fu Jen Catholic University
New Taipei City
Taiwan

Editorial Office
MDPI
Grosspeteranlage 5
4052 Basel, Switzerland

This is a reprint of articles from the Special Issue published online in the open access journal *Algorithms* (ISSN 1999-4893) (available at: https://www.mdpi.com/journal/algorithms/special_issues/L1262KVU3T).

For citation purposes, cite each article independently as indicated on the article page online and as indicated below:

Lastname, A.A.; Lastname, B.B. Article Title. *Journal Name* **Year**, *Volume Number*, Page Range.

ISBN 978-3-7258-1591-3 (Hbk)
ISBN 978-3-7258-1592-0 (PDF)
doi.org/10.3390/books978-3-7258-1592-0

© 2024 by the authors. Articles in this book are Open Access and distributed under the Creative Commons Attribution (CC BY) license. The book as a whole is distributed by MDPI under the terms and conditions of the Creative Commons Attribution-NonCommercial-NoDerivs (CC BY-NC-ND) license.

Contents

Frank Nielsen
Generalizing the Alpha-Divergences and the Oriented Kullback–Leibler Divergences with Quasi-Arithmetic Means
Reprinted from: *Algorithms* **2022**, *15*, 435, doi:10.3390/a15110435 1

Steve de Rose, Philippe Meyer and Frédéric Bertrand
Human Body Shapes Anomaly Detection and Classification Using Persistent Homology
Reprinted from: *Algorithms* **2023**, *16*, 161, doi:10.3390/a16030161 26

Taiki Arakane and Takeshi Saitoh
Efficient DNN Model for Word Lip-Reading
Reprinted from: *Algorithms* **2023**, *16*, 269, doi:10.3390/a16060269 44

Paolo Massimo Buscema, Giulia Massini, Giovanbattista Raimondi, Giuseppe Caporaso, Marco Breda and Riccardo Petritoli
A Pattern Recognition Analysis of Vessel Trajectories
Reprinted from: *Algorithms* **2023**, *16*, 414, doi:10.3390/a16090414 59

Thomas P. Oghalai, Ryan Long, Wihan Kim, Brian E. Applegate and John S. Oghalai
Automated Segmentation of Optical Coherence Tomography Images of the Human Tympanic Membrane Using Deep Learning
Reprinted from: *Algorithms* **2023**, *16*, 445, doi:10.3390/a16090445 71

Mikhail Zotov, Dmitry Anzhiganov, Aleksandr Kryazhenkov, Dario Barghini, Matteo Battisti, Alexander Belov, et al.
Neural Network Based Approach to Recognition of Meteor Tracksin the Mini-EUSO Telescope Data
Reprinted from: *Algorithms* **2023**, *16*, 448, doi:10.3390/a16090448 84

Yun-Wei Lin, Yuh-Hwan Liu, Yi-Bing Lin and Jian-Chang Hong
FenceTalk: Exploring False Negatives in Moving Object Detection
Reprinted from: *Algorithms* **2023**, *16*, 481, doi:10.3390/a16100481 98

Sai Bharadwaj Appakaya, Ruchira Pratihar and Ravi Sankar
Parkinson's Disease Classification Framework Using Vocal Dynamics in Connected Speech
Reprinted from: *Algorithms* **2023**, *16*, 509, doi:10.3390/a16110509 116

Li-Na Wang, Guoqiang Zhong, Yaxin Shi and Mohamed Cheriet
Relational Fisher Analysis: Dimensionality Reduction in Relational Data with Global Convergence
Reprinted from: *Algorithms* **2023**, *16*, 522, doi:10.3390/a16110522 135

Jorge Juarez-Lucero, Maria Guevara-Villa, Anabel Sanchez-Sanchez, Raquel Diaz-Hernandez and Leopoldo Altamirano-Robles
A New Algorithm for Detecting GPN Protein Expression and Overexpression of IDC and ILC Her2+ Subtypes on Polyacrylamide Gels Associated with Breast Cancer
Reprinted from: *Algorithms* **2024**, *17*, 149, doi:10.3390/a17040149 157

Jiachen Guo and Wenjie Luo
Point-Sim: A Lightweight Network for 3D Point Cloud Classification
Reprinted from: *Algorithms* **2024**, *17*, 158, doi:10.3390/a17040158 177

Article

Generalizing the Alpha-Divergences and the Oriented Kullback–Leibler Divergences with Quasi-Arithmetic Means

Frank Nielsen

Sony Computer Science Laboratories, Tokyo 141-0022, Japan; frank.nielsen.x@gmail.com

Abstract: The family of α-divergences including the oriented forward and reverse Kullback–Leibler divergences is often used in signal processing, pattern recognition, and machine learning, among others. Choosing a suitable α-divergence can either be done beforehand according to some prior knowledge of the application domains or directly learned from data sets. In this work, we generalize the α-divergences using a pair of strictly comparable weighted means. Our generalization allows us to obtain in the limit case $\alpha \to 1$ the 1-divergence, which provides a generalization of the forward Kullback–Leibler divergence, and in the limit case $\alpha \to 0$, the 0-divergence, which corresponds to a generalization of the reverse Kullback–Leibler divergence. We then analyze the condition for a pair of weighted quasi-arithmetic means to be strictly comparable and describe the family of quasi-arithmetic α-divergences including its subfamily of power homogeneous α-divergences. In particular, we study the generalized quasi-arithmetic 1-divergences and 0-divergences and show that these counterpart generalizations of the oriented Kullback–Leibler divergences can be rewritten as equivalent conformal Bregman divergences using strictly monotone embeddings. Finally, we discuss the applications of these novel divergences to k-means clustering by studying the robustness property of the centroids.

Keywords: Kullback–Leibler divergence; α-divergences; comparable weighted means; weighted quasi-arithmetic means; information geometry; conformal divergences; k-means clustering

1. Introduction

1.1. Statistical Divergences and α-Divergences

Consider a measurable space [1] $(\mathcal{X}, \mathcal{F})$ where $\mathcal{F}$ denotes a finite σ-algebra and $\mathcal{X}$ the sample space, and let μ denotes a positive measure on $(\mathcal{X}, \mathcal{F})$, usually chosen as the Lebesgue measure or the counting measure. The notion of statistical dissimilarities [2–4] $D(P:Q)$ between two distributions P and Q is at the core of many algorithms in signal processing, pattern recognition, information fusion, data analysis, and machine learning, among others. A dissimilarity may be oriented, i.e., asymmetric: $D(P:Q) \neq D(Q:P)$, where the colon mark ":" between the arguments of the dissimilarities represents the asymmetric property of the division operation. When the arbitrary probability measures P and Q are dominated by a measure μ (e.g., one can always choose $\mu = \frac{P+Q}{2}$), we consider their Radon–Nikodym (RN) densities $p_\mu = \frac{dP}{d\mu}$ and $q_\mu = \frac{dQ}{d\mu}$ with respect to μ, and define $D(P:Q)$ as $D_\mu(p_\mu : q_\mu)$. A good dissimilarity measure shall be invariant of the chosen dominating measure so that we can write $D(P:Q) = D_\mu(p_\mu : q_\mu)$ [5]. When those statistical dissimilarities are smooth, they are called divergences [6] in information geometry, as they induce a dualistic geometric structure [7].

The most renowned statistical divergence rooted in information theory [8] is the Kullback–Leibler divergence (KLD, also called relative entropy):

$$\mathrm{KL}_\mu(p_\mu : q_\mu) := \int_{\mathcal{X}} p_\mu(x) \log \frac{p_\mu(x)}{q_\mu(x)} \mathrm{d}\mu(x). \qquad (1)$$

Since the KLD is independent of the reference measure μ, i.e., $\mathrm{KL}_\mu(p_\mu : q_\mu) = \mathrm{KL}_\nu(p_\nu : q_\nu)$ for $p_\mu = \frac{\mathrm{d}P}{\mathrm{d}\mu}$ and $q_\mu = \frac{\mathrm{d}Q}{\mathrm{d}\mu}$, and $p_\nu = \frac{\mathrm{d}P}{\mathrm{d}\nu}$ and $q_\nu = \frac{\mathrm{d}Q}{\mathrm{d}\nu}$ are the RN derivatives with respect to another positive measure ν, we write concisely in the remainder:

$$\mathrm{KL}(p : q) = \int p \log \frac{p}{q} \mathrm{d}\mu, \qquad (2)$$

instead of $\mathrm{KL}_\mu(p_\mu : q_\mu)$.

The KLD belongs to a parametric family of α-divergences [9] $I_\alpha(p:q)$ for $\alpha \in \mathbb{R}$:

$$I_\alpha(p:q) := \begin{cases} \frac{1}{\alpha(1-\alpha)}(1 - \int p^\alpha q^{1-\alpha} \mathrm{d}\mu), & \alpha \in \mathbb{R}\backslash\{0,1\} \\ I_1(p:q) = \mathrm{KL}(p:q), & \alpha = 1 \\ I_0(p:q) = \mathrm{KL}(q:p), & \alpha = 0 \end{cases} \qquad (3)$$

The α-divergences extended to positive densities [10] (not necessarily normalized densities) play a central role in information geometry [6]:

$$I_\alpha^+(p:q) := \begin{cases} \frac{1}{\alpha(1-\alpha)} \int (\alpha p + (1-\alpha)q - p^\alpha q^{1-\alpha}) \mathrm{d}\mu, & \alpha \in \mathbb{R}\backslash\{0,1\} \\ I_1^+(p:q) = \mathrm{KL}^+(p:q), & \alpha = 1 \\ I_0^+(p:q) = \mathrm{KL}^+(q:p), & \alpha = 0 \end{cases}, \qquad (4)$$

where KL^+ denotes the Kullback–Leibler divergence extended to positive measures:

$$\mathrm{KL}^+(p:q) := \int \left(p \log \frac{p}{q} + q - p \right) \mathrm{d}\mu. \qquad (5)$$

The α-divergences are asymmetric for $\alpha \neq \frac{1}{2}$ (i.e., $I_\alpha(p:q) \neq I_\alpha(q:p)$ for $\alpha \neq \frac{1}{2}$) but exhibit the following reference duality [11]:

$$I_\alpha(q:p) = I_{1-\alpha}(p:q) =: I_\alpha^*(p:q), \qquad (6)$$

where we denoted by $D^*(p:q) := D(q:p)$, the reverse divergence for an arbitrary divergence $D(p:q)$ (e.g., $I_\alpha^*(p:q) := I_\alpha(q:p) = I_{1-\alpha}(p:q)$). The α-divergences have been extensively used in many applications [12], and the parameter α may not be necessarily fixed beforehand but can also be learned from data sets in applications [13,14]. When $\alpha = \frac{1}{2}$, the α-divergence is symmetric and called the squared Hellinger divergence [15]:

$$I_{\frac{1}{2}}(p:q) := 4\left(1 - \int \sqrt{pq} \mathrm{d}\mu\right) = 2\int (\sqrt{p} - \sqrt{q})^2 \mathrm{d}\mu. \qquad (7)$$

The α-divergences belong to the family of Ali–Silvey–Csizár's f-divergences [16,17] which are defined for a convex function $f(u)$ satisfying $f(1) = 0$ and strictly convex at 1:

$$I_f(p:q) := \int p f\left(\frac{q}{p}\right) \mathrm{d}\mu. \qquad (8)$$

We have

$$I_\alpha(p:q) = I_{f_\alpha}(p:q), \qquad (9)$$

with the following class of f-generators:

$$f_\alpha(u) := \begin{cases} \frac{1}{\alpha(1-\alpha)}(\alpha + (1-\alpha)u - u^{1-\alpha}), & \alpha \in \alpha \in \mathbb{R}\backslash\{0,1\} \\ u - 1 - \log u, & \alpha = 1 \\ 1 - u + u \log u, & \alpha = 0 \end{cases} \qquad (10)$$

In information geometry, α-divergences and more generally f-divergences are called invariant divergences [6], since they are provably the only statistical divergences which are invariant under invertible smooth transformations of the sample space. That is, let $Y = m(X)$ be a smooth invertible transformation and let $\mathcal{Y} = m(\mathcal{X})$ denote the transformed sample space. Denote by $p_Y(y)$ and $p_{Y'}(y)$ the densities with respect to y corresponding to $p_X(x)$ and $p_{X'}(x)$, respectively. Then, we have $I_f(p_X : p_{X'}) = I_f(p_Y : p_{Y'})$ [18]. The dualistic information-geometric structures induced by these invariant f-divergences between densities of a same parametric family $\{p_\theta(x) : \theta \in \Theta\}$ of statistical models yield the Fisher information metric and the dual $\pm\alpha$-connections for $\alpha = 3 + 2\frac{f'''(1)}{f''(1)}$, see [6] for details. It is customary to rewrite the α-divergences in information geometry using rescaled parameter $\alpha_A = 1 - 2\alpha$ (i.e., $\alpha = \frac{1-\alpha_A}{2}$). Thus, the extended α_A-divergence in information geometry is defined as follows:

$$\hat{I}^+_{\alpha_A}(p:q) = \begin{cases} \frac{4}{1-\alpha_A^2} \int \left(\frac{1-\alpha_A}{2}p + \frac{1+\alpha_A}{2}q - p^{\frac{1-\alpha_A}{2}} q^{\frac{1+\alpha_A}{2}} \right) d\mu, & \alpha_A \in \mathbb{R}\setminus\{-1,1\} \\ \hat{I}_1(p:q) = \mathrm{KL}^+(p:q), & \alpha_A = 1 \\ \hat{I}_{-1}(p:q) = \mathrm{KL}^+(q:p), & \alpha_A = -1 \end{cases}, \quad (11)$$

and the reference duality is expressed by $\hat{I}^+_{\alpha_A}(q:p) = \hat{I}^+_{-\alpha_A}(p:q)$.

A statistical divergence $D(\cdot : \cdot)$ when evaluated on densities belonging to a given parametric family $\mathcal{P} = \{p_\theta : \theta \in \Theta\}$ of densities is equivalent to a corresponding contrast function $D_\mathcal{P}$ [7]:

$$D_\mathcal{P}(\theta_1 : \theta_2) := D(p_{\theta_1} : p_{\theta_2}). \quad (12)$$

Remark 1. *Although quite confusing, those contrast functions [7] have also been called divergences in the literature [6]. Any smooth parameter divergence $D(\theta_1 : \theta_2)$ (contrast function [7]) induces a dualistic structure in information geometry [6]. For example, the KLD on the family Δ of probability mass functions defined on a finite alphabet $\mathcal{X}$ is equivalent to a Bregman divergence, and thus induces a dually flat space [6]. More generally, the α_A-divergences on the probability simplex Δ induce the α_A-geometry in information geometry [6].*

We refer the reader to [3] for a richly annotated bibliography of many common statistical divergences investigated in signal processing and statistics. Building and studying novel statistical/parameter divergences from first principles is an active research area. For example, Li [19,20] recently introduced some new divergence functionals based on the framework of transport information geometry [21], which considers information entropy functionals in Wasserstein spaces. Li defined (i) the transport information Hessian distances [20] between univariate densities supported on a compact, which are symmetric distances satisfying the triangle inequality, and obtained the counterpart of the Hellinger distance on the L^2-Wasserstein space by choosing the Shannon information entropy, and (ii) asymmetric transport Bregman divergences (including the transport Kullback–Leibler divergence) between densities defined on a multivariate compact smooth support in [19].

The α-divergences are widely used in information sciences, see [22–27] just to cite a few applications. The singly parametric α-divergences have also been generalized to biparametric families of divergences such as the (α, β)-divergences [6] or the $\alpha\beta$-divergences [28].

In this work, based on the observation that the term $\alpha p + (1 - \alpha)q - p^\alpha q^{1-\alpha}$ in the extended $I^+_\alpha(p:q)$ divergence for $\alpha \in (0,1)$ of Equation (4) is a difference between a weighted arithmetic mean $A_{1-\alpha}(p,q) := \alpha p + (1-\alpha)q$ and a weighted geometric mean $G_{1-\alpha}(p,q) := p^\alpha q^{1-\alpha}$, we investigate a generalization of α-divergences with respect to a generic pair of strictly comparable weighted means [29]. In particular, we consider the class of quasi-arithmetic weighted means [30], analyze the condition for two quasi-arithmetic means to be strictly comparable, and report their induced α-divergences with limit KL type divergences when $\alpha \to 1$ and $\alpha \to 0$.

1.2. Divergences and Decomposable Divergences

A statistical divergence $D(p:q)$ shall satisfy the following two basic axioms:

D1 (Non-negativity). $D(p:q) \geq 0$ for all densities p and q,

D2 (Identity of indiscernibles). $D(p:q) = 0$ if and only if $p = q$ μ-almost everywhere.

These axioms are a subset of the metric axioms, since we do not consider the symmetry axiom nor the triangle inequality axiom of metric distances. See [31,32] for some common examples of probability metrics (e.g., total variation distance or Wasserstein metrics).

A divergence $D(p:q)$ is said decomposable [6] when it can be written as a definite integral of a scalar divergence $d(\cdot, \cdot)$:

$$D(p:q) = \int d(p(x):q(x))\mathrm{d}\mu(x), \qquad (13)$$

or $D(p:q) = \int d(p:q)\mathrm{d}\mu$ for short, where $d(a,b)$ is a scalar divergence between $a > 0$ and $b > 0$ (hence one-dimensional parameter divergence).

The α-divergences are decomposable divergences since we have

$$I_\alpha^+(p:q) = \int i_\alpha(p(x):q(x))\mathrm{d}\mu \qquad (14)$$

with the following scalar α-divergence:

$$i_\alpha(a:b) := \begin{cases} \frac{1}{\alpha(1-\alpha)}\left(\alpha a + (1-\alpha)b - a^\alpha b^{1-\alpha}\right), & \alpha \in \mathbb{R}\setminus\{0,1\} \\ i_1(a:b) = a\log\frac{a}{b} + b - a & \alpha = 1 \\ i_0(a:b) = i_1(b:a), & \alpha = 0 \end{cases} \qquad (15)$$

1.3. Contributions and Paper Outline

The outline of the paper and its main contributions are summarized as follows:

We first define for two families of strictly comparable means (Definition 1) their generic induced α-divergences in Section 2 (Definition 2). Then, Section 2.2 reports a closed-form formula (Theorem 3) for the quasi-arithmetic α-divergences induced by two strictly comparable quasi-arithmetic means with monotonically increasing generators f and g such that $f \circ g^{-1}$ is strictly convex and differentiable (Theorem 1). In Section 2.3, we study the divergences I_0^+ and I_1^+ obtained in the limit cases when $\alpha \to 0$ and $\alpha \to 1$, respectively, (Theorem 2). We obtain generalized counterparts of the Kullback–Leibler divergence when $\alpha \to 1$ and generalized counterparts of the reverse Kullback–Leibler divergence when $\alpha \to 0$. Moreover, these generalized KLDs can be rewritten as generalized cross-entropies minus entropies. In Section 2.4, we show how to express these generalized I_1-divergences and I_0-divergences as conformal Bregman representational divergences, and briefly explain their induced conformally flat statistical manifolds (Theorem 4). Section 3 introduces the subfamily of bipower homogeneous α-divergences (Definition 2) which belong to the family of Ali–Silvey–Csiszár f-divergences [16,17]. In Section 4, we consider k-means clustering [33] and k-means++ seeding [34] for the generic class of extended α-divergences: we first study the robustness of quasi-arithmetic means in Section 4.1 and then the robustness of the newly class of generalized Kullback–Leibler centroids in Section 4.2. Finally, Section 5 summarizes the results obtained in this work and discusses perspectives for future research.

2. The α-Divergences Induced by a Pair of Strictly Comparable Weighted Means

2.1. The (M, N) α-Divergences

The point of departure for generalizing the α-divergences is to rewrite Equation (4) for $\alpha \in \mathbb{R}\setminus\{0,1\}$ as

$$I_\alpha^+(p:q) = \frac{1}{\alpha(1-\alpha)}\int (A_{1-\alpha}(p,q) - G_{1-\alpha}(p,q))\mathrm{d}\mu, \qquad (16)$$

where A_λ and G_λ for $\lambda \in (0,1)$ stands for the weighted arithmetic mean and the weighted geometric mean, respectively:

$$A_\lambda(x,y) = (1-\lambda)x + \lambda y,$$
$$G_\lambda(x,y) = x^{1-\lambda} y^\lambda.$$

For a weighted mean $M_\lambda(a,b)$, we choose the (geometric) convention $M_0(x,y) = x$ and $M_1(x,y) = 1$ so that $\{M_\lambda(x,y)\}_{\lambda \in [0,1]}$ smoothly interpolates between x ($\lambda = 0$) and y ($\lambda = 1$). For the converse convention, we simply define $M'_\lambda(a,b) = M_{1-\lambda}(a,b)$ and get the conventional definition of $I_\alpha^+(p:q) = \frac{1}{\alpha(1-\alpha)} \int (A'_\alpha(p,q) - G'_\alpha(p,q)) d\mu$.

In general, a mean $M(x,y)$ aggregates two values x and y of an interval $I \subset \mathbb{R}$ to produce an intermediate quantity which satisfies the innerness property [35,36]:

$$\min\{x,y\} \leq M(x,y) \leq \max\{x,y\}, \quad \forall x,y \in I. \tag{17}$$

This in-between property of means (Equation (17)) was postulated by Cauchy [37] in 1821. A mean is said strict if the inequalities of Equation (17) are strict whenever $x \neq y$. A mean M is said reflexive iff $M(x,x) = x$ for all $x \in I$. The reflexive property of means was postulated by Chisini [38] in 1929.

In the remainder, we consider $I = (0, \infty)$. By using the unique dyadic representation of any real $\lambda \in (0,1)$ (i.e., $\lambda = \sum_{i=1}^\infty \frac{d_i}{2^i}$ with $d_i \in \{0,1\}$ the binary digit expansion of λ), one can build a weighted mean M_λ from any given mean M; see [29] for such a construction.

In the remainder, we drop the "+" notation to emphasize that the divergences are defined between positive measures. By analogy to the α-divergences, let us define the (decomposable) (M,N) α-divergences between two positive densities p and q for a pair of weighted means $M_{1-\alpha}$ and $N_{1-\alpha}$ for $\alpha \in (0,1)$ as

$$I_\alpha^{M,N}(p:q) := \frac{1}{\alpha(1-\alpha)} \int (M_{1-\alpha}(p,q) - N_{1-\alpha}(p,q)) d\mu. \tag{18}$$

The ordinary α-divergences for $\alpha \in (0,1)$ are recovered as the (A,G) α-divergences:

$$I_\alpha^{A,G}(p:q) = \frac{1}{\alpha(1-\alpha)} \int (A_{1-\alpha}(p,q) - G_{1-\alpha}(p,q)) d\mu, \tag{19}$$
$$= I_{1-\alpha}(p:q) = I_\alpha(q:p) = I_\alpha^*(p:q). \tag{20}$$

In order to define generalized α-divergences satisfying axioms D1 and D2 of proper divergences, we need to characterize the class of acceptable means. We give a definition strengthening the notion of comparable means in [29]:

Definition 1 (Strictly comparable weighted means). *A pair (M,N) of means are said strictly comparable whenever $M_\lambda(x,y) \geq N_\lambda(x,y)$ for all $x,y \in (0,\infty)$ with equality if and only if $x = y$, and for all $\lambda \in (0,1)$.*

Example 1. *For example, the inequality of the arithmetic and geometric means states that $A(x,y) \geq G(x,y)$ implies means A and G are comparable, denoted by $A \geq G$. Furthermore, the arithmetic and geometric weighted means are distinct whenever $x \neq y$. Indeed, consider the equation $(1-\alpha)x + \alpha y = x^{1-\alpha} y^\alpha$ for $x,y > 0$ and $x \neq y$. By taking the logarithm on both sides, we get*

$$\log((1-\alpha)x + \alpha y) = (1-\alpha)\log x + \alpha \log y. \tag{21}$$

Since the logarithm is a strictly convex function, the only solution is $x = y$. Thus, (A,G) is a pair of strictly comparable weighted means.

For a weighted mean M, define $M'_\lambda(x,y) := M_{1-\lambda}(x,y)$. We are ready to state the definition of generalized α-divergences:

Definition 2 ((M, N) α-divergences). *The (M, N) α-divergences $I_\alpha^{M,N}(p:q)$ between two positive densities p and q for $\alpha \in (0,1)$ is defined for a pair of strictly comparable weighted means M_α and N_α with $M_\alpha \geq N_\alpha$ by:*

$$I_\alpha^{M,N}(p:q) := \frac{1}{\alpha(1-\alpha)} \int (M_{1-\alpha}(p,q) - N_{1-\alpha}(p,q)) d\mu, \quad \alpha \in (0,1) \tag{22}$$

$$= \frac{1}{\alpha(1-\alpha)} \int (M'_\alpha(p,q) - N'_\alpha(p,q)) d\mu, \quad \alpha \in (0,1). \tag{23}$$

Using $\alpha = \frac{1-\alpha_A}{2}$, we can rewrite this α-divergence as

$$\hat{I}_{\alpha_A}^{M,N}(p:q) := \frac{4}{1-\alpha_A^2} \int \left(M_{\frac{1+\alpha_A}{2}}(p,q) - N_{\frac{1+\alpha_A}{2}}(p,q) \right) d\mu, \quad \alpha_A \in (-1,1) \tag{24}$$

$$= \frac{4}{1-\alpha_A^2} \int \left(M'_{\frac{1-\alpha_A}{2}}(p,q) - N'_{\frac{1-\alpha_A}{2}}(p,q) \right) d\mu, \quad \alpha_A \in (-1,1). \tag{25}$$

It is important to check the conditions on the weighted means M_α and N_α which ensures the law of the indiscernibles of a divergence $D(p:q)$, namely, $D(p:q) = 0$ iff $p = q$ almost μ-everywhere. This condition rewrites as $\int M_\alpha(p,q) d\mu = \int N_\alpha(p,q) d\mu$ if and only if $p(x) = q(x)$ μ-almost everywhere. A sufficient condition is to ensure that $M_\alpha(x,y) \neq N_\alpha(x,y)$ for $x \neq y$. In particular, this condition holds if the weighted means M_α and N_α are strictly comparable weighted means.

Instead of taking the difference $M_{1-\alpha}(x:y) - N_{1-\alpha}(x:y)$ between two weighted means, we may also measure the gap logarithmically, and thus define the family of $\log \frac{M}{N}$ α-divergences as follows:

Definition 3 ($\log \frac{M}{N}$ α-divergence). *The $\log \frac{M}{N}$ α-divergences $L_\alpha^{M,N}(p:q)$ between two positive densities p and q for $\alpha \in (0,1)$ is defined for a pair of strictly comparable weighted means M_α and N_α with $M_\alpha \geq N_\alpha$ by:*

$$L_\alpha^{M,N}(p:q) := \int \left(\log \frac{M_{1-\alpha}(p,q)}{N_{1-\alpha}(p,q)} \right) d\mu, \tag{26}$$

$$= -\int \left(\log \frac{N_{1-\alpha}(p,q)}{M_{1-\alpha}(p,q)} \right) d\mu. \tag{27}$$

Note that this definition is different from the skewed Bhattacharyya type distance [39,40], which rather measures

$$B_\alpha^{M,N}(p:q) := \log \frac{\int M_{1-\alpha}(p,q) d\mu}{\int N_{1-\alpha}(p,q) d\mu}, \tag{28}$$

$$= -\log \frac{\int N_{1-\alpha}(p,q) d\mu}{\int M_{1-\alpha}(p,q) d\mu}. \tag{29}$$

The ordinary α-skewed Bhattacharyya distance [39] is recovered when $N_\alpha = G_\alpha$ (weighted geometric mean) and $M_\alpha = A_\alpha$ the arithmetic mean since $\int A_{1-\alpha}(p,q) d\mu = 1$. The Bhattacharyya type divergences $B_\alpha^{M,N}$ were introduced in [41] in order to upper bound the probability of error in Bayesian hypothesis testing.

A weighted mean M_α is said symmetric if and only if $M_\alpha(x,y) = M_{1-\alpha}(y,x)$. When both the weighted means M and N are symmetric, we have the following reference duality [11]:

$$I_\alpha^{M,N}(p:q) = I_{1-\alpha}^{M,N}(q:p). \tag{30}$$

We consider symmetric weighted means in the remainder.

In the limit cases of $\alpha \to 0$ or $\alpha \to 1$, we define the 0-divergence $I_0^{M,N}(p:q)$ and the 1-divergence $I_1^{M,N}(p:q)$, respectively, by

$$I_0^{M,N}(p:q) = \lim_{\alpha \to 0} I_\alpha^{M,N}(p:q), \tag{31}$$

$$I_1^{M,N}(p:q) = \lim_{\alpha \to 1} I_\alpha^{M,N}(p:q) = I_0^{M,N}(q:p), \tag{32}$$

provided that those limits exist.

Notice that the ordinary α-divergences are defined for any $\alpha \in \mathbb{R}$ but our generic quasi-arithmetic α-divergences are defined in general on $(0,1)$. However, when the weighted means M_α and N_α admit weighted extrapolations (e.g., the arithmetic mean A_α or the geometric mean G_α) the quasi-arithmetic α-divergences can be extended to $\mathbb{R}\setminus\{0,1\}$. Furthermore, when the limits of quasi-arithmetic α-divergences exist for $\alpha \in \{0,1\}$, the quasi-arithmetic α-divergences may be defined on the full range of $\alpha \in \mathbb{R}$. To demonstrate the restricted range $(0,1)$, consider the weighted harmonic mean for $x,y > 0$ with $x \neq y$:

$$H_\lambda(x,y) = \frac{1}{(1-\lambda)\frac{1}{x} + \lambda\frac{1}{y}} = \frac{xy}{\lambda x + (1-\lambda)y} = \frac{xy}{y + \lambda(x-y)}. \tag{33}$$

Clearly, the denominator may become zero when $\lambda = \frac{y}{y-x}$ and even possibly negative. Thus, to avoid this issue, we restrict the range of α to $(0,1)$ for defining quasi-arithmetic α-divergences.

2.2. The Quasi-Arithmetic α-Divergences

A quasi-arithmetic mean (QAM) is defined for a continuous and strictly monotonic function $f : I \subset \mathbb{R}_+ \to J \subset \mathbb{R}_+$ as:

$$M^f(x,y) := f^{-1}\left(\frac{f(x) + f(y)}{2}\right). \tag{34}$$

Function f is called the generator of the quasi-arithmetic mean. These strict and reflexive quasi-arithmetic means are also called Kolmogorov means [30], Nagumo means [42] de Finetti means [43], or quasi-linear means [44] in the literature. These means are called quasi-arithmetic means because they can be interpreted as arithmetic means on the arguments $f(x)$ and $f(y)$:

$$f(M^f(x,y)) = \frac{f(x) + f(y)}{2} = A(f(x), f(y)). \tag{35}$$

QAMs are strict, reflexive, and symmetric means.

Without loss of generality, we may assume strictly increasing functions f instead of monotonic functions since $M^{-f} = M^f$. Indeed, $M^{-f}(x,y) = (-f)^{-1}(-f(M^f(x,y)))$ and $((-f)^{-1} \circ (-f))(u) = u$, the identity function. Notice that the composition $f_1 \circ f_2$ of two strictly monotonic increasing functions f_1 and f_2 is a strictly monotonic increasing function. Furthermore, we consider $I = J = (0, \infty)$ in the remainder since we apply these means on positive densities. Two quasi-arithmetic means M^f and M^g coincide if and only if $f(u) = ag(u) + b$ for some $a > 0$ and $b \in \mathbb{R}$, see [44]. The quasi-arithmetic means were considered in the axiomatization of the entropies by Rényi to define the α-entropies (see Equation (2).11 of [45]).

By choosing $f_A(u) = u$, $f_G(u) = \log u$, or $f_H(u) = \frac{1}{u}$, we obtain the Pythagorean's arithmetic A, geometric G, and harmonic H means, respectively:

- the arithmetic mean (A): $A(x,y) = \frac{x+y}{2} = M^{f_A}(x,y)$,
- the geometric mean (G): $G(x,y) = \sqrt{xy} = M^{f_G}(x,y)$, and
- the harmonic mean (H): $H(x,y) = \frac{2}{\frac{1}{x}+\frac{1}{y}} = \frac{2xy}{x+y} = M^{f_H}(x,y)$.

More generally, choosing $f_{P_r}(u) = u^r$, we obtain the parametric family of power means also called Hölder means [46] or binary means [47]:

$$P_r(x,y) = \left(\frac{x^r + y^r}{2}\right)^{\frac{1}{r}} = M^{f_{P_r}}(x,y), \quad r \in \mathbb{R}\setminus\{0\}. \tag{36}$$

In order to get a smooth family of power means, we define the geometric mean as the limit case of $r \to 0$:

$$P_0(x,y) = \lim_{r \to 0} P_r(x,y) = G(x,y) = \sqrt{xy}. \tag{37}$$

A mean M is positively homogeneous if and only if $M(ta,tb) = tM(a,b)$ for any $t > 0$. It is known that the only positively homogeneous quasi-arithmetic means coincide exactly with the family of power means [44]. The weighted QAMs are given by

$$\begin{align}
M_\alpha^f(p,q) &= f^{-1}((1-\alpha)f(p) + \alpha f(q))), \tag{38}\\
&= f^{-1}(f(p) + \alpha(f(q) - f(p))) = M_{1-\alpha}^f(q,p). \tag{39}
\end{align}$$

Let us remark that QAMs were generalized to complex-valued generators in [48] and to probability measures defined on a compact support in [49].

Notice that there exist other positively homogeneous means which are not quasi-arithmetic means. For example, the logarithmic mean [50,51] $L(x,y)$ for $x > 0$ and $y > 0$:

$$L(x,y) = \frac{y-x}{\log y - \log x} \tag{40}$$

is an example of a homogeneous mean (i.e., $L(tx,ty) = tL(x,y)$ for any $t > 0$) that is not a QAM. Besides the family of QAMs, there exist many other families of means [35]. For example, let us mention the Lagrangian means [52], which intersect with the QAMs only for the arithmetic mean, or a generalization of the QAMs called the Bajraktarević means [53].

Let us now strengthen a recent theorem (Theorem 1 of [54], 2010):

Theorem 1 (Strictly comparable weighted QAMs). *The pair (M^f, M^g) of quasi-arithmetic means obtained for two strictly increasing generators f and g is strictly comparable provided that function $f \circ g^{-1}$ is strictly convex, where $\circ$ denotes the function composition.*

Proof. Since $f \circ g^{-1}$ is strictly convex, it is convex, and therefore it follows from Theorem 1 of [54] that $M_\alpha^f \geq M_\alpha^g$ for all $\alpha \in [0,1]$. Thus, the very nice property of QAMs is that $M^f \geq M^g$ implies that $M_\alpha^f \geq M_\alpha^g$ for any $\alpha \in [0,1]$. Now, let us consider the equation $M_\alpha^f(p,q) = M_\alpha^g(p,q)$ for $p \neq q$:

$$f^{-1}((1-\alpha)f(p) + \alpha f(q)) = g^{-1}((1-\alpha)g(p) + \alpha g(q)). \tag{41}$$

Since $f \circ g^{-1}$ is assumed strictly convex, and g is strictly increasing, we have $g(p) \neq g(q)$ for $p \neq q$, and we reach the following contradiction:

$$\begin{align}
(1-\alpha)f(p) + \alpha f(q) &= (f \circ g^{-1})((1-\alpha)g(p) + \alpha g(q)), \tag{42}\\
&< (1-\alpha)(f \circ g^{-1})(g(p)) + \alpha(f \circ g^{-1})(g(q)), \tag{43}\\
&< (1-\alpha)f(p) + \alpha f(q). \tag{44}
\end{align}$$

Thus, $M_\alpha^f(p,q) \neq M_\alpha^g(p,q)$ for $p \neq q$, and $M_\alpha^f(p,q) = M_\alpha^g(p,q)$ for $p = q$. □

Thus, we can define the quasi-arithmetic α-divergences as follows:

Definition 4 (Quasi-arithmetic α-divergences). *The (f, g) α-divergences $I_\alpha^{f,g}(p:q) := I_\alpha^{M^f,M^g}(p:q)$ between two positive densities p and q for $\alpha \in (0,1)$ are defined for two strictly increasing and differentiable functions f and g such that $f \circ g^{-1}$ is strictly convex by:*

$$I_\alpha^{f,g}(p:q) := \frac{1}{\alpha(1-\alpha)} \int \left(M_{1-\alpha}^f(p,q) - M_{1-\alpha}^g(p,q) \right) d\mu, \quad (45)$$

where M_λ^f and M_λ^g are the weighted quasi-arithmetic means induced by f and g, respectively.

We have the following corollary:

Corollary 1 (Proper quasi-arithmetic α-divergences). *Let (M^f, M^g) be a pair of quasi-arithmetic means with $f \circ g^{-1}$ strictly convex, then the (M^f, M^g) α-divergences are proper divergences for $\alpha \in (0,1)$.*

Proof. Consider p and q with $p(x) \neq q(x)$ μ-almost everywhere. Since $f \circ g^{-1}$ is strictly convex, we have $M^f(x,y) - M^g(x,y) \geq 0$ with strict inequality when $x \neq y$. Thus, $\int M^f(p,q)d\mu - \int M^g(p,q)d\mu > 0$ and $I_\alpha^{f,g}(p:q) > 0$. Therefore the quasi-arithmetic α-divergences $I_\alpha^{f,g}$ satisfy the law of the indiscernibles for $\alpha \in (0,1)$. □

Note that the (A, G) α-divergences (i.e., the ordinary α-divergences) are proper divergences satisfying both the properties D1 and D2 because $f_A(u) = u$ and $f_G(u) = \log u$, and hence $(f_A \circ f_G^{-1})(u) = \exp(u)$ is strictly convex on $(0, \infty)$.

Let us denote by $I_\alpha^{f,g}(p:q) := I_\alpha^{M^f,M^g}(p:q)$ the quasi-arithmetic α-divergences. Since the QAMs are symmetric means, we have $I_\alpha^{f,g}(p:q) = I_{1-\alpha}^{f,g}(q:p)$.

Remark 2. *Let us notice that Zhang [55] in their study of divergences under monotone embeddings also defined the following family of related divergences (Equation (71) of [55]):*

$$\hat{I}_{\alpha_A}^{f,g}(p:q) = \frac{4}{1-\alpha_A^2} \int \left(M_{\frac{1+\alpha_A}{2}}^f(p,q) - M_{\frac{1+\alpha_A}{2}}^g(p,q) \right) d\mu. \quad (46)$$

However, Zhang did not study the limit case divergences $\hat{I}_{\alpha_A}^{f,g}(p:q)$ when $\alpha_A \to \pm 1$.

2.3. Limit Cases of 1-Divergences and 0-Divergences

We seek a closed-form formula of the limit divergence $\lim_{\alpha \to 0} I_\alpha^{f,g}(p:q)$ when $\alpha \to 0$.

Lemma 1. *A first-order Taylor approximation of the quasi-arithmetic mean [56] M_α^f for a C_1 strictly increasing generator f when $\alpha \simeq 0$ yields*

$$M_\alpha^f(p,q) = p + \frac{\alpha(f(q) - f(p))}{f'(p)} + o(\alpha(f(q) - f(p))). \quad (47)$$

Proof. By taking the first-order Taylor expansion of $f^{-1}(x)$ at x_0 (i.e., Taylor polynomial of order 1), we get:

$$f^{-1}(x) = f^{-1}(x_0) + (x - x_0)(f^{-1})'(x_0) + o(x - x_0). \quad (48)$$

Using the property of the derivative of an inverse function

$$(f^{-1})'(x) = \frac{1}{(f'(f^{-1})(x))'} \quad (49)$$

it follows that the first-order Taylor expansion of $f^{-1}(x)$ is:

$$f^{-1}(x) = f^{-1}(x_0) + (x - x_0)\frac{1}{(f'(f^{-1})(x_0))} + o(x - x_0). \tag{50}$$

Plugging $x_0 = f(p)$ and $x = f(p) + \alpha(f(q) - f(p))$, we get a first-order approximation of the weighted quasi-arithmetic mean M_α^f when $\alpha \to 0$:

$$M_\alpha^f(p, q) = p + \frac{\alpha(f(q) - f(p))}{f'(p)} + o(\alpha(f(q) - f(p))). \tag{51}$$

□

Let us introduce the following bivariate function:

$$E_f(p, q) := \frac{f(q) - f(p)}{f'(p)}. \tag{52}$$

Remark 3. *Notice that $E_f(p,q) = E_{-f}(p,q)$ matches the fact that $M_\alpha^f(p,q) = M_\alpha^{-f}(p,q)$. That is, we may either consider a strictly increasing differentiable generator f, or equivalently a strictly decreasing differentiable generator $-f$.*

Thus, we obtain closed-form formulas for the I_1-divergence and I_0-divergence:

Theorem 2 (Quasi-arithmetic I_1-divergence and reverse I_0-divergence). *The quasi-arithmetic I_1-divergence induced by two strictly increasing and differentiable functions f and g such that $f \circ g^{-1}$ is strictly convex is*

$$I_1^{f,g}(p:q) := \lim_{\alpha \to 1} I_\alpha^{f,g}(p:q) = \int \left(E_f(p,q) - E_g(p,q) \right) d\mu \geq 0, \tag{53}$$

$$= \int \left(\frac{f(q) - f(p)}{f'(p)} - \frac{g(q) - g(p)}{g'(p)} \right) d\mu. \tag{54}$$

Furthermore, we have $I_0^{f,g}(p:q) = I_1^{f,g}(q:p) = (I_1^{f,g})^*(p:q)$, the reverse divergence.

Proof. Let us prove that $I_1^{f,g}$ is a proper divergence satisfying axioms D1 and D2. Note that a sufficient condition for $I_1^{f,g}(p:q) \geq 0$ is to check that

$$E_f(p,q) \geq E_g(p,q), \tag{55}$$

$$\frac{f(q) - f(p)}{f'(p)} \geq \frac{g(q) - g(p)}{g'(p)}. \tag{56}$$

If $p = q$ μ-almost everywhere then clearly $I_1^{f,g}(p:q) = 0$. Consider $p \neq q$ (i.e., at some observation x: $p(x) \neq q(x)$).

We use the following property of a strictly convex and differentiable function h for $x < y$ (sometimes called the chordal slope lemma, see [29]):

$$h'(x) \leq \frac{h(y) - h(x)}{y - x} \leq h'(y). \tag{57}$$

We consider $h(x) = (f \circ g^{-1})(x)$ so that $h'(x) = \frac{f'(g^{-1}(x))}{g'(g^{-1}(x))}$. There are two cases to consider:

- $p < q$ and therefore $g(p) < g(q)$. Let $y = g(q)$ and $x = g(p)$ in Equation (57). We have $h'(x) = \frac{f'(p)}{g'(p)}$ and $h'(y) = \frac{f'(q)}{g'(q)}$, and the double inequality of Equation (57) becomes

$$\frac{f'(p)}{g'(p)} \leq \frac{f(q) - f(p)}{g(q) - g(p)} \leq \frac{f'(q)}{g'(q)}.$$

Since $g(q) - g(p) > 0$, $g'(p) > 0$, and $f'(p) > 0$, we get

$$\frac{g(q) - g(p)}{g'(p)} \leq \frac{f(q) - f(p)}{f'(p)}.$$

- $q < p$ and therefore $g(p) > g(q)$. Then, the double inequality of Equation (57) becomes

$$\frac{f'(q)}{g'(q)} \leq \frac{f(q) - f(p)}{g(q) - g(p)} \leq \frac{f'(p)}{g'(p)}$$

That is,

$$\frac{f(q) - f(p)}{f'(p)} \geq \frac{g(q) - g(p)}{g'(p)},$$

since $g(q) - g(p) < 0$.

Thus, in both cases, we checked that $E_f(p(x), q(x)) \geq E_g(p(x), q(x))$. Therefore, $I_1^{f,g}(p:q) \geq 0$, and since the QAMs are distinct, $I_1^{f,g}(p:q) = 0$ iff $p(x) = q(x)$ μ-a.e. □

We can interpret the I_1 divergences as generalized KL divergences and define generalized notions of cross-entropies and entropies. Since the KL divergence can be written as the cross-entropy minus the entropy, we can also decompose the I_1 divergences as follows:

$$I_1^{f,g}(p:q) = \int \left(\frac{f(q)}{f'(p)} - \frac{g(q)}{g'(p)} \right) d\mu - \int \left(\frac{f(p)}{f'(p)} - \frac{g(p)}{g'(p)} \right) d\mu, \qquad (58)$$

$$= h_\times^{f,g}(p:q) - h^{f,g}(p), \qquad (59)$$

where $h_\times^{f,g}(p:q)$ denotes the (f,g)-cross-entropy (for a constant $c \in \mathbb{R}$):

$$h_\times^{f,g}(p:q) = \int \left(\frac{f(q)}{f'(p)} - \frac{g(q)}{g'(p)} \right) d\mu + c, \qquad (60)$$

and $h^{f,g}(p)$ stands for the (f,g)-entropy (self cross-entropy):

$$h^{f,g}(p) = h_\times^{f,g}(p:p) = \int \left(\frac{f(p)}{f'(p)} - \frac{g(p)}{g'(p)} \right) d\mu + c. \qquad (61)$$

Notice that we recover the Shannon entropy for $f(x) = x$ and $g(x) = \log(x)$ with $f \circ g^{-1})(x) = \exp(x)$ (strictly convex) and $c = -1$ to annihilate the $\int p d\mu = 1$ term:

$$h^{\mathrm{id},\log}(p) = \int (p - p \log p) d\mu - 1 = -\int p \log p d\mu. \qquad (62)$$

We define the generalized (f,g)-Kullback–Leibler divergence or generalized (f,g)-relative entropies:

$$\mathrm{KL}_{f,g}(p:q) := h_\times^{f,g}(p:q) - h^{f,g}(p). \qquad (63)$$

When $f=f_A$ and $g=f_G$, we resolve the constant to $c=0$, and recover the ordinary Shannon cross-entropy and entropy:

$$h_\times^{f_A,f_G}(p:q) = \int (q - p\log q)d\mu = h_\times(p:q), \tag{64}$$

$$h^{f_A,f_G}(p:q) = h_\times^{f_A,f_G}(p:p) = \int (p - p\log p)d\mu = h(p), \tag{65}$$

and we have the (f_A, f_G)-Kullback–Leibler divergence that is the extended Kullback–Leibler divergence:

$$\mathrm{KL}_{f_A,f_G}(p:q) = \mathrm{KL}^+(p:q) = h_\times(p:q) - h(p) = \int \left(p\log\frac{p}{q} + q - p\right)d\mu. \tag{66}$$

Thus, we have the (f,g)-cross-entropy and (f,g)-entropy expressed as

$$h_\times^{f,g}(p:q) = \int \left(\frac{f(q)}{f'(p)} - \frac{g(q)}{g'(p)}\right)d\mu, \tag{67}$$

$$h^{f,g}(p) = \int \left(\frac{f(p)}{f'(p)} - \frac{g(p)}{g'(p)}\right)d\mu. \tag{68}$$

In general, we can define the (f,g)-Jeffreys divergence as:

$$J^{f,g}(p:q) = \mathrm{KL}^{f,g}(p:q) + \mathrm{KL}^{f,g}(q:p). \tag{69}$$

Thus, we define the quasi-arithmetic mean α-divergences as follows:

Theorem 3 (Quasi-arithmetic α-divergences). *Let f and g be two strictly continuously increasing and differentiable functions on $(0,\infty)$ such that $f \circ g^{-1}$ is strictly convex. Then, the quasi-arithmetic α-divergences induced by (f,g) for $\alpha \in [0,1]$ is*

$$I_\alpha^{f,g}(p:q) = \begin{cases} \frac{1}{\alpha(1-\alpha)}\int \left(M_{1-\alpha}^f(p,q) - M_{1-\alpha}^g(p,q)\right)d\mu, & \alpha \in \mathbb{R}\setminus\{0,1\}, \\ I_1^{f,g}(p:q) = \int \left(\frac{f(q)-f(p)}{f'(p)} - \frac{g(q)-g(p)}{g'(p)}\right)d\mu & \alpha = 1, \\ I_0^{f,g}(p:q) = \int \left(\frac{f(p)-f(q)}{f'(q)} - \frac{g(p)-g(q)}{g'(q)}\right)d\mu, & \alpha = 0. \end{cases} \tag{70}$$

When $f(u) = f_A(u) = u$ ($M^f = A$) and $g(u) = f_G(u) = \log u$ ($M^g = G$), we get

$$I_1^{A,G}(p:q) = \int \left(q - p - p\log\frac{q}{p}\right)d\mu = \mathrm{KL}^+(p:q) = I_1(p:q), \tag{71}$$

the Kullback–Leibler divergence (KLD) extended to positive densities, and $I_0 = \mathrm{KL}^{+*}$ the reverse extended KLD.

Let $\mathcal{M}$ denote the class of strictly increasing and differentiable real-valued univariate functions. An interesting question is to study the class of pairs of functions $(f,g) \in \mathcal{M} \times \mathcal{M}$ such that $I_1^{f,g}(p:q) = \mathrm{KL}(p:q)$. This involves solving integral-based functional equations [57].

We can rewrite the α-divergence $I_\alpha^{f,g}(p:q)$ for $\alpha \in (0,1)$ as

$$I_\alpha^{f,g}(p:q) = \frac{1}{\alpha(1-\alpha)}\left(S_{1-\alpha}^f(p,q) - S_{1-\alpha}^g(p,q)\right), \tag{72}$$

where

$$S_\lambda^h(p,q) := \int M_\lambda^h(p,q)d\mu. \tag{73}$$

Zhang [11] (pp. 188–189) considered the (A, M^ρ) α_A-divergences:

$$D_\alpha^\rho(p:q) := \frac{4}{1-\alpha^2}\int \left(\frac{1-\alpha}{2}p + \frac{1+\alpha}{2}q - \rho^{-1}\left(\frac{1-\alpha}{2}\rho(p) + \frac{1+\alpha}{2}\rho(q)\right)\right)\mathrm{d}\mu. \quad (74)$$

Zhang obtained for $D_{\pm 1}^\rho(p:q)$ the following formula:

$$D_1^\rho(p:q) = \int \left(p - q - \left(\rho^{-1}\right)'(\rho(q))(\rho(p) - \rho(q))\right)\mathrm{d}\mu = D_{-1}^\rho(q:p), \quad (75)$$

which is in accordance with our generic formula of Equation (53) since $(\rho^{-1}(x))' = \frac{1}{\rho'(\rho^{-1}(x))}$. Notice that $A_\alpha \geq P_\alpha^r$ for $r \leq 1$; the arithmetic weighted mean dominates the weighted power means P^r when $r \leq 1$.

Furthermore, by imposing the homogeneity condition $I_\alpha^{A,M^\rho}(tp:tq) = t I_\alpha^{A,M^\rho}(p:q)$ for $t > 0$, Zhang [11] obtained the class of (α_A, β_A)-divergences for $(\alpha_A, \beta_A) \in [-1,1]^2$:

$$D_{\alpha_A,\beta_A}(p:q) := \frac{4}{1-\alpha_A^2}\frac{2}{1+\beta_A}\int \left(\frac{1-\alpha_A}{2}p + \frac{1+\alpha_A}{2}q\right.$$
$$\left. - \left(\frac{1-\alpha_A}{2}p^{\frac{1-\beta_A}{2}} + \frac{1+\alpha_A}{2}q^{\frac{1-\beta_A}{2}}\right)^{\frac{2}{1-\beta_A}}\right)\mathrm{d}\mu. \quad (76)$$

2.4. Generalized KL Divergences as Conformal Bregman Divergences on Monotone Embeddings

Let us rewrite the generalized KLDs $I_1^{f,g}$ as a conformal Bregman representational divergence [58–60] as follows:

Theorem 4. *The generalized KLDs $I_1^{f,g}$ divergences are conformal Bregman representational divergences*

$$I_1^{f,g}(p:q) = \int \frac{1}{f'(p)} B_F(g(q):g(p))\mathrm{d}\mu, \quad (77)$$

with $F = f \circ g^{-1}$ a strictly convex and differentiable Bregman convex generator defining the scalar Bregman divergence [61] B_F:

$$B_F(a:b) = F(a) - F(b) - (a-b)F'(b).$$

Proof. For the Bregman strictly convex and differentiable generator $F = f \circ g^{-1}$, we expand the following conformal divergence

$$\frac{1}{f'(p)}B_F(g(q):g(p)) = \frac{1}{f'(p)}(F(g(q)) - F(g(p)) - (g(q) - g(p))F'(g(p))), \quad (78)$$
$$= \frac{1}{f'(p)}\left((f(q) - f(p)) - (g(q) - g(p))\frac{f'(p)}{g'(p)}\right), \quad (79)$$

since $(g^{-1} \circ g)(x) = x$ and $F'(g(x)) = \frac{f'(x)}{g'(x)}$. It follows that

$$\frac{1}{f'(p)}B_F(g(q):g(p)) = \frac{f(q) - f(p)}{f'(p)} - \frac{g(q) - g(p)}{g'(p)}, \quad (80)$$
$$= E_f(p,q) - E_g(p,q) = I_1^{f,g}(p:q). \quad (81)$$

Hence, we easily check that $I_1^{f,g}(p:q) = \int \frac{1}{f'(p)}B_F(g(q):g(p))\mathrm{d}\mu \geq 0$ since $f'(p) > 0$ and $B_F \geq 0$. □

In general, for a functional generator f and a strictly monotonic representational function r (also called monotone embedding [62] in information geometry), we can define the representational Bregman divergence [63] $B_{f \circ r^{-1}}(r(p) : r(q))$ provided that $F = f \circ r^{-1}$ is a Bregman generator (i.e., strictly convex and differentiable).

The Itakura–Saito divergence [64] (IS) between two densities p and q is defined by:

$$D_{\text{IS}}(p:q) = \int \left(\frac{p}{q} - \log \frac{p}{q} - 1\right) d\mu, \tag{82}$$

$$= \int D_{\text{IS}}(p(x) : q(x)) d\mu(x), \tag{83}$$

where $D_{\text{IS}}(x:y) = \frac{x}{y} - \log \frac{x}{y} - 1$ is the scalar IS divergence. This divergence was originally designed in sound processing for measuring the discrepancy between two speech power spectra. Observe that the IS divergence is invariant by rescaling: $D_{\text{IS}}(tp:tq) = D_{\text{IS}}(p:q)$ for any $t > 0$. The IS divergence is a Bregman divergence [61] obtained for the Burg information generator (i.e., negative Burg entropy): $F_{\text{Burg}}(u) = -\log u$ with $F'_{\text{Burg}}(u) = -\frac{1}{u}$. It follows that we have

$$I_1^f(p:q) = \int p B_f(q:p) d\mu, \tag{84}$$

The Itakura–Saito divergence may further be extended to a family of α-Itakura–Saito divergences (see [6], Equation (10).45 of Theorem 10.1):

$$D_{\text{IS},\alpha}(p:q) = \begin{cases} \int \frac{1}{\alpha^2}\left(\left(\frac{p}{q}\right)^\alpha - \alpha \log \frac{p}{q} - 1\right) d\mu & \alpha \neq 0 \\ \frac{1}{2}\int (\log q - \log p)^2 d\mu & \alpha = 0. \end{cases} \tag{85}$$

In [56], a generalization of the Bregman divergences was obtained using the comparative convexity induced by two abstract means M and N to define (M, N)-Bregman divergences as limit of scaled (M, N)-Jensen divergences. The skew (M, N)-Jensen divergences are defined for $\alpha \in (0, 1)$ by:

$$J_{F,\alpha}^{M,N}(p:q) = \frac{1}{\alpha(1-\alpha)} (N_\alpha(F(p), F(q))) - F(M_\alpha(p,q))), \tag{86}$$

where M_α and N_α are weighted means that should be regular [56] (i.e., homogeneous, symmetric, continuous, and increasing in each variable). Then, we can define the (M, N)-Bregman divergence as

$$B_F^{M,N}(p:q) = \lim_{\alpha \to 1^-} J_{F,\alpha}^{M,N}(p:q), \tag{87}$$

$$= \lim_{\alpha \to 1^-} \frac{1}{\alpha(1-\alpha)} (N_\alpha(F(p), F(q))) - F(M_\alpha(p,q))). \tag{88}$$

The formula obtained in [56] for the quasi-arithmetic means M^f and M^g and a functional generator F that is (M^f, M^g)-convex is:

$$B_F^{f,g}(p:q) = \frac{g(F(p)) - g(F(q))}{g'(F(q))} - \frac{f(p) - f(q)}{f'(q)} F'(q), \tag{89}$$

$$= \frac{1}{f'(F(q))} B_{g \circ F \circ f^{-1}}(f(p) : f(q)) \geq 0. \tag{90}$$

This is a conformal divergence [58] that can be written using the E_f terms as:

$$B_F^{f,g}(p:q) = E_g(F(q), F(p)) - E_f(q,p) F'(q). \tag{91}$$

A function F is (M^f, M^g)-convex iff $g \circ F \circ f^{-1}$ is (ordinary) convex [56].

The information geometry induced by a Bregman divergence (or equivalently by its convex generator) is a dually flat space [6]. The dualistic structure induced by a conformal Bregman representational divergence is related to conformal flattening [59,60]. The notion of conformal structures was first introduced in information geometry by Okamoto et al. [65].

Following the work of Ohara [59,60,66], the Kurose geometric divergence $\rho(p, r)$ [67] (a contrast function in affine differential geometry) induced by a pair (L, M) of strictly monotone smooth functions between two distributions p and r of the d-dimensional probability simplex Δ_d is defined by (Equation (28) in [59]):

$$\rho(p:r) = \frac{1}{\Lambda(r)} \sum_{i=1}^{d+1} \frac{L(p_i) - L(r_i)}{L'(r_i)} = \frac{1}{\Lambda(r)} \sum_{i=1}^{d+1} E_L(r_i, p_i), \qquad (92)$$

where $\Lambda(r) = \sum_{i=1}^{d+1} \frac{1}{L'(p_i)} p_i$. Affine immersions [67] can be interpreted as special embeddings.

Let ρ be a divergence (contrast function) and $(^\rho g, ^\rho \nabla, ^\rho \nabla^*)$ be the induced statistical manifold structure with

$$^\rho g_{ij}(p) := -(\partial_i)_p (\partial_j)_p \, \rho(p,q)|_{q=p}, \qquad (93)$$
$$\Gamma_{ij,k}(p) := -(\partial_i)_p (\partial_j)_p (\partial_k)_q \, \rho(p,q)|_{q=p}, \qquad (94)$$
$$\Gamma^*_{ij,k}(p) := -(\partial_i)_p (\partial_j)_q (\partial_k)_q \, \rho(p,q)|_{q=p}, \qquad (95)$$

where $(\partial_i)_s$ denotes the tangent vector at s of a vector field ∂_i.

Consider a conformal divergence $\rho_\kappa(p:q) = \kappa(q)\rho(p:q)$ for a positive function $\kappa(q) > 0$, called the conformal factor. Then, the induced statistical manifold [6,7] $(^{\rho_\kappa} g, ^{\rho_\kappa}\nabla, ^{\rho_\kappa}\nabla^*)$ is 1-conformally equivalent to $(^\rho g, ^\rho \nabla, ^\rho \nabla^*)$ and we have

$$^{\rho_\kappa} g = \kappa \, ^\rho g, \qquad (96)$$
$$^\rho g(^{\rho_\kappa}\nabla_X Y, Z) = {}^\rho g(^\rho \nabla_X Y, Z) - d(\log \kappa)(Z) {}^\rho g(X, Y). \qquad (97)$$

The dual affine connections $^{\rho_\kappa}\nabla^*$ and $^\rho \nabla^*$ are projectively equivalent [67] (and $^\rho \nabla^*$ is said -1-conformally flat).

Conformal flattening [59,60] consists of choosing the conformal factor κ such that $(^{\rho_\kappa} g, ^{\rho_\kappa}\nabla, ^{\rho_\kappa}\nabla)$ becomes a dually flat space [6] equipped with a canonical Bregman divergence.

Therefore, it follows that the statistical manifolds induced by the 1-divergence $I_1^{f,g}$ is a representational 1-conformally flat statistical manifold. Figure 1 gives an overview of the interplay of divergences with information-geometric structures. The logarithmic divergence [68] $L_{G,\alpha}$ is defined for $\alpha > 0$ and an α-exponentially concave generator G by:

$$L_{G,\alpha}(\theta_1 : \theta_2) = \frac{1}{\alpha} \log\left(1 + \alpha \nabla G(\theta_2)^\top (\theta_1 - \theta_2)\right) + G(\theta_2) - G(\theta_1). \qquad (98)$$

When $\alpha \to 0$, we have $L_{G,\alpha}(\theta_1 : \theta_2) \to B_{-G}(\theta_1 : \theta_2)$, where B_F is the Bregman divergence [61] induced by a strictly convex and smooth function F:

$$B_F(\theta_1 : \theta_2) = F(\theta_1) - F(\theta_2) - (\theta_1 - \theta_2)^\top \nabla F(\theta_2).$$

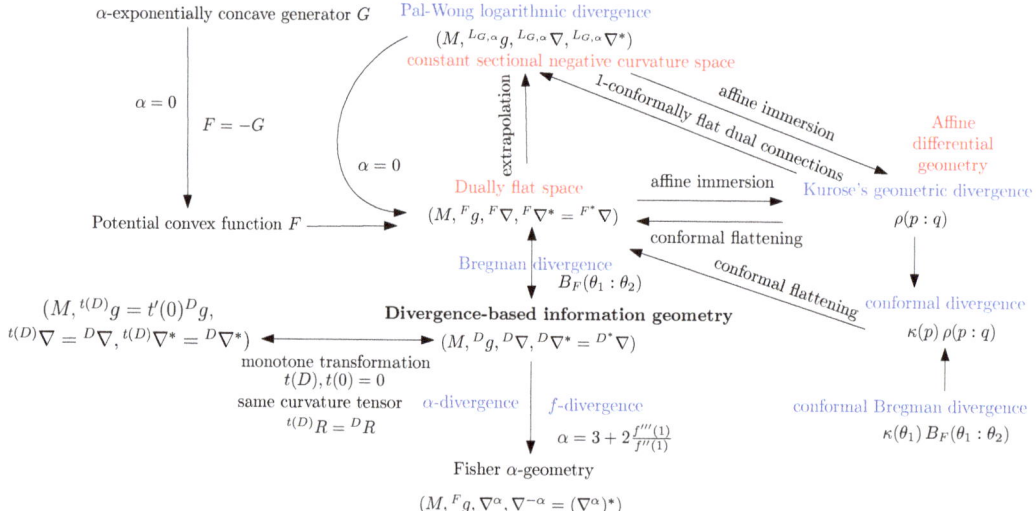

Figure 1. Interplay of divergences and their information-geometric structures: Bregman divergences are canonical divergences of dually flat structures, and the α-logarithmic divergences are canonical divergences of 1-conformally flat statistical manifolds. When $\alpha \to 0$, the logarithmic divergence $L_{F,\alpha}$ tends to the Bregman divergence B_F.

3. The Subfamily of Homogeneous (r,s)-Power α-Divergences for $r > s$

In particular, we can define the (r,s)-power α-divergences from two power means $P_r = M^{\text{pow}_r}$ and $P_s = M^{\text{pow}_s}$ with $r > s$ (and $P_r \geq P_s$) with the family of generators $\text{pow}_l(u) := u^l$. Indeed, we check that $f_{rs}(u) := \text{pow}_r \circ \text{pow}_s^{-1}(u) = u^{\frac{r}{s}}$ is strictly convex on $(0, \infty)$ since $f''_{rs}(u) = \frac{r}{s}(\frac{r}{s} - 1) u^{\frac{r}{s} - 2} > 0$ for $r > s$. Thus, P_r and P_s are two QAMs which are both comparable and distinct. Table 1 lists the expressions of $E_r(p,q) := E_{\text{pow}_r}(p,q)$ obtained from the power mean generators $\text{pow}_r(u) = u^r$.

Table 1. Expressions of the terms E_r for the family of power means P_r, $r \in \mathbb{R}$.

Power Mean	$E_r(p,q)$
$P_r (r \in \mathbb{R}\backslash\{0\})$	$\frac{q^r - p^r}{rp^{r-1}}$
$Q(r=2)$	$\frac{q^2 - p^2}{2p}$
$A(r=1)$	$q - p$
$G(r=0)$	$p \log \frac{q}{p}$
$H(r=-1)$	$-p^2\left(\frac{1}{q} - \frac{1}{p}\right) = p - \frac{p^2}{q}$

We conclude with the definition of the (r,s)-power α-divergences:

Corollary 2 (power α-divergences). *Given $r > s$, the α-power divergences are defined for $r > s$ and $r, s \neq 0$ by*

$$I_\alpha^{r,s}(p:q) = \begin{cases} \frac{1}{\alpha(1-\alpha)} \int \left((\alpha p^r + (1-\alpha) q^r)^{\frac{1}{r}} - (\alpha p^s + (1-\alpha) q^s)^{\frac{1}{s}} \right) d\mu, & \alpha \in \mathbb{R}\backslash\{0,1\}. \\ I_1^{r,s}(p:q) = \int \left(\frac{q^r - p^r}{rp^{r-1}} - \frac{q^s - p^s}{sp^{s-1}} \right) d\mu & \alpha = 1, \\ I_0^{r,s}(p:q) = I_1^{r,s}(q:p) & \alpha = 0. \end{cases} \quad (99)$$

When $r = 0$, we get the following power α-divergences for $s < 0$:

$$I_\alpha^{0,s}(p:q) = \begin{cases} \frac{1}{\alpha(1-\alpha)} \int \left(p^\alpha q^{1-\alpha} - (\alpha p^s + (1-\alpha)q^s)^{\frac{1}{s}} \right) d\mu, & \alpha \in \mathbb{R}\backslash\{0,1\}. \\ I_1^{0,s}(p:q) = \int \left(p \log \frac{q}{p} - \frac{q^s - p^s}{sp^{s-1}} \right) d\mu & \alpha = 1, \\ I_0^{0,s}(p:q) = I_1^{r,s}(q:p) & \alpha = 0. \end{cases} \quad (100)$$

When $s = 0$, we get the following power α-divergences for $r > 0$:

$$I_\alpha^{r,0}(p:q) = \begin{cases} \frac{1}{\alpha(1-\alpha)} \int \left((\alpha p^r + (1-\alpha)q^r)^{\frac{1}{r}} - p^\alpha q^{1-\alpha} \right) d\mu, & \alpha \in \mathbb{R}\backslash\{0,1\}. \\ I_1^{r,0}(p:q) = \int \left(\frac{q^r - p^r}{rp^{r-1}} - p \log \frac{q}{p} \right) d\mu & \alpha = 1, \\ I_0^{r,0}(p:q) = I_1^{r,s}(q:p) & \alpha = 0. \end{cases} \quad (101)$$

In particular, we get the following family of (A, H) α-divergences

$$I_\alpha^{A,H}(p:q) = I_\alpha^{1,-1}(p:q) = \begin{cases} \frac{1}{\alpha(1-\alpha)} \int \left(\alpha p + (1-\alpha)q - \frac{pq}{\alpha q + (1-\alpha)p} \right) d\mu, & \alpha \in \mathbb{R}\backslash\{0,1\}. \\ I_1^{1,-1}(p:q) = \int \left(q - 2p + \frac{p^2}{q} \right) d\mu & \alpha = 1, \\ I_0^{1,-1}(p:q) = I_1^{1,-1}(q:p) & \alpha = 0. \end{cases}, \quad (102)$$

and the family of (G, H) α-divergences:

$$I_\alpha^{G,H}(p:q) = I_\alpha^{0,-1}(p:q) = \begin{cases} \frac{1}{\alpha(1-\alpha)} \int \left(p^\alpha q^{1-\alpha} - \frac{pq}{\alpha q + (1-\alpha)p} \right) d\mu, & \alpha \in \mathbb{R}\backslash\{0,1\}. \\ I_1^{0,-1}(p:q) = \int \left(p \log \frac{q}{p} - p + \frac{p^2}{q} \right) d\mu & \alpha = 1, \\ I_0^{0,-1}(p:q) = I_1^{0,-1}(q:p) & \alpha = 0. \end{cases} \quad (103)$$

The (r,s)-power α-divergences for $r, s \neq 0$ yield homogeneous divergences: $I_\alpha^{r,s}(tp : tq) = t\, I_\alpha^{r,s}(p:q)$ for any $t > 0$ because the power means are homogeneous: $P_\alpha^r(tx, ty) = tP_\alpha^r(x,y) = txP_\alpha^r(1, \frac{y}{x})$. Thus, the $I_\alpha^{r,s}$-divergences are Csiszár f-divergences [17]

$$I_\alpha^{r,s}(p:q) = \int p(x) f_{r,s}\left(\frac{q(x)}{p(x)} \right) d\mu \quad (104)$$

for the generator

$$f_{r,s}(u) = \frac{1}{\alpha(1-\alpha)} (P_\alpha^r(1,u) - P^s(1,u)). \quad (105)$$

Thus, the family of (r,s)-power α-divergences are homogeneous divergences:

$$I_\alpha^{r,s}(tp:tq) = t\, I_\alpha^{r,s}(p:q), \quad \forall t > 0. \quad (106)$$

4. Applications to Center-Based Clustering

Clustering is a class of unsupervised learning algorithms which partitions a given d-dimensional point set $\mathcal{P} = \{p_1, \ldots, p_n\}$ into clusters such that data points falling into a same cluster tend to be more similar to data points belonging to different clusters. The celebrated k-means clustering [69] is a center-based method for clustering $\mathcal{P}$ into k clusters $\mathcal{C}_1, \ldots, \mathcal{C}_k$ (with $\mathcal{P} = \cup_{i=1}^k \mathcal{C}_i$), by minimizing the following k-means objective function

$$L(\mathcal{P}, \mathcal{C}) = \frac{1}{n} \sum_{i=1}^n \min_{j \in \{1,\ldots,k\}} \|p_i - c_j\|^2, \quad (107)$$

where the c_j's denote the cluster representatives. Let $\mathcal{C} = \{c_1, \ldots, c_k\}$ denote the set of cluster centers. The cluster $\mathcal{C}_j$ is defined as the points of $\mathcal{P}$ closer to cluster representative c_j than any other c_i for $i \neq j$:

$$\mathcal{C}_j = \{p \in \mathcal{P} : \|p - c_j\|^2 \leq \|p - c_l\|^2, \forall l \in \{1, \ldots, k\}\}.$$

When $k = 1$, it can be shown that the centroid of the point set $\mathcal{P}$ is the unique best cluster representative:

$$\arg\min_{c_1} L(\mathcal{P}, \{c_1\}) \Rightarrow c_1 = \frac{1}{n}\sum_{i=1}^{n} p_i.$$

When $d > 1$ and $k > 1$, finding a best partition $\mathcal{P} = \cup_{j=1}^{k} \mathcal{C}_j$ which minimizes the objective function of Equation (107) is NP-hard [70]. When $d = 1$, k-means clustering can be solved efficiently using dynamic programming [71] in subcubic $O(n^3)$ time.

The k-means objective function can be generalized to any arbitrary (potentially asymmetric) divergence $D(\cdot : \cdot)$ by considering the following objective function:

$$L_D(\mathcal{P}, \mathcal{C}) := \frac{1}{n}\sum_{i=1}^{n} \min_{j \in \{1, \ldots, k\}} D(p_i : c_j). \tag{108}$$

Thus, when $D(p : q) = \|p - q\|^2$, one recovers the ordinary k-means clustering [69]. When $D(p : q) = B_F(p : q)$ is chosen as a Bregman divergence, one gets the right-sided Bregman k-means clustering [72] as the minimization of the cluster centers are defined on the right-sided arguments of D in Equation (108). When $F(x) = \|x\|_2^2$, Bregman k-means clustering (i.e., $D(p : q) = B_F(p : q)$ in Equation (108)) amounts to the ordinary k-means clustering. The right-sided Bregman centroid for $k = 1$ coincides with the center of mass and is independent of the Bregman generator F:

$$\arg\min_{c_1} L_{B_F}(\mathcal{P}, \{c_1\}) \Rightarrow c_1 = \frac{1}{n}\sum_{i=1}^{n} p_i.$$

The left-sided Bregman k-means clustering is obtained by considering the right-sided Bregman centroid for the reverse Bregman divergence $(B_F)^*(p : q) = B_F(q : p)$, and the left-sided Bregman centroid [73] can be expressed as a multivariate generalization of the quasi-arithmetic mean:

$$c_1 = (\nabla F)^{-1}\left(\frac{1}{n}\sum_{i=1}^{n} \nabla F(p_i)\right).$$

In order to study the robustness of k-means clustering with respect to our novel family of divergences $I_\alpha^{f,g}$, we first study the robustness of the left-sided Bregman centroids to outliers.

4.1. Robustness of the Left-Sided Bregman Centroids

Consider two d-dimensional points $p = (p_1, \ldots, p_d)$ and $p' = (p'_1, \ldots, p'_d)$ of a domain $\Theta \subset \mathbb{R}^d$. The centroid of p and p' with respect to any arbitrary divergence $D(\cdot : \cdot)$ is by definition the minimizer of

$$L_D(c) = \frac{1}{2}D(p : c) + \frac{1}{2}D(p' : c),$$

provided that the minimizer $\min_{c \in \Theta} L_D(c)$ is unique. Assume a separable Bregman divergence induced by the generator $F(p) = \sum_{i=1}^{d} F(p_i)$. The left-sided Bregman centroid [73] of p and p' is given by the following separable quasi-arithmetic centroid:

$$c = (c_1, \ldots, c_d),$$

with
$$c_i = M^f(p_i, p'_i) = f^{-1}\left(\frac{f(p_i) + f(p'_i)}{2}\right),$$

where $f(x) = F'(x)$ denotes the derivative of the Bregman generator $F(x)$.

Now, fix p (say, $p = (1, \ldots, 1) \in \Theta$), and let the coordinates p'_i of p' all tend to infinity: That is, point p' plays the role of an outlier data point. We use the general framework of influence functions [74] in statistics to study the robustness of divergence-based centroids. Consider the r-power mean, a quasi-arithmetic mean induced by $\text{pow}_r(x) = x^r$ for $r \neq 0$ and by extension $\text{pow}_0(x) = \log x$ when $r = 0$ (geometric mean).

When $r < 0$, we check that

$$\lim_{p'_i \to +\infty} M^{\text{pow}_r}(p_i, p'_i) = \lim_{p'_i \to +\infty} \left(\frac{1 + p'^r_i}{2}\right)^{\frac{1}{r}}, \tag{109}$$

$$= \left(\frac{1}{2}\right)^{\frac{1}{r}} < \infty. \tag{110}$$

That is, the r-power mean is robust to an outlier data point when $r < 0$ (see Figure 2). Note that if instead of considering the centroid, we consider the barycenter with w denoting the weight of point p and $1 - w$ denoting the weight of the outlier p' for $w \in (0, 1)$, then the power r-mean falls in a square box of side $w^{\frac{1}{r}}$ when $r < 0$.

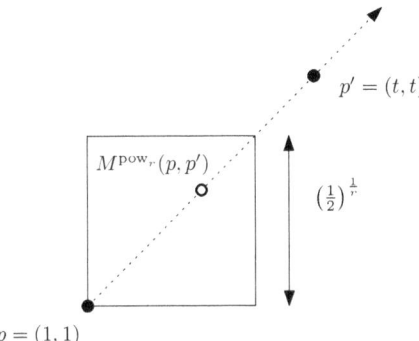

$p = (1, 1)$

Figure 2. Illustration of the robustness property of the r-power mean $M^{\text{pow}_r}(p, p')$ when $r < 0$ for two points: a prescribed point $p = (1, 1)$ and an outlier point $p' = (t, t)$. When $t \to +\infty$, the r-power mean of p and p' for $r < 0$ (e.g., coordinatewise harmonic mean when $r = -1$) is contained inside the box anchored at p of size length $\left(\frac{1}{2}\right)^{\frac{1}{r}}$. The r-power mean can be interpreted as a left-sided Bregman centroid for $F'(x) = -x^r$, i.e., $F(x) = -\frac{1}{r}x^{r+1}$ when $r < -1$ and $F(x) = -\log x$ when $r = -1$.

On the contrary, when $r > 0$ or $r = 0$, we have $\lim_{p'_i \to +\infty} M^{\text{pow}_r}(p_i, p'_i) = \infty$, and the r-power mean diverges to infinity.

Thus, when $r < 0$, the quasi-arithmetic centroid of $p = (1, \ldots, 1)$ and p' is contained in a bounding box of length $\left(\frac{1}{2}\right)^{\frac{1}{r}}$ with left corner $(1, \ldots, 1)$, and the left-sided Bregman power centroid minimizing

$$\frac{1}{2}B_F(c : p) + \frac{1}{2}B_F(c : p')$$

is robust to outlier p'.

To contrast with this result, notice that the right-sided Bregman centroid [72] is always the center of mass (arithmetic mean), and therefore not robust to outliers as a single outlier data point may potentially drag the centroid to infinity.

Example 2. Since $M^f = M^{-f}$ for any strictly smooth increasing function f, we deduce that the quasi-arithmetic left-sided Bregman centroid induced by $F(x) = -\log x$ with $f(x) = F'(x) = -x^{-1} = -\frac{1}{x}$ for $x > 0$ is the harmonic mean which is robust to outliers. The corresponding Bregman divergence is the Itakura–Saito divergence [72].

Notice that it is enough to consider without loss of generality two points p and p': Indeed, the case of the quasi-arithmetic mean of $\mathcal{P} = \{p_1, \ldots, p_n\}$ and p' can be rewritten as an equivalent weighted quasi-arithmetic mean of two points $\bar{p} = M^f(p_1, \ldots, p_n)$ with weight $w = \frac{n}{n+1}$ and p' of weight $\frac{1}{n+1}$ using the replacement property of quasi-arithmetic means:

$$M^f(p_1, \ldots, p_k, p_{k+1}, \ldots, p_n) = M^f(\bar{p}, \ldots, \bar{p}, p_{k+1}, p_n)$$

where $\bar{p} = M^f(p_1, \ldots, p_k)$.

4.2. Robustness of Generalized Kullback–Leibler Centroids

The fact that the generalized KLDs are conformal representational Bregman divergences can be used to design efficient algorithms in computational geometry [60]. For example, let us consider the centroid (or barycenter) of a finite set of weighted probability measures $P_1, \ldots, P_n \ll \mu$ (with RN derivatives $p_1, \ldots, p_n$) defined as the minimizer of

$$\min \sum_{i=1}^{n} w_i I_1^{f,g}(p_i : c),$$

where the w_i's are positive weights summing up to one ($\sum_{i=1}^{n} w_i = 1$). The divergences $I_1^{f,g}(p_i : c)$ are separable. Thus, consider without loss of generality, the scalar-generalized KLDs so that we have

$$I_1^{f,g}(p : q) = \frac{1}{f'(p)} B_F(g(q) : g(p)),$$

where p and q are scalars.

Since the Bregman centroid is unique and always coincide with the center of mass [72]

$$c^* = \arg\min w_i \sum_{i=1}^{n} B_F(p_i : c) = \sum_{i=1}^{n} w_i p_i,$$

for positive weights w_i's summing up to one, we deduce that the right-sided generalized KLD centroid

$$\arg\min_c \frac{1}{n} \sum_{i=1}^{n} I_1^{f,g}(p_i : c) = \arg\min_c \frac{1}{n} \sum_{i=1}^{n} \frac{1}{f'(p_i)} B_F(g(c) : g(p_i))$$

amounts to a left-sided Bregman centroid with un-normalized positive weights $W_i = \frac{1}{f'(p_i)}$ for the scalar Bregman generator $F(x) = f(g^{-1}(x))$ with $F'(x) = \frac{f'(g^{-1}(x))}{g'(g^{-1}(x))}$. Therefore, the right-sided generalized KLD centroid c^* is calculated for normalized weights $w_i = \frac{W_i}{\sum_{j=1}^{n} W_j}$ as:

$$c^* = (F')^{-1}\left(\sum_{i=1}^{n} w_i F'(g(p_i))\right), \qquad (111)$$

$$= (F')^{-1}\left(\sum_{i=1}^{n} \frac{1}{f'(p_i) \sum_{j=1}^{n} \frac{1}{f'(p_j)}} \frac{f'(p_i)}{g'(p_i)}\right), \qquad (112)$$

$$= (F')^{-1}\left(\sum_{i=1}^{n} \frac{1}{g'(p_i) \sum_{j=1}^{n} \frac{1}{f'(p_j)}}\right). \qquad (113)$$

Thus, we obtain a closed-form formula when $(F')^{-1}$ is computationally tractable. For example, consider the (r,s)-power KLD (with $r > s$). We have $f'(x) = rx^{r-1}, g'(x) = sx^{s-1}$, $F(x) = x^{\frac{r}{s}}$, $F'(x) = \frac{r}{s} x^{\frac{r-s}{s}}$ and therefore, we get $F'^{-1}(x) = \left(\frac{s}{r}x\right)^{\frac{s}{r-s}}$. Thus, we get a closed-form formula for the right-sided (r,s)-power Kullback–Leibler centroid using Equation (113).

Overall, we can design a k-means-type algorithm with respect to our generalized KLDs following [72]. Moreover, we can initialize probabilistically k-means with a fast k-means++ seeding [34] described in Algorithm 1. The performance of the k-means++ seeding (i.e., the ratio $\frac{L_D(\mathcal{P},\mathcal{C})}{\min_{\mathcal{C}} L_D(\mathcal{P},\mathcal{C})}$) is $O(\log k)$ when $D(p:q) = \|p - q\|^2$, and the analysis has been extended to arbitrary divergences in [75]. The merit of using the k-means++ seeding is that we do not need to iteratively update the cluster representatives using Lloyd's heuristic [69] and we can thus bypass the calculations of centroids and merely choose the cluster representatives from the source data points $\mathcal{P}$ as described in Algorithm 1.

Algorithm 1 Generic seeding of k-means with divergence-based k-means++.

input: A finite set $\mathcal{P} = \{p_1, \ldots, p_n\}$ of n points, the number of cluster representatives $k \geq 1$, and an arbitrary divergence $D(\cdot : \cdot)$
Output: Set of initial cluster centers $\mathcal{C} = \{c_1, \ldots, c_k\}$
Choose $c_1 \leftarrow p_i$ with uniform probability and $\mathcal{C} = \{c_1\}$;
for $i \leftarrow 2$ **to** k **do**
  Pick at random $c_i = p_j \in \mathcal{P}$ with probability

$$\pi(p_j) = \frac{D(p_j : \mathcal{C})}{\sum_{p \in \mathcal{P}} D(p : \mathcal{C})}$$

  where $D(p : \mathcal{C}) := \min_{c \in \mathcal{C}} D(p : c)$;
  $\mathcal{C} \leftarrow \mathcal{C} \cup \{c_i\}$;
end
return $\mathcal{C}$;

The advantage of using a conformal Bregman divergence such as a total Bregman divergence [33] or $I_1^{f,g}$ is to potentially ensure robustness to outliers (e.g., see Theorem III.2 of [33]). Robustness property of these novel $I_1^{f,g}$ divergences can also be studied for statistical inference tasks based on minimum divergence methods [4,76].

5. Conclusions and Discussion

For two comparable strict means [35] $M(p,q) \geq N(p,q)$ (with equality holding if and only if $p = q$), one can define their (M,N)-divergence as

$$I^{M,N}(p:q) := 4 \int (M(p,q) - N(p,q)) d\mu. \tag{114}$$

When the property of strict comparable means extend to their induced weighted means $M_\alpha(p,q)$ and $N_\alpha(p,q)$ (i.e., $M_\alpha(p,q) \geq N_\alpha(p,q)$), one can further define the family of (M,N) α-divergences for $\alpha \in (0,1)$:

$$I_\alpha^{M,N}(p:q) := \frac{1}{\alpha(1-\alpha)} \int (M_{1-\alpha}(p,q) - N_{1-\alpha}(p,q)) d\mu, \tag{115}$$

so that $I^{M,N}(p:q) = I_{\frac{1}{2}}^{M,N}(p:q)$. When the weighted means are symmetric, the reference duality holds (i.e., $I_\alpha^{M,N}(q:p) = I_{1-\alpha}^{M,N}(p:q)$), and we can define the (M,N)-equivalent of the Kullback–Leibler divergence, i.e., the (M,N) 1-divergence, as the limit case (when it exists): $I_1^{M,N}(p:q) = \lim_{\alpha \to 1} I_\alpha^{M,N}(p:q)$. Similarly, the (M,N)-equivalent of the reverse Kullback–Leibler divergence is obtained as $I_0^{M,N}(p:q) = \lim_{\alpha \to 0} I_\alpha^{M,N}(p:q)$.

We proved that the quasi-arithmetic weighted means [30] M_α^f and M_α^g were strictly comparable whenever $f \circ g^{-1}$ was strictly convex. In the limit cases of $\alpha \to 0$ and $\alpha \to 1$, we reported a closed-form formula for the equivalent of the forward and the reverse Kullback–Leibler divergences. We reported closed-form formulas for the quasi-arithmetic α-divergences $I_\alpha^{f,g}(p:q) := I_\alpha^{M^f,M^g}(p:q)$ for $\alpha \in [0,1]$ (Theorem 3) and for the subfamily of homogeneous (r,s)-power α-divergences $I_\alpha^{r,s}(p:q) := I_\alpha^{M^{\text{pow}_r},M^{\text{pow}_s}}(p:q)$ induced by power means (Corollary 2). The ordinary (A,G) α-divergences [12], the (A,H) α-divergences, and the (G,H) α-divergences are examples of (r,s)-power α-divergences obtained for $(r,s) = (1,0)$, $(r,s) = (1,-1)$ and $(r,s) = (0,-1)$, respectively.

Generalized α-divergences may prove useful in reporting a closed-form formula between densities of a parametric family $\{p_\theta\}$. For example, consider the ordinary α-divergences between two scale Cauchy densities $p_1(x) = \frac{1}{\pi}\frac{s_1}{x^2+s_1^2}$ and $p_2(x) = \frac{1}{\pi}\frac{s_2}{x^2+s_2^2}$; there is no obvious closed-form for the ordinary α-divergences, but we can report a closed-form for the (A,H) α-divergences following the calculus reported in [41]:

$$I_\alpha^{A,H}(p_1:p_2) = \frac{1}{\alpha(1-\alpha)}\left(1 - \int H_{1-\alpha}(p_1(x),p_2(x))\mathrm{d}\mu(x)\right), \quad (116)$$

$$= \frac{1}{\alpha(1-\alpha)}\left(1 - \frac{s_1 s_2}{(\alpha s_1 + (1-\alpha)s_2)s_{1-\alpha}}\right), \quad (117)$$

with $s_\alpha = \sqrt{\frac{\alpha s_1 s_2^2 + (1-\alpha)s_2 s_1^2}{\alpha s_1 + (1-\alpha)s_2}}$. For probability distributions p_{θ_1} and p_{θ_2} belonging to the same exponential family [77] with cumulant function F, the ordinary α-divergences admit the following closed-form solution:

$$I_\alpha(p_{\theta_1}:p_{\theta_2}) = \begin{cases} \frac{1}{\alpha(1-\alpha)}(1 - \exp(F(\alpha\theta_1 + (1-\alpha)\theta_2) - (\alpha F(\theta_1) + (1-\alpha)F(\theta_2)))), & \alpha \in (0,1) \\ I_1(p_{\theta_1}:p_{\theta_2}) = \mathrm{KL}(p_{\theta_1}:p_{\theta_2}) = B_F(\theta_2:\theta_1), & \alpha = 1 \\ I_0(p_{\theta_1}:p_{\theta_2}) = \mathrm{KL}(p_{\theta_2}:p_{\theta_1}) = B_F(\theta_1:\theta_2) & \alpha = 0 \end{cases} \quad (118)$$

where B_F is the Bregman divergence: $B_F(\theta_2:\theta_1) = F(\theta_2) - F(\theta_1) - (\theta_2 - \theta_1)^\top \nabla F(\theta_1)$.

Instead of considering ordinary α-divergences in applications, one may consider the (r,s)-power α-divergences, and tune the three scalar parameters (r,s,α) according to the various tasks (say, by cross-validation in supervised machine learning tasks, see [13]). For the limit cases of $\alpha \to 0$ or of $\alpha \to 1$, we further proved that the limit KL type divergences amounted to conformal Bregman divergences on strictly monotone embeddings and explained the connection of conformal divergences with conformal flattening [60], which allows one to build fast algorithms for centroid-based k-means clustering [72], Voronoi diagrams, and proximity data-structures [60,63]. Some ideas left for future directions is to study the properties of these new (M,N) α-divergences for statistical inference [2,4,76].

Funding: This research received no external funding

Conflicts of Interest: The authors declare no conflict of interest.

References

1. Keener, R.W. *Theoretical Statistics: Topics for a Core Course*; Springer: Berlin/Heidelberg, Germany, 2011.
2. Basu, A.; Shioya, H.; Park, C. *Statistical Inference: The Minimum Distance Approach*; CRC Press: Boca Raton, FL, USA, 2011.
3. Basseville, M. Divergence measures for statistical data processing — An annotated bibliography. *Signal Process.* **2013**, *93*, 621–633. [CrossRef]
4. Pardo, L. *Statistical Inference Based on Divergence Measures*; CRC Press: Boca Raton, FL, USA, 2018.

5. Oller, J.M. Some geometrical aspects of data analysis and statistics. In *Statistical Data Analysis and Inference*; Elsevier: Amsterdam, The Netherlands, 1989; pp. 41–58.
6. Amari, S. *Information Geometry and Its Applications*; Applied Mathematical Sciences; Springer: Tokyo, Japan, 2016.
7. Eguchi, S. Geometry of minimum contrast. *Hiroshima Math. J.* **1992**, *22*, 631–647. [CrossRef]
8. Cover, T.M.; Thomas, J.A. *Elements of Information Theory*; John Wiley & Sons: Hoboken, NJ, USA, 2012.
9. Cichocki, A.; Amari, S.i. Families of alpha-beta-and gamma-divergences: Flexible and robust measures of similarities. *Entropy* **2010**, *12*, 1532–1568. [CrossRef]
10. Amari, S.i. α-Divergence is Unique, belonging to Both f-divergence and Bregman Divergence Classes. *IEEE Trans. Inf. Theory* **2009**, *55*, 4925–4931. [CrossRef]
11. Zhang, J. Divergence function, duality, and convex analysis. *Neural Comput.* **2004**, *16*, 159–195. [CrossRef]
12. Hero, A.O.; Ma, B.; Michel, O.; Gorman, J. *Alpha-Divergence for Classification, Indexing and Retrieval*; Technical Report CSPL-328; Communication and Signal Processing Laboratory, University of Michigan: Ann Arbor, MI, USA, 2001.
13. Dikmen, O.; Yang, Z.; Oja, E. Learning the information divergence. *IEEE Trans. Pattern Anal. Mach. Intell.* **2014**, *37*, 1442–1454. [CrossRef]
14. Liu, W.; Yuan, K.; Ye, D. On α-divergence based nonnegative matrix factorization for clustering cancer gene expression data. *Artif. Intell. Med.* **2008**, *44*, 1–5. [CrossRef]
15. Hellinger, E. Neue Begründung der Theorie Quadratischer Formen von Unendlichvielen Veränderlichen. *J. Für Die Reine Und Angew. Math.* **1909**, *1909*, 210–271. [CrossRef]
16. Ali, S.M.; Silvey, S.D. A general class of coefficients of divergence of one distribution from another. *J. R. Stat. Soc. Ser. B* **1966**, *28*, 131–142. [CrossRef]
17. Csiszár, I. Information-type measures of difference of probability distributions and indirect observation. *Stud. Sci. Math. Hung.* **1967**, *2*, 229–318.
18. Qiao, Y.; Minematsu, N. A study on invariance of f-divergence and its application to speech recognition. *IEEE Trans. Signal Process.* **2010**, *58*, 3884–3890. [CrossRef]
19. Li, W. Transport information Bregman divergences. *Inf. Geom.* **2021**, *4*, 435–470. [CrossRef]
20. Li, W. Transport information Hessian distances. In Proceedings of the International Conference on Geometric Science of Information (GSI), Paris, France, 21–23 July 2021; Springer: Berlin/Heidelberg, Germany, 2021; pp. 808–817.
21. Li, W. Transport information geometry: Riemannian calculus on probability simplex. *Inf. Geom.* **2022**, *5*, 161–207. [CrossRef]
22. Amari, S.i. Integration of stochastic models by minimizing α-divergence. *Neural Comput.* **2007**, *19*, 2780–2796. [CrossRef] [PubMed]
23. Cichocki, A.; Lee, H.; Kim, Y.D.; Choi, S. Non-negative matrix factorization with α-divergence. *Pattern Recognit. Lett.* **2008**, *29*, 1433–1440. [CrossRef]
24. Wada, J.; Kamahara, Y. Studying malapportionment using α-divergence. *Math. Soc. Sci.* **2018**, *93*, 77–89. [CrossRef]
25. Maruyama, Y.; Matsuda, T.; Ohnishi, T. Harmonic Bayesian prediction under α-divergence. *IEEE Trans. Inf. Theory* **2019**, *65*, 5352–5366. [CrossRef]
26. Iqbal, A.; Seghouane, A.K. An α-Divergence-Based Approach for Robust Dictionary Learning. *IEEE Trans. Image Process.* **2019**, *28*, 5729–5739. [CrossRef]
27. Ahrari, V.; Habibirad, A.; Baratpour, S. Exponentiality test based on alpha-divergence and gamma-divergence. *Commun. Stat.-Simul. Comput.* **2019**, *48*, 1138–1152. [CrossRef]
28. Sarmiento, A.; Fondón, I.; Durán-Díaz, I.; Cruces, S. Centroid-based clustering with $\alpha\beta$-divergences. *Entropy* **2019**, *21*, 196. [CrossRef]
29. Niculescu, C.P.; Persson, L.E. *Convex Functions and Their Applications: A Contemporary Approach*, 1st ed.; Springer Science & Business Media: Berlin/Heidelberg, Germany, 2006.
30. Kolmogorov, A.N. Sur la notion de moyenne. *Acad. Naz. Lincei Mem. Cl. Sci. His. Mat. Natur. Sez.* **1930**, *12*, 388–391.
31. Gibbs, A.L.; Su, F.E. On choosing and bounding probability metrics. *Int. Stat. Rev.* **2002**, *70*, 419–435. [CrossRef]
32. Rachev, S.T.; Klebanov, L.B.; Stoyanov, S.V.; Fabozzi, F. *The Methods of Distances in the Theory of Probability and Statistics*; Springer: Berlin/Heidelberg, Germany, 2013; Volume 10.
33. Vemuri, B.C.; Liu, M.; Amari, S.I.; Nielsen, F. Total Bregman divergence and its applications to DTI analysis. *IEEE Trans. Med Imaging* **2010**, *30*, 475–483. [CrossRef]
34. Arthur, D.; Vassilvitskii, S. k-means++: The advantages of careful seeding. In Proceedings of the SODA '07: Proceedings of the Eighteenth Annual ACM-SIAM Symposium on Discrete Algorithms, New Orleans, LA, USA, 7–9 January 2007; Society for Industrial and Applied Mathematics: Philadelphia, PA, USA, 2007; pp. 1027–1035.
35. Bullen, P.S.; Mitrinovic, D.S.; Vasic, M. *Means and Their Inequalities*; Springer Science & Business Media: Berlin/Heidelberg, Germany, 2013; Volume 31.
36. Toader, G.; Costin, I. *Means in Mathematical Analysis: Bivariate Means*; Academic Press: Cambridge, MA, USA, 2017.
37. Cauchy, A.L.B. *Cours d'analyse de l'École Royale Polytechnique*; Debure frères: Paris, France 1821.
38. Chisini, O. Sul concetto di media. *Period. Di Mat.* **1929**, *4*, 106–116.
39. Bhattacharyya, A. On a measure of divergence between two statistical populations defined by their probability distributions. *Bull. Calcutta Math. Soc.* **1943**, *35*, 99–109.

40. Nielsen, F.; Boltz, S. The Burbea-Rao and Bhattacharyya centroids. *IEEE Trans. Inf. Theory* **2011**, *57*, 5455–5466. [CrossRef]
41. Nielsen, F. Generalized Bhattacharyya and Chernoff upper bounds on Bayes error using quasi-arithmetic means. *Pattern Recognit. Lett.* **2014**, *42*, 25–34. [CrossRef]
42. Nagumo, M. Über eine klasse der mittelwerte. *Jpn. J. Math. Trans. Abstr.* **1930**, *7*, 71–79. [CrossRef]
43. De Finetti, B. Sul concetto di media. *Ist. Ital. Degli Attuari* **1931**, *3*, 369–396.
44. Hardy, G.; Littlewood, J.; Pólya, G. *Inequalities*; Cambridge Mathematical Library, Cambridge University Press: Cambridge, UK, 1988.
45. Rényi, A. On measures of entropy and information. In Proceedings of the Fourth Berkeley Symposium on Mathematical Statistics and Probability, Berkeley, CA, USA, 20 June–30 July 1960; The Regents of the University of California: Oakland, CA, USA, 1961; Volume 1: Contributions to the Theory of Statistics; .
46. Holder, O.L. Über einen Mittelwertssatz. *Nachr. Akad. Wiss. Gottingen Math.-Phys. Kl.* **1889**, *44*, 38–47.
47. Bhatia, R. The Riemannian mean of positive matrices. In *Matrix Information Geometry*; Springer: Berlin/Heidelberg, Germany, 2013; pp. 35–51.
48. Akaoka, Y.; Okamura, K.; Otobe, Y. Bahadur efficiency of the maximum likelihood estimator and one-step estimator for quasi-arithmetic means of the Cauchy distribution. *Ann. Inst. Stat. Math.* **2022**, *74*, 1–29. [CrossRef]
49. Kim, S. The quasi-arithmetic means and Cartan barycenters of compactly supported measures. *Forum Math. Gruyter* **2018**, *30*, 753–765. [CrossRef]
50. Carlson, B.C. The logarithmic mean. *Am. Math. Mon.* **1972**, *79*, 615–618. [CrossRef]
51. Stolarsky, K.B. Generalizations of the logarithmic mean. *Math. Mag.* **1975**, *48*, 87–92. [CrossRef]
52. Jarczyk, J. When Lagrangean and quasi-arithmetic means coincide. *J. Inequal. Pure Appl. Math.* **2007**, *8*, 71.
53. Páles, Z.; Zakaria, A. On the Equality of Bajraktarević Means to Quasi-Arithmetic Means. *Results Math.* **2020**, *75*, 19. [CrossRef]
54. Maksa, G.; Páles, Z. Remarks on the comparison of weighted quasi-arithmetic means. *Colloq. Math.* **2010**, *120*, 77–84. [CrossRef]
55. Zhang, J. Nonparametric information geometry: From divergence function to referential-representational biduality on statistical manifolds. *Entropy* **2013**, *15*, 5384–5418. [CrossRef]
56. Nielsen, F.; Nock, R. Generalizing Skew Jensen Divergences and Bregman Divergences with Comparative Convexity. *IEEE Signal Process. Lett.* **2017**, *24*, 1123–1127. [CrossRef]
57. Kuczma, M. *An Introduction to the Theory of Functional Equations and Inequalities: Cauchy's Equation and Jensen's Inequality*; Springer Science & Business Media: Berlin/Heidelberg, Germany, 2009.
58. Nock, R.; Nielsen, F.; Amari, S.i. On conformal divergences and their population minimizers. *IEEE Trans. Inf. Theory* **2015**, *62*, 527–538. [CrossRef]
59. Ohara, A. Conformal flattening for deformed information geometries on the probability simplex. *Entropy* **2018**, *20*, 186. [CrossRef]
60. Ohara, A. Conformal Flattening on the Probability Simplex and Its Applications to Voronoi Partitions and Centroids. In *Geometric Structures of Information*; Springer: Berlin/Heidelberg, Germany, 2019; pp. 51–68.
61. Bregman, L.M. The relaxation method of finding the common point of convex sets and its application to the solution of problems in convex programming. *USSR Comput. Math. Math. Phys.* **1967**, *7*, 200–217. [CrossRef]
62. Zhang, J. On monotone embedding in information geometry. *Entropy* **2015**, *17*, 4485–4499. [CrossRef]
63. Nielsen, F.; Nock, R. The dual Voronoi diagrams with respect to representational Bregman divergences. In Proceedings of the Sixth International Symposium on Voronoi Diagrams (ISVD), Copenhagen, Denmark, 23–26 June 2009; IEEE: Piscataway, NJ, USA, 2009; pp. 71–78.
64. Itakura, F.; Saito, S. Analysis synthesis telephony based on the maximum likelihood method. In Proceedings of the 6th International Congress on Acoustics, Tokyo, Japan, 21–28 August 1968; pp. 280–292.
65. Okamoto, I.; Amari, S.I.; Takeuchi, K. Asymptotic theory of sequential estimation: Differential geometrical approach. *Ann. Stat.* **1991**, *19*, 961–981. [CrossRef]
66. Ohara, A.; Matsuzoe, H.; Amari, S.I. Conformal geometry of escort probability and its applications. *Mod. Phys. Lett. B* **2012**, *26*, 1250063. [CrossRef]
67. Kurose, T. On the divergences of 1-conformally flat statistical manifolds. *Tohoku Math. J. Second Ser.* **1994**, *46*, 427–433. [CrossRef]
68. Pal, S.; Wong, T.K.L. The geometry of relative arbitrage. *Math. Financ. Econ.* **2016**, *10*, 263–293. [CrossRef]
69. Lloyd, S. Least squares quantization in PCM. *IEEE Trans. Inf. Theory* **1982**, *28*, 129–137. [CrossRef]
70. Mahajan, M.; Nimbhorkar, P.; Varadarajan, K. The planar k-means problem is NP-hard. *Theor. Comput. Sci.* **2012**, *442*, 13–21. [CrossRef]
71. Wang, H.; Song, M. Ckmeans.1d.dp: Optimal k-means clustering in one dimension by dynamic programming. *R J.* **2011**, *3*, 29. [CrossRef]
72. Banerjee, A.; Merugu, S.; Dhillon, I.S.; Ghosh, J.; Lafferty, J. Clustering with Bregman divergences. *J. Mach. Learn. Res.* **2005**, *6*, 1705–1749.
73. Nielsen, F.; Nock, R. Sided and symmetrized Bregman centroids. *IEEE Trans. Inf. Theory* **2009**, *55*, 2882–2904. [CrossRef]
74. Ronchetti, E.M.; Huber, P.J. *Robust Statistics*; John Wiley & Sons: Hoboken, NJ, USA, 2009.
75. Nielsen, F.; Nock, R. Total Jensen divergences: Definition, properties and clustering. In Proceedings of the 2015 IEEE International Conference on Acoustics, Speech and Signal Processing (ICASSP), South Brisbane, Australia, 19–24 April 2015; IEEE: Piscataway, NJ, USA, 2015; pp. 2016–2020.

76. Eguchi, S.; Komori, O. *Minimum Divergence Methods in Statistical Machine Learning*; Springer: Berlin/Heidelberg, Germany, 2022.
77. Kailath, T. The divergence and Bhattacharyya distance measures in signal selection. *IEEE Trans. Commun. Technol.* **1967**, *15*, 52–60. [CrossRef]

Article

Human Body Shapes Anomaly Detection and Classification Using Persistent Homology

Steve de Rose [1,2], Philippe Meyer [1,*] and Frédéric Bertrand [1]

1 Computer Science and Digital Society Laboratory (LIST3N), Université de Technologie de Troyes, 10004 Troyes Cedex, France; steve.de-rose@etu.unistra.fr (S.d.R.); frederic.bertrand1@utt.fr (F.B.)
2 Institut de Recherche Mathématique Avancée (IRMA), CNRS UMR 7501, Université de Strasbourg, 67404 Strasbourg Cedex, France
* Correspondence: philippe.meyer@utt.fr

Abstract: Accurate sizing systems of a population permit the minimization of the production costs of the textile apparel industry and allow firms to satisfy their customers. Hence, information about human body shapes needs to be extracted in order to examine, compare and classify human morphologies. In this paper, we use topological data analysis to study human body shapes. Persistence theory applied to anthropometric point clouds together with clustering algorithms show that relevant information about shapes is extracted by persistent homology. In particular, the homologies of human body points have interesting interpretations in terms of human anatomy. In the first place, anomalies of scans are detected using complete-linkage hierarchical clusterings. Then, a discrimination index shows which type of clustering separates gender accurately and if it is worth restricting to body trunks or not. Finally, Ward-linkage hierarchical clusterings with Davies–Bouldin, Dunn and Silhouette indices are used to define eight male morphotypes and seven female morphotypes, which are different in terms of weight classes and ratios between bust, waist and hip circumferences. The techniques used in this work permit us to classify human bodies and detect scan anomalies directly on the full human body point clouds rather than the usual methods involving the extraction of body measurements from individuals or their scans.

Keywords: topological data analysis; machine learning; persistent homology; clustering; anomaly detection; morphotype

Citation: de Rose, S.; Meyer, P.; Bertrand, F. Human Body Shapes Anomaly Detection and Classification Using Persistent Homology. *Algorithms* **2023**, *16*, 161. https://doi.org/10.3390/a16030161

Academic Editors: Chih-Lung Lin, Bor-Jiunn Hwang, Shaou-Gang Miaou and Yuan-Kai Wang

Received: 7 February 2023
Revised: 8 March 2023
Accepted: 10 March 2023
Published: 15 March 2023

Copyright: © 2023 by the authors. Licensee MDPI, Basel, Switzerland. This article is an open access article distributed under the terms and conditions of the Creative Commons Attribution (CC BY) license (https:// creativecommons.org/licenses/by/ 4.0/).

1. Introduction

The separation of human bodies into groups of morphologies is a common issue for garment industries. Rather than targeting a single standard body shape, the discrimination of morphologies helps to improve sizing systems and can reduce production costs for apparel manufacturing. Among the classifications already established, there is one from [1] particularly used by industries, where the authors obtained nine types of female body shapes such as triangle, inverted triangle, hourglass, oval, etc. This is the first work where mathematical criteria, together with the help of experts, have been used to define these groups.

In terms of data science, we can approach this problem by clustering algorithms. To this end, different types of data can be extracted from a body such as measurements or anthropometric point clouds. Body measurements can be directly represented in a Euclidean space to use methods from data analysis. In [2], principal component and K-means cluster analyses are performed on measurements and key body locations, and three female lower body shape groups are obtained. This representation in a vector space to perform the clustering is straightforward but has disadvantages. For example, it is not clear that it is appropriate to compare with Euclidean metric measurements of different types such as body lengths, circumferences or individual weight. On the other hand, this requires the choice of the set of measurements extracted from the body, and key morphological characteristics may be omitted. The use of 3D representations of the bodies is suitable for these issues but becomes difficult to implement since we need a way to extract information from anthropometric point clouds and compare them. For example, in [3], the authors use control points and correlation strength principal component analysis of trunks. Reinterpreting these components by averaged shape figures and combining factor loading maps, five female trunk shape groups are defined by a Ward-linkage hierarchical clustering. Different methods from data science have been used to classify human body shapes; see for example [4–7].

Topological data analysis [8,9] is a powerful tool to study and understand the shape of data, and thus it naturally applies in this context. In particular, persistent homology [10,11] can be used to extract relevant topological information from data and point clouds. These extracted features are encoded by diagrams and have stability properties relative to specific distances [12,13]. Several applications of this theory have been established in different contexts such as time-series data analysis [14], object recognition [15], complex network analysis [16], molecular biology data exploration [17], biomedicine [18], geographical information science [19] and environmental science [20]. Feature extraction for classification is an active research topic in pattern recognition and machine learning; see for example [21,22] or [23].

In this work, we use persistent theory applied on human point clouds in order to perform the following:

- Extract information from human bodies with interpretation in terms of human anatomy;
- Detect scans anomalies;
- Identify and separate human point clouds by gender;
- Classify male and female morphotypes.

More precisely, we compute the persistence diagrams, Wasserstein distance and associated silhouettes on the human point clouds of the CAESAR database [24]. Using graph theory, among other things, approaches by homological degree allow us to interpret persistent homologies and identify them to body areas and limbs. To define morphotypes independently of individuals' height, we normalize the point clouds using three-dimensional homotheties. Then, we show that anomalies of scans are naturally isolated clusters when performing complete-linkage hierarchical clustering on the persistence diagrams of the point clouds using the Wasserstein distance. Then, a gender discrimination index is defined to study which hierarchical clustering linkage is interesting to separate males and females accurately. We compare the performance of these clustering algorithms on persistence diagrams, on silhouettes, and whether point clouds are restricted to trunks or not. Finally, Ward-linkage hierarchical clusterings on the silhouettes of the persistence diagrams of the point clouds, together with a mix of different clustering criteria such as Davies–Bouldin, Dunn and Silhouette indices are used to obtain eight male morphotypes and seven female morphotypes. Then, we study the properties of these clusters, and their medoids are computed and considered as representatives of the groups.

The paper is organized as follows. In Section 2, we introduce the tools of persistence theory that we use. In Section 3, we detect scan anomalies. In Section 4, we study which type of clustering accurately separates males and females. Finally, we classify morphotypes in Section 5.

2. Methodology

2.1. Dataset

The CAESAR (Civilian American and European Surface Anthropometry Resource) 3D Anthropometric Database is composed of 3D body scans of thousands of men and women aged from 18 to 65 and originated from various NATO countries: the United States of America, Canada, the Netherlands and Italy.

In this paper, we are using the dataset of [24], which is derived from the CAESAR dataset and is composed of 1517 male and 1531 female meshes, registered as OBJ files. Each mesh has 12,500 vertices (Figure 1a) and 25,000 faces (Figure 1b), and we extract and consider only the underlying point clouds of all the meshes. In the figures, the meshes and point clouds are presented headless for confidentiality. The individuals are numbered discontinuously from Spring0001 to Spring4800, and for convenience we refer to SpringXXXX by SXXXX.

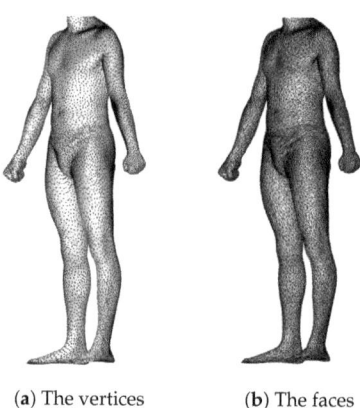

(a) The vertices   (b) The faces

Figure 1. The mesh of the individual S0013.

2.2. Persistence Diagrams, Landscapes, Silhouettes and Distances

Persistent homology is a tool used to efficiently compute and encode the multidimensional homological features of topological spaces associated to a dataset. To compute these homological invariants, we have to build topological structures on the data such as filtered simplicial complexes.

A simplex is a notion generalizing points, line segments, triangles and tetrahedrons to any dimension and composed of faces that are also simplices of lower dimension. A simplicial complex K is a collection of simplices satisfying two properties: each face of a simplex of K is in K and the non-empty intersection of two simplices of K is a face of both of them. Given a body point cloud X in $\mathbb{R}^3$, several types of simplicial complexes can be constructed on X, such as the Vietoris–Rips and the Čech complexes. We center three-dimensional balls of radius ϵ on each data point, and we vary ϵ from 0 to $+\infty$. The data points are considered as 0-simplices, and when $n+1$ balls intersect, we add an n-dimensional face between them. The result is called a Čech complex. For each fixed ϵ, we count the homological features of the associated topological space. Since the underlying vector space is of dimension 3, we have three types of homological classes to consider:

- H_0: The connected components;
- H_1: The non-homotopic loops;
- H_2: The two-dimensional voids.

Thus, we represent each homological feature by a point in $\mathbb{R}^2$, where its abscissa is the birth time of the feature and its ordinate is the death time. The set of points obtained in this way is the persistence diagram of X. The persistence barcode represents each homology class with a bar defined by its birth time, when the topological feature appears, and a death time, when the topological feature disappears. In order not to have too many points due to the creation and death of small homological features, a minimal persistence is fixed.

For example, in Figure 2, the persistence barcode and diagram of the individual S0013 (Figure 1) are given.

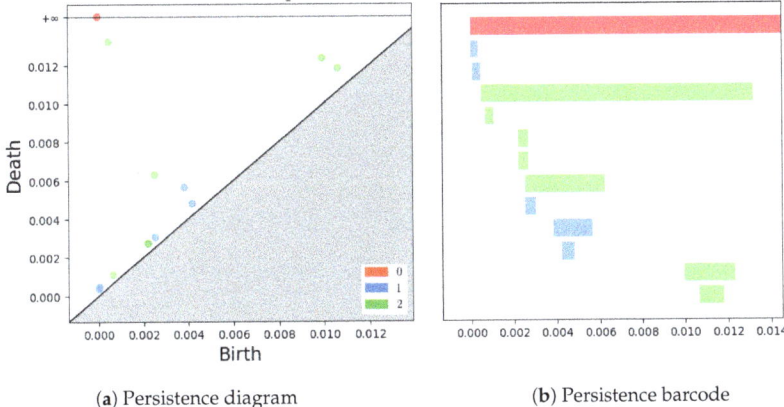

(a) Persistence diagram

(b) Persistence barcode

Figure 2. The persistence diagram and barcode of S0013.

It is possible to compare persistence diagrams using the Wasserstein distance. Let D and D' be persistence diagrams. A perfect matching between D and D' is a subset $\phi \subseteq D \times D'$ such that every point of D and D' is exactly one time in ϕ, completing with the diagonal if necessary in order to ignore cardinality mismatches. The (p,q)-Wasserstein distance between D and D' is defined by

$$W_{p,q}(D, D') = \inf_{\phi \in \Phi} \Big(\sum_{x \in D} ||x - \phi(x)||_q^p \Big)^{1/p}, \tag{1}$$

where $||x||_q$ is the q-norm of x defined by

$$||x||_q = \Big(\sum |x_i|^q \Big)^{1/q}. \tag{2}$$

We exclusively use the $(2,2)$-Wasserstein distance. For precise definitions and details, see [8,9].

Persistence landscapes are an encoding of persistence diagrams by series of piecewise continuous linear functions [25,26]; see Figure 3. This allows us to perform statistics on them, the absence of which was a disadvantage of persistence diagrams. In particular, it is possible to calculate unique averages of landscapes. While a persistence landscape has a corresponding persistence diagram, an average of persistence landscapes does not.

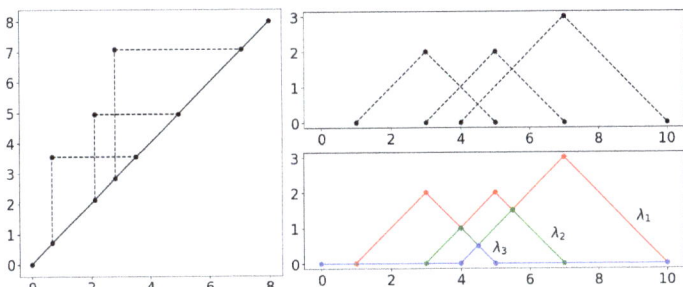

Figure 3. Visual explanation of persistence landscapes. The persistence diagram (**left**) is tilted so that the diagonal becomes the new horizontal axis (**top right**). The λ_i are the piecewise linear functions (**bottom right**).

A persistence silhouette is computed by taking a weighted average of the collection of 1D-piecewise-linear functions given by the persistence landscapes and then by evenly sampling this average on a given range. Finally, the corresponding vector of samples is returned; see Figure 4. For the implementation of clustering, we choose to make a vector consisting of 25 points of the silhouette of H_0 homologies, 250 points equidistant from the silhouette of H_1 homologies and 250 points equidistant from the silhouette of H_2 homologies for each persistence diagram. The points are the values of the silhouette equally spaced. Hence, each individual is represented by a vector in a real vector space of dimension 525 together with the Euclidean distance.

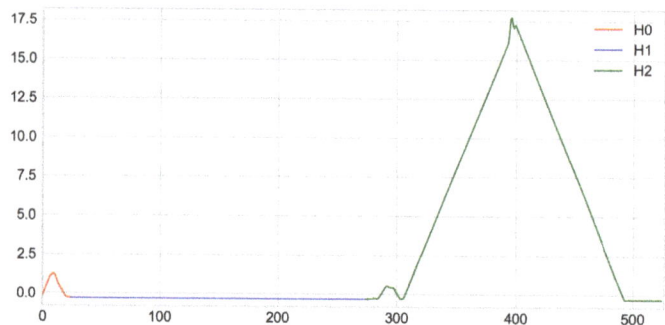

Figure 4. Representation of a vector obtained by persistence silhouette.

2.3. Interpretation of Persistent Homology

The persistence diagram of a body point cloud is composed of three types of homologies (see Figure 2). Since the points are distant from each other at an equivalent distance, all the balls are rapidly connected, thus giving a single connected component. Several H_1 and H_2 homologies representing the internal body cavities appear and disappear when the radius ϵ of the balls varies to $+\infty$. We now explain our approach to interpret and identify these homological features in terms of human anatomy. Since displaying the homologies in their entirety is too costly, we thought of other approaches for each degree.

For each homology, we know the radii of the balls at their birth and death. A simplex tree represents abstract simplicial complexes of any dimension. All faces of the simplicial complex are explicitly stored in a tree whose nodes are in bijection with the faces of the complex. This data structure allows us to efficiently implement a large range of basic operations on simplicial complexes. Using the simplex tree of a set of points, we know the values of the radii when pairs of points, triangles and tetrahedra are covered. The approach is slightly different depending on the dimension:

- Dimension 0: All H_0 homologies are born when the radius of the balls is zero. For each homology H_0, we choose to display the second point of the pair covered at the birth of the homology as its representative.
- Dimension 1: First, we make an undirected graph containing all the points of a set, where each time a pair of points is covered, as the radius of the balls increases, we connect these points by an edge with a weight equal to the radius of the balls. At the birth of a homology H_1, before adding the edge to our graph, we compute the shortest path connecting these two points, which we display by closing it with the segment connecting these points. The lace displayed is a likely representative of this homology. At the death of this homology, we recover the information of the triangle covered by the balls, and we add it to the display to give a general idea of the evolution of our homology.
- Dimension 2: For each homology H_2, we simply display the triangle covered at its birth and the tetrahedron covered at its death.

For example, in the persistence diagram of the individual S0013 given in Figure 2, there are 13 different homologies numbered in the persistence barcode from 0 to 12. With this approach, we display each homology in Figure 5 and we can interpret them as follows:

- n°0: H_2 corresponding to the left part of the torso,
- n°1: H_2 corresponding to the right part of the torso,
- n°2: H_1 corresponding to a loop between legs at foot level,
- n°3: H_1 corresponding to a loop between legs from ankles to calves,
- n°4: H_1 corresponding to a loop between legs from knees to calves,
- n°5: H_2 corresponding to the head,
- n°6: H_2 corresponding to the right calf,
- n°7: H_2 corresponding to the left calf,
- n°8: H_2 corresponding to the right foot,
- n°9: H_2 corresponding to the whole body,
- n°10: H_1 corresponding to a loop around the right foot,
- n°11: H_1 corresponding to a loop around the left foot,
- n°12: H_0 of all the connected balls.

We remark that the arms and the left foot do not appear on the diagram. This is caused by the minimal persistence and the facts that the arms are too thin and that the scan of the left foot is more flat and deformed compared to the right one. Homology n°9 is particularly distinguished, and we call it the principal H_2-homology. It corresponds to the aggregation of the parts and limbs of the body, thus forming the inner cavity of the body point cloud.

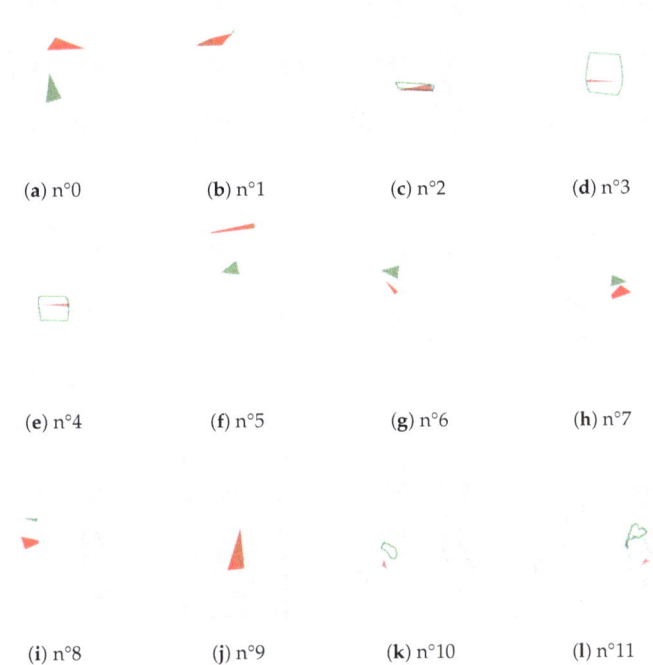

Figure 5. All non H_0 homologies of the persistence diagram of S0013.

2.4. Normalization of Point Clouds by Homothety

We want morphotypes to be independent of the size of the individuals in order to propose a sizing system associated to each morphotype. For this purpose, we apply a homothety on each point cloud so that each individual is the same height: 1 m 70 cm. This affects the distances between them and individuals with similar morphology, but different heights become closer (Figure 6).

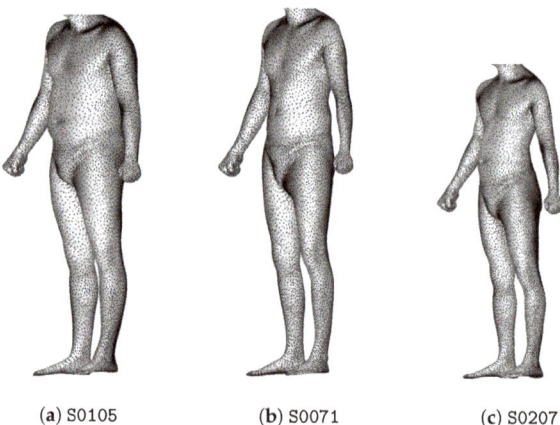

(a) S0105   (b) S0071   (c) S0207

Figure 6. Individuals (**a**–**c**) are 1.89 m, 1.93 m and 1.65 m tall, respectively. Among them, the couple (**a**,**b**) is the closest before normalization, and the couple (**b**,**c**) is the closest after normalization.

3. Anomaly Detection

Among the data, there are anomalies of scans. We have found five anomalies for men and four for women. It turns out that they are encoded and detected by the persistence diagrams, Wasserstein distance and clustering algorithms. More precisely, we perform complete-linkage hierarchical clusterings on the persistence diagrams of the point clouds together with the Wasserstein distance (with $p = q = 2$), separately for men and women. Analyzing corresponding truncated dendrograms, we remark that anomalies are very often isolated individuals agglomerating late. To find the best truncation of the dendrogram, we use as criteria the mean between the percentage of isolated individuals that are anomalies and the percentage of anomalies isolated in this way.

For men, the best truncation range is $[21, 46]$, where the criteria show that 90%: 100% of isolated individuals are anomalies and 80% of anomalies are detected. Figure 7 shows the dendrogram for male point clouds truncated at 21 clusters, where the 4 isolated individuals are anomalies as shown in Figure 8.

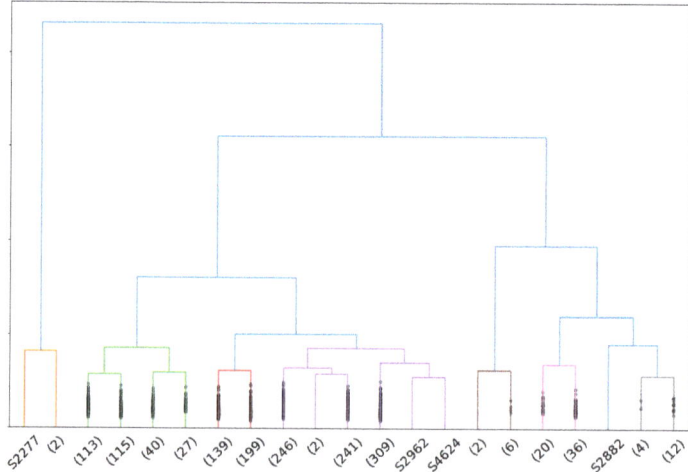

Figure 7. Dendrogram associated to a complete-linkage hierarchical clustering of the persistence diagrams of male point clouds with the Wasserstein distance. Clusters composed of one individual are presented without parentheses.

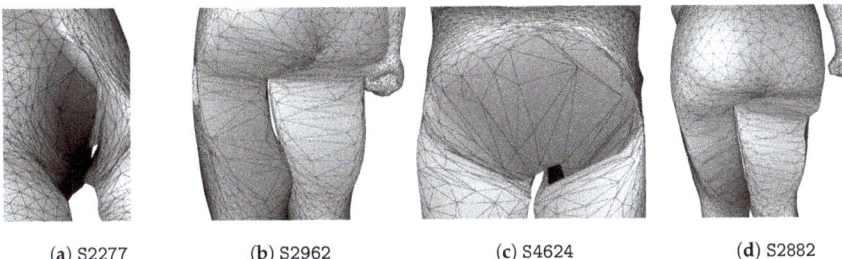

(a) S2277 (b) S2962 (c) S4624 (d) S2882

Figure 8. Anomalies of men scans detected by complete-linkage hierarchical clustering of persistence diagrams.

To illustrate that anomalies are detected by persistence, we analyze the persistence diagram of Figure 9, which corresponds to the individual S2962.

Its three H_2 homologies $n°3, 5, 6$ are particularly distinguished and can be seen at birth and death in Figure 10. Homologies H_2 numbers 3 and 5 correspond to the right and left leg, respectively, while the number 6 corresponds to the torso and is the principal H_2-homology.

For normal scans, the principal H_2-homology also aggregates legs. Because of the misplaced points and the holes on the point cloud, leg homologies are separated from the principal H_2-homology which starts later than in the usual case.

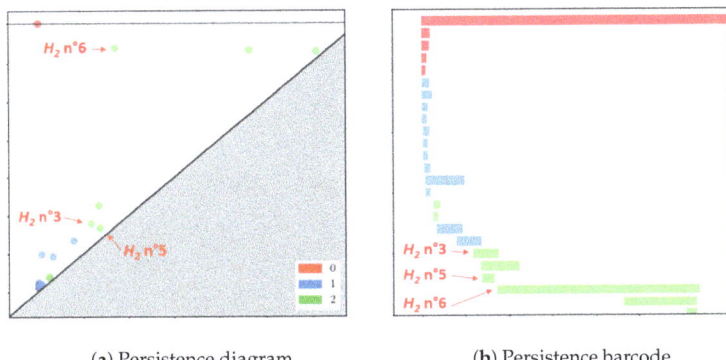

(a) Persistence diagram (b) Persistence barcode

Figure 9. Persistence diagram and barcode of the anomaly of scan S2962. Three particular homologies reflecting the anomaly are highlighted.

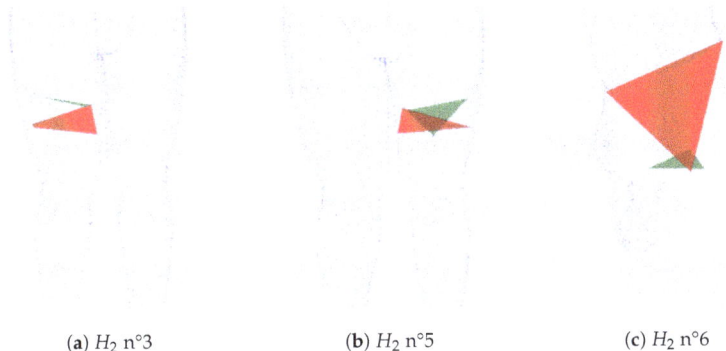

(a) H_2 n°3 (b) H_2 n°5 (c) H_2 n°6

Figure 10. Three abnormal homologies at birth and death of the defective scan S2962.

For women, the best truncation range is [23, 37], where the criteria show that 87.5%:100% of isolated individuals are anomalies and 75% of anomalies are detected. Figure 11 shows the dendrogram for female point clouds truncated at 23 clusters, where the 3 isolated individuals are anomalies as shown in Figure 12.

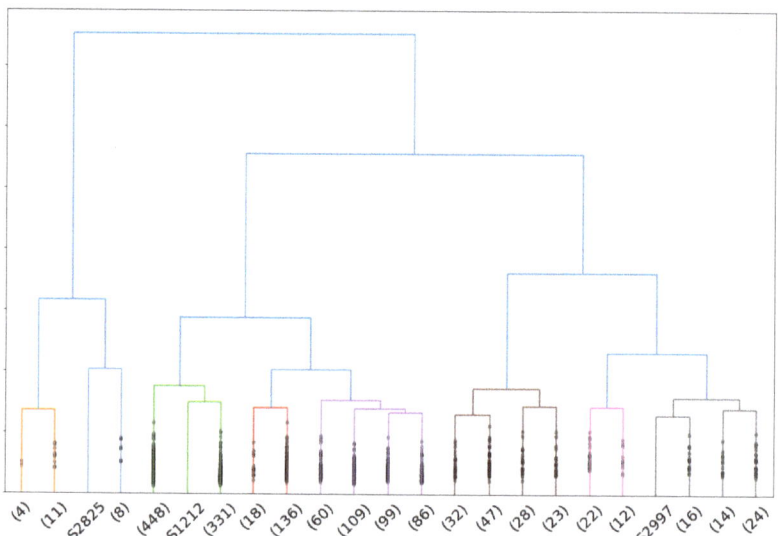

Figure 11. Dendrogram associated to a complete-linkage hierarchical clustering of the persistence diagrams of female point clouds with the Wasserstein distance.

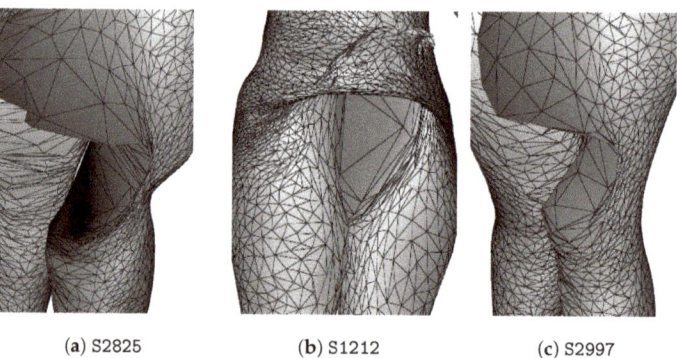

(a) S2825   (b) S1212   (c) S2997

Figure 12. Anomalies of female scans detected by complete-linkage hierarchical clustering of persistence diagrams.

4. Gender Discrimination Index

In this section, we analyze if clustering algorithms on persistence diagrams and silhouettes give groups separating men from women scans by changing the number of clusters. To this end, we use persistence diagrams or silhouettes, restricted to trunks of point clouds or not.

Let $P_m(C)$ and $P_f(C)$ be respectively the proportions of men and women in a cluster C. We have

$$P_m(C) = \frac{n_m(C)}{s(C)}, \qquad P_f(C) = \frac{n_f(C)}{s(C)} \qquad (3)$$

where $n_m(C)$ is the number of men in C, $n_f(C)$ is the number of women in C and $s(C)$ is the size of C. To measure the quality of a clustering $\mathcal{C}$ of a set of mixed male and female diagrams or silhouettes D_{MF}, we introduce a gender discrimination index (GDI) defined by

$$GDI(\mathcal{C}) = \frac{2}{s(D_{MF})} \sum_{k=1}^{K} s(C_k) \left| P_m(C_k) - \frac{1}{2} \right| \qquad (4)$$

where K is the number of clusters of $\mathcal{C}$, C_k are the clusters of $\mathcal{C}$ and $s(D_{MF})$ is the number of elements in D_{MF}. Thus, the better the clustering $\mathcal{C}$ separates men from women, the closer $GDI(\mathcal{C})$ is to 1, and the worse it is, the closer $GDI(\mathcal{C})$ is to 0. We can consider that a clustering is satisfactory to separate men from women if its GDI is greater or equal to $\frac{1}{2}$.

4.1. Evolution of the GDI Score as a Function of the Number of Clusters

In this section, we observe the ability of different clustering methods to separate male from female persistence diagrams or silhouettes.

We use a matrix of Wasserstein distances between diagrams to perform hierarchical clustering with complete and Ward's linkage methods [27] as well as K-Medoids clustering with the PAM (Partitioning Around Medoids) algorithm [28]. The notion of a barycenter between persistence diagrams is delicate [29,30], but we can use the Ward-linkage method with the Lance–Williams algorithm [31].

As shown in Figure 13, hierarchical clustering with the complete-linkage method does not differentiate correctly between female and male scans. However, the K-Medoids clustering has a correct GDI score for more than 10 clusters and becomes good on some occasions for more than 13 clusters. The Ward-linkage hierarchical clustering has a correct GDI score for more than 12 clusters and becomes good for more than 19 clusters.

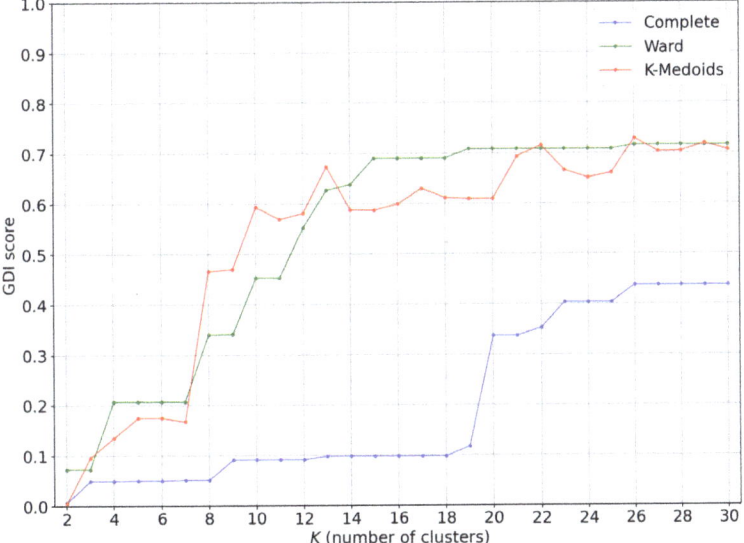

Figure 13. GDI score evolution of various clustering algorithms on the persistence diagrams with Wasserstein distance.

We now use vectors obtained from the silhouettes associated to the persistence diagrams of scans on which we perform a Ward-linkage hierarchical clustering as well as a K-Means clustering and a K-Medoids clustering with the PAM algorithm. This time, these three clustering algorithms give very good GDI scores; see Figure 14.

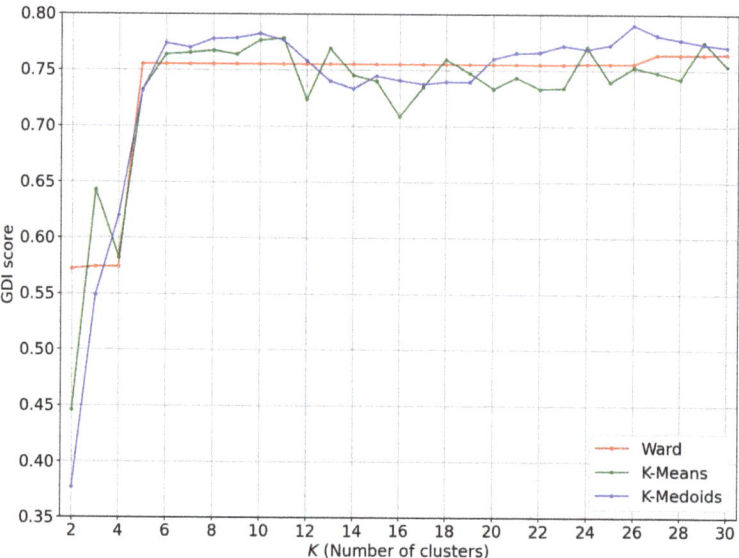

Figure 14. GDI score evolution of various clustering algorithms on the persistence silhouettes.

4.2. Restriction to Trunks

When constructing the silhouettes, we used a weighting that tended to favor the H_2 homologies corresponding to the trunks of the subjects, so the question then arises as to whether we would obtain better results by using only the points corresponding to the trunk of the body. To this end, we have developed an algorithm to isolate the points corresponding to the trunk of an individual which we now describe.

Let X be a normalized body point cloud at 1.70 m. We rotate and translate the scan such that the individual is standing along the height axis z and is at the minimal height of 0. Then, we isolate points located in the range $[66.5, 146.5]$ cm to exclude points corresponding to the legs and head. We compute the director and intercept coefficients of two linear equations delimiting the trunk, taking into account the mean width of the individual. More precisely, we compute the lines $x = a_1 z + b_1$ and $x = a_2 z + b_2$, which intersect at the height 107.5 cm. Projecting the points on the plane (x, z), we obtain a set of points X_1 located between the first line, its symmetric with respect to the axis $x = 0$ and below 107.5 cm and a set of points X_2 located between the second line, its symmetric with respect to the axis $x = 0$ and above 107.5 cm. The union X' of X_1 and X_2 is composed of points of the individual's trunk. In Figure 15, a body point cloud and the trunk point cloud isolated by this process are represented.

We now compare the clustering results using a Wasserstein distance matrix applied to the whole body and applied to the trunk.

From the curves in Figure 16 and the average GDI scores of Table 1, it appears that for clustering algorithms based on Wasserstein distances between persistence diagrams, it is not worth restricting these to trunk points.

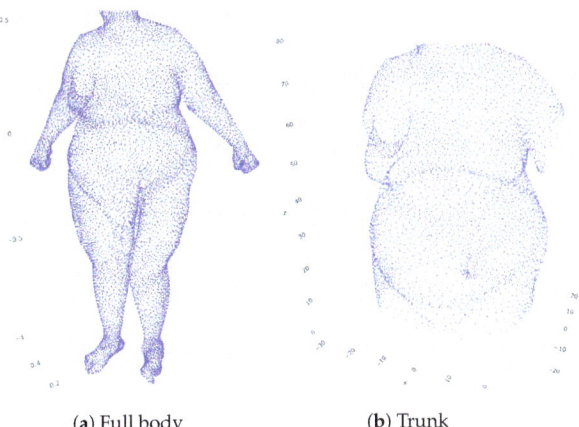

(a) Full body   (b) Trunk

Figure 15. An individual and its isolated trunk.

Table 1. Average GDI scores on the persistence diagrams of the whole body and the trunk with the Wasserstein distance.

	Complete	Ward	K-Medoids
Body	0.2	0.54	0.526
Trunk	0.21	0.553	0.582

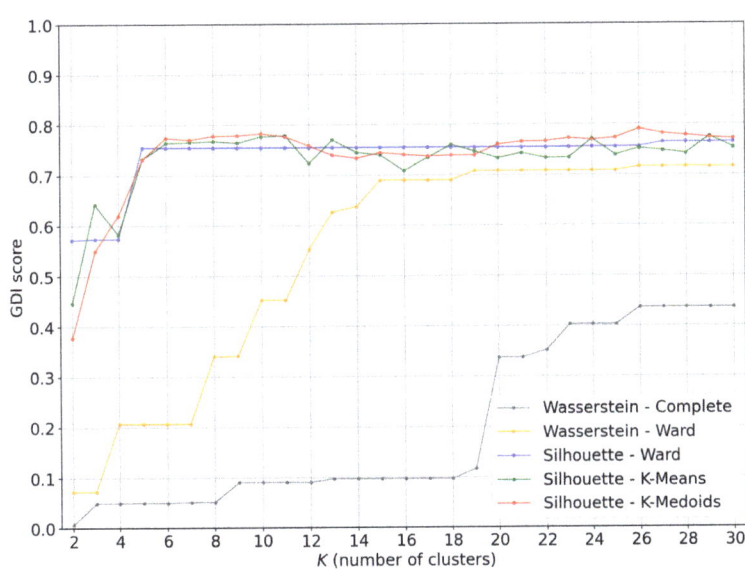

Figure 16. Comparison of GDI score on the persistence diagrams of the whole body and the trunk with the Wasserstein distance.

We now compare the results of clustering algorithms using the vectors obtained from the persistence silhouettes applied to the whole body and applied to the trunk.

From the curves of Figure 17 and the average GDI scores of Table 2, it appears that for clustering based on silhouette persistence vectors, it is worth restricting these to trunk points, particularly for K-Medoids clustering.

Table 2. Average GDI scores on the persistence silhouettes of the whole body and the trunk.

	Ward	K-Means	K-Medoids
Body	0.738	0.73	0.737
Trunk	0.765	0.767	0.827

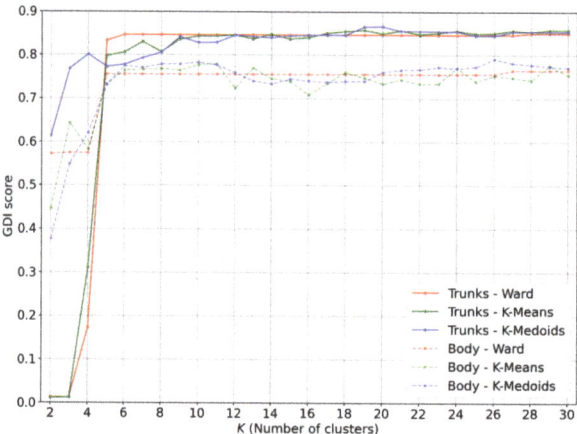

Figure 17. Comparison of GDI score on the persistence silhouettes of the whole body and the trunk.

5. Human Body Shapes Classification

5.1. Male Morphotypes

To define morphotypes of men's body shapes, we perform a Ward-linkage hierarchical clustering on silhouettes of the persistence diagrams of the men's point clouds together with the euclidean distance. The associated dendrogram is given in Figure 18.

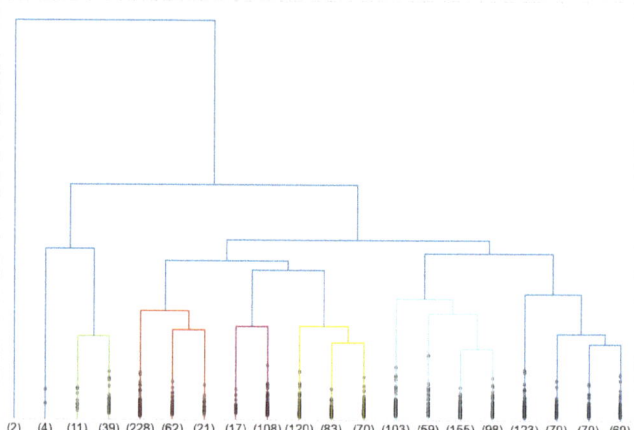

Figure 18. Dendrogram associated to a Ward-linkage hierarchical clustering of the silhouettes of the persistence diagrams of male point clouds.

To find a correct truncation of the dendrogram, we use the following clustering quality indices:
- The Elbow method;
- The Davies–Bouldin index [32];

- The Silhouette index [33];
- The Dunn index [34].

Since there is a continuity between human body shapes, there is no distinguished point common to all these indices. However, the Davies–Bouldin and Dunn indices both suggest to truncate at eight clusters. Information about size, mean distance of all pairs, diameter, mean distance to the mean and distance between the mean and the medoid of each cluster is given in Table 3.

Table 3. Clustering of male body shapes.

Cluster	C_1	C_2	C_3	C_4	C_5	C_6	C_7	C_8
Size	2	4	50	311	125	273	415	332
Proportion (in percent)	0.1	0.3	3	21	8	18	27	22
Mean distance	79.6	39.4	28.6	17.4	20.3	16.3	18.9	19.4
Diameter	79.6	53.9	62	49.6	59.2	54.6	67.1	69.4
Distance to the mean	39.8	24.4	19.6	12.1	14.2	11.5	13.1	13.6
Distance mean–medoid	39.8	20.1	9	3.4	5.9	6.9	5.1	6.1

The first cluster is only composed of two individuals who are extremely overweight, and their meshes are shown in Figure 19. The four men in the second cluster are also extremely overweight.

The medoid is the element minimizing the distance with other elements of the cluster. It can be considered as a representative, and we show in Figure 20 the medoids associated to every cluster, except for the first cluster.

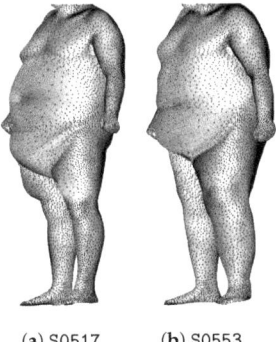

(a) S0517   (b) S0553

Figure 19. The two individuals of cluster C_1.

Since we do not have measurements associated with the individuals of the CAESAR database, in each group, we have to look at all the individuals in order to identify the predominant morphological features. It turns out that the clusters C_3 and C_7 are composed of overweight individuals of different categories, while the thinnest men are located in cluster C_6. It turns out that individuals of clusters C_4, C_5 and C_8 have a standard morphotype but that men of C_8 have a shorter torso than in C_4 and C_8 and that men of C_4 are more corpulent that in the two others.

5.2. Female Morphotypes

Similarly, to define morphotypes of women's body shapes, we perform a Ward-linkage hierarchical clustering on silhouettes of the persistence diagrams of the women's point clouds together with the euclidean distance. The associated dendrogram is given in Figure 21.

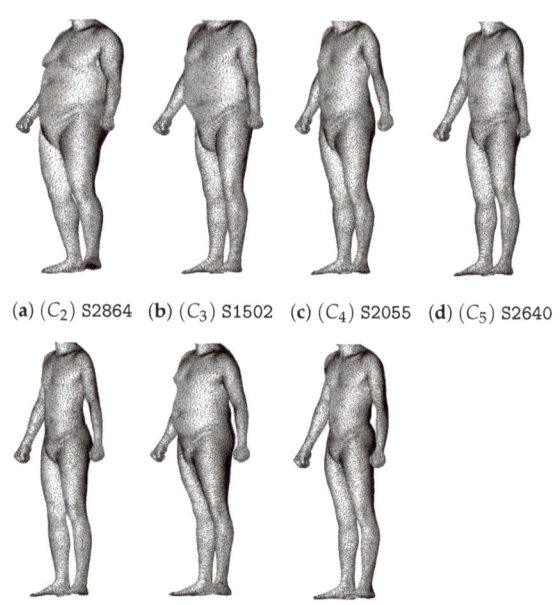

(a) (C_2) S2864 (b) (C_3) S1502 (c) (C_4) S2055 (d) (C_5) S2640

(e) (C_6) S4286 (f) (C_7) S2982 (g) (C_8) S4505

Figure 20. Medoids of clusters C_2 to C_8 of the Ward-linkage hierarchical clustering.

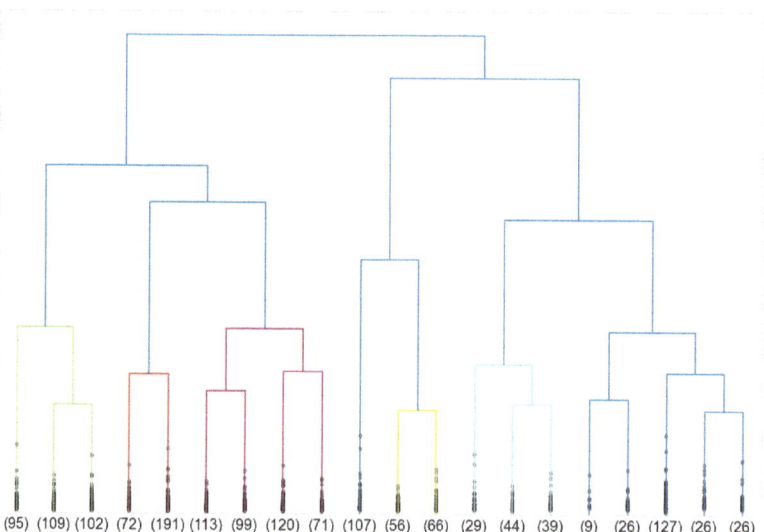

Figure 21. Dendrogram associated to a Ward-linkage hierarchical clustering of the silhouettes of the persistence diagrams of female point clouds.

This time, the Silhouette and Dunn indices suggest truncating at seven clusters. Information about size, mean distance of all pairs, diameter, mean distance to the mean and distance between the mean and the medoid of each cluster is given in Table 4. Remark that clusters of women are more compact than clusters of men since the mean distance of pairs and diameter are much smaller.

Table 4. Clustering of female body shapes.

Clusters	C_1	C_2	C_3	C_4	C_5	C_6	C_7
Size	306	263	403	107	122	112	214
Proportion (in percent)	20	17	27	7	8	7	14
Mean distance	14.2	12.7	14.1	14.3	14	19	21.3
Diameter	36.6	32	32.1	31.9	38.2	59.3	62.4
Distance to the mean	10.1	9	10.1	10.1	9.9	13.3	14.8
Distance mean–medoid	3.4	3.4	4.5	3.3	3.6	5.9	5.2

We show in Figure 22 the medoids associated to the seven clusters.

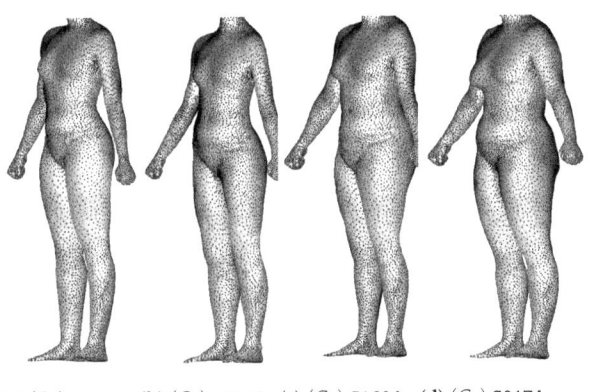

(**a**) (C_1) S0522   (**b**) (C_2) S2018   (**c**) (C_3) S1604   (**d**) (C_4) S0174

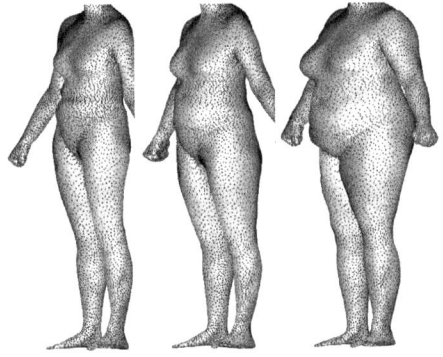

(**e**) (C_5) S1076   (**f**) (C_6) S4507   (**g**) (C_7) S1174

Figure 22. Medoids of the seven clusters of the Ward-linkage hierarchical clustering.

The first two clusters are composed of thin women, but in the first one, they have a shorter torso with a waist circumference that is more pronounced. The clusters C_6 and C_7 are composed of overweight individuals of different categories. The women of the clusters C_3 have a straight body without much difference between waist, hip and chest circumferences. Individuals of C_4 and C_5 have a larger hip circumference compared to the waist circumference, but women of C_4 have a stronger lower body while women of C_5 have a shorter torso.

6. Discussion

The research conducted in this paper demonstrates that the tools of topological data analysis and persistence theory permit us to extract pertinent information about the shape of anthropometric point clouds. The homologies of the persistence diagram of human body points have interesting interpretations in terms of human anatomy. Hence, most of the scan anomalies are correctly detected by clustering algorithms. The gender discrimination index shows that it is worth restricting our search to trunk body points to separate men from women and that the Ward-linkage hierarchical clustering and the K-Medoids clustering give better results than the complete-linkage hierarchical clustering. Finally, we obtain eight morphotypes of men and seven morphotypes of women's body shapes with Ward-linkage hierarchical clusterings. The clusters are composed of individuals of similar weight classes, and the groups can be distinguished by their ratios between bust, waist and hip circumferences or by their torso sizes or their lower body shapes. It is worth noting that the female clusters have better proportions and smaller diameters than the male clusters.

The proposed approach is promising for anomaly detection and classification and should be applied to other types of point clouds in different contexts. The method can also be extended to other problems related to human bodies, such as measurement extraction with supervised machine learning algorithms.

Author Contributions: Conceptualization, S.d.R., P.M. and F.B.; methodology, S.d.R., P.M. and F.B.; formal analysis, S.d.R. and P.M.; software, S.d.R. and P.M. ; writing—original draft preparation, S.d.R. and P.M.; writing—review and editing, S.d.R. and P.M.; visualization, S.d.R. and P.M.; supervision, P.M. and F.B. All authors have read and agreed to the published version of the manuscript.

Funding: This work was supported by Labcom-DiTeX, a joint research group in Textile Data Innovation between Institut Français du Textile et de l'Habillement (IFTH) and Université de Technologie de Troyes (UTT).

Institutional Review Board Statement: Not applicable.

Informed Consent Statement: Not applicable.

Data Availability Statement: The data used in this article come from [24].

Conflicts of Interest: The authors declare no conflict of interest.

References

1. Simmons, K.; Istook, C.; Devarajan, P. Female Figure Identification Technique (FFIT) for apparel part I: Describing female shapes. *J. Text. Appar. Technol. Manag.* **2004**, *4*, 1–16.
2. Song, H.K.; Ashdown, S. Categorization of lower body shapes for adult females based on multiple view analysis. *Text. Res. J.* **2011**, *81*, 914–931. [CrossRef]
3. Nakamura, K.; Kurokawa, T. Analysis and classification of three-dimensional trunk shape of women by using the human body shape model. *Int. J. Comput. Appl. Technol.* **2009**, *34*, 278–284. [CrossRef]
4. Cottle, F.S. *Statistical Human Body Form Classification: Methodology Development and Application*; Auburn University: Auburn, AL, USA, 2012.
5. Hamad, M.; Thomassey, S.; Bruniaux, P. A new sizing system based on 3D morphology clustering. *Comput. Ind. Eng.* **2017**, *113*, 683–692. [CrossRef]
6. Naveed, T.; Zhong, Y.; Hussain, A.; Babar, A.A.; Naeem, A.; Iqbal, A.; Saleemi, S. Female Body Shape Classifications and Their Significant Impact on Fabric Utilization. *Fibers Polym.* **2018**, *19*, 2642–2656. [CrossRef]
7. Pei, J.; Park, H.; Ashdown, S.P. Female breast shape categorization based on analysis of CAESAR 3D body scan data. *Text. Res. J.* **2019**, *89*, 590–611. [CrossRef]

8. Chazal, F.; Michel, B. An Introduction to Topological Data Analysis: Fundamental and Practical Aspects for Data Scientists. *Front. Artif. Intell.* **2021**, *4*, 667963. [CrossRef] [PubMed]
9. Munch, E. A User's Guide to Topological Data Analysis. *J. Learn. Anal.* **2017**, *4*, 47–61. [CrossRef]
10. Edelsbrunner, H.; Letscher, D.; Zomorodian, A. Topological Persistence and Simplification. *Discret. Comput. Geom.* **2002**, *28*, 511–533. [CrossRef]
11. Zomorodian, A.; Carlsson, G. Computing Persistent Homology. *Discret. Comput. Geom.* **2005**, *33*, 249–274. [CrossRef]
12. Cohen-Steiner, D.; Edelsbrunner, H.; Harer, J. Stability of Persistence Diagrams. In Proceedings of the SCG '05: Twenty-First Annual Symposium on Computational Geometry, Pisa, Italy, 6–8 June 2005; Association for Computing Machinery: New York, NY, USA, 2005; pp. 263–271. [CrossRef]
13. Chazal, F.; Cohen-Steiner, D.; Glisse, M.; Guibas, L.J.; Oudot, S.Y. Proximity of Persistence Modules and Their Diagrams. In Proceedings of the SCG '09: Twenty-Fifth Annual Symposium on Computational Geometry, Aarhus, Denmark, 8–10 June 2009; Association for Computing Machinery: New York, NY, USA, 2009; pp. 237–246. [CrossRef]
14. Umeda, Y.; Kaneko, J.; Kikuchi, H. Topological data analysis and its application to time-series data analysis. *Fujitsu Sci. Tech. J.* **2019**, *55*, 65–71.
15. Li, C.; Ovsjanikov, M.; Chazal, F. Persistence-Based Structural Recognition. In Proceedings of the 2014 IEEE Conference on Computer Vision and Pattern Recognition, Columbus, OH, USA, 23–28 June 2014; pp. 2003–2010. [CrossRef]
16. Horak, D.; Maletić, S.; Rajković, M. Persistent homology of complex networks. *J. Stat. Mech. Theory Exp.* **2009**, *2009*, P03034. [CrossRef]
17. Yao, Y.; Sun, J.; Huang, X.; Bowman, G.R.; Singh, G.; Lesnick, M.; Guibas, L.J.; Pande, V.S.; Carlsson, G. Topological methods for exploring low-density states in biomolecular folding pathways. *J. Chem. Phys.* **2009**, *130*, 144115. [CrossRef]
18. Skaf, Y.; Laubenbacher, R. Topological data analysis in biomedicine: A review. *J. Biomed. Inform.* **2022**, *130*, 104082. [CrossRef]
19. Corcoran, P.; Jones, C.B. Topological data analysis for geographical information science using persistent homology. *Int. J. Geogr. Inf. Sci.* **2023**, *37*, 712–745. [CrossRef]
20. Ver Hoef, L.; Adams, H.; King, E.J.; Ebert-Uphoff, I. A Primer on Topological Data Analysis to Support Image Analysis Tasks in Environmental Science. *Artif. Intell. Earth Syst.* **2023**, *2*, e220039. [CrossRef]
21. Mahmmod, B.M.; Abdulhussain, S.H.; Suk, T.; Hussain, A. Fast computation of Hahn polynomials for high order moments. *IEEE Access* **2022**, *10*, 48719–48732. [CrossRef]
22. Jassim, W.A.; Raveendran, P.; Mukundan, R. New orthogonal polynomials for speech signal and image processing. *IET Signal Process.* **2012**, *6*, 713–723. [CrossRef]
23. Abdulhussain, S.H.; Mahmmod, B.M.; Baker, T.; Al-Jumeily, D. Fast and accurate computation of high-order Tchebichef polynomials. *Concurr. Comput. Pract. Exp.* **2022**, *34*, e7311. [CrossRef]
24. Yang, Y.; Yu, Y.; Zhou, Y.; Du, S.; Davis, J.; Yang, R. Semantic Parametric Reshaping of Human Body Models. In Proceedings of the 2014 2nd International Conference on 3D Vision, Tokyo, Japan, 8–11 December 2014; Volume 2, pp. 41–48. [CrossRef]
25. Chazal, F.; Fasy, B.T.; Lecci, F.; Rinaldo, A.; Wasserman, L. Stochastic Convergence of Persistence Landscapes and Silhouettes. In Proceedings of the SOCG'14: Thirtieth Annual Symposium on Computational Geometry, Kyoto, Japan, 8–11 June 2014; Association for Computing Machinery: New York, NY, USA, 2014; pp. 474–483. [CrossRef]
26. Bubenik, P. Statistical Topological Data Analysis Using Persistence Landscapes. *J. Mach. Learn. Res.* **2015**, *16*, 77–102.
27. Ward, J.H.W., Jr. Hierarchical Grouping to Optimize an Objective Function. *J. Am. Stat. Assoc.* **1963**, *58*, 236–244. [CrossRef]
28. Kaufman, L.; Rousseeuw, P.J. Partitioning around Medoids (Program PAM). In *Finding Groups in Data*; John Wiley & Sons, Ltd.: Hoboken, NJ, USA, 1990; Chapter 2, pp. 68–125. [CrossRef]
29. Mileyko, Y.; Mukherjee, S.; Harer, J. Probability measures on the space of persistence diagrams. *Inverse Probl.* **2011**, *27*, 124007. [CrossRef]
30. Turner, K.; Mileyko, Y.; Mukherjee, S.; Harer, J. Frechet Means for Distributions of Persistence Diagrams. *Discret. Comput. Geom.* **2014**, *52*, 44–70. [CrossRef]
31. Lance, G.N.; Williams, W.T. A general theory of classificatory sorting strategies: II. Clustering systems. *Comput. J.* **1967**, *10*, 271–277. [CrossRef]
32. Davies, D.L.; Bouldin, D.W. A Cluster Separation Measure. *IEEE Trans. Pattern Anal. Mach. Intell.* **1979**, *PAMI-1*, 224–227. [CrossRef]
33. Rousseeuw, P.J. Silhouettes: A graphical aid to the interpretation and validation of cluster analysis. *J. Comput. Appl. Math.* **1987**, *20*, 53–65. [CrossRef]
34. Dunn, J.C. A Fuzzy Relative of the ISODATA Process and Its Use in Detecting Compact Well-Separated Clusters. *J. Cybern.* **1973**, *3*, 32–57. [CrossRef]

Disclaimer/Publisher's Note: The statements, opinions and data contained in all publications are solely those of the individual author(s) and contributor(s) and not of MDPI and/or the editor(s). MDPI and/or the editor(s) disclaim responsibility for any injury to people or property resulting from any ideas, methods, instructions or products referred to in the content.

Article

Efficient DNN Model for Word Lip-Reading

Taiki Arakane and Takeshi Saitoh *

Department of Artificial Intelligence, Kyushu Institute of Technology, Fukuoka 820-8502, Japan
* Correspondence: saitoh@ai.kyutech.ac.jp

Abstract: This paper studies various deep learning models for word-level lip-reading technology, one of the tasks in the supervised learning of video classification. Several public datasets have been published in the lip-reading research field. However, few studies have investigated lip-reading techniques using multiple datasets. This paper evaluates deep learning models using four publicly available datasets, namely Lip Reading in the Wild (LRW), OuluVS, CUAVE, and Speech Scene by Smart Device (SSSD), which are representative datasets in this field. LRW is one of the large-scale public datasets and targets 500 English words released in 2016. Initially, the recognition accuracy of LRW was 66.1%, but many research groups have been working on it. The current state of the art (SOTA) has achieved 94.1% by 3D-Conv + ResNet18 + {DC-TCN, MS-TCN, BGRU} + knowledge distillation + word boundary. Regarding the SOTA model, in this paper, we combine existing models such as ResNet, WideResNet, WideResNet, EfficientNet, MS-TCN, Transformer, ViT, and ViViT, and investigate the effective models for word lip-reading tasks using six deep learning models with modified feature extractors and classifiers. Through recognition experiments, we show that similar model structures of 3D-Conv + ResNet18 for feature extraction and MS-TCN model for inference are valid for four datasets with different scales.

Keywords: lip-reading; word recognition; deep neural network; LRW; OuluVS; CUAVE; SSSD; 3D convolutional layer; ResNet; WideResNet; EfficientNet; transformer; ViT; ViViT; MS-TCN

Citation: Arakane, T.; Saitoh, T. Efficient DNN Model for Word Lip-Reading. *Algorithms* **2023**, *16*, 269. https://doi.org/10.3390/a16060269

Academic Editors: Chih-Lung Lin, Bor-Jiunn Hwang, Shaou-Gang Miaou and Yuan-Kai Wang

Received: 28 April 2023
Revised: 24 May 2023
Accepted: 26 May 2023
Published: 27 May 2023

Copyright: © 2023 by the authors. Licensee MDPI, Basel, Switzerland. This article is an open access article distributed under the terms and conditions of the Creative Commons Attribution (CC BY) license (https://creativecommons.org/licenses/by/4.0/).

1. Introduction

This paper focuses on word lip-reading technology that estimates the utterance content from visual information only without audio information. This paper uses "words," but more precisely, it includes both words and short phrases. This technology is expected to be used in the following cases where it is difficult to use audio-based speech recognition: it is used in noisy environments where it is difficult to obtain speech, in public places where it is difficult to speak, and used by people with disabilities who cannot speak due to laryngectomy. Since various problems can be solved using lip-reading technology, it is expected to be one of the next-generation communication tools.

As an academic framework, lip-reading technology is classified as supervised learning for video data. There are several topics for research on lip-reading; shooting directions such as frontal and side [1], recognition targets such as single sound [2,3], word [4–9], and sentence [10–12]. Word recognition is an active research topic, and various algorithms have been proposed.

Word lip-reading has been studied since the early days of lip-reading technology. However, research has become active with the release of datasets such as OuluVS [13], CUAVE [14], SSSD [15], LRW [4], CAS-VSR-W1k (LRW-1000) [16], and RUSAVIC [17], and the introduction of deep learning. In particular, research groups using LRW, one of the large-scale datasets, have been competing for several percent accuracies in recent years (https://paperswithcode.com/sota/lipreading-on-lip-reading-in-the-wild, accessed on 26 April 2023).

This paper explores various deep learning models and their effectiveness in word lip-reading. While many papers use one or two datasets, this paper conducts experiments on four publicly available datasets; LRW, OuluVS, CUAVE, and SSSD.

This paper is organized as follows. Section 2 describes the related research. Section 3 summarizes the basic model of the deep learning model considered in this paper. Section 4 introduces the deep learning model investigated in this paper. Section 5 shows recognition experiments on four datasets, and Section 6 concludes this paper.

2. Related Research

There are many studies on word lip-reading techniques. Here, this paper focuses on research targeting LRW.

LRW is a dataset published by Chung et al. in 2016 [4] (https://www.robots.ox.ac.uk/~vgg/data/lip_reading/lrw1.html, accessed on 26 April 2023) and is a large-scale lip-reading research dataset containing 500 English words. LRW is used as a benchmark in the lip-reading field. LRW contains utterance scenes clipped from news and discussion programs broadcast from 2010 to 2016 by the British Broadcasting Corporation (BBC), which collectively manages radio and television in the United Kingdom. While most datasets are utterance scenes in which the speaker was recording in a roughly specified posture, LRW has the feature of recording utterance scenes in a natural posture even though it is a TV program. The number of speakers is more than 1000. LRW consists of three types of data: training data, validation data, and test data, and provides video data containing 488,766 scenes, 25,000 scenes, and 25,000 scenes, respectively. The train data contain 800–1000 scenes per word, and the validation and test data contain 50 scenes per word. All scenes are extracted face images with 256×256 pixels, the frame rate is 25 fps, the scene length is 1.16 s, and the number of frames is 29. The total recording time of the three types is about 173 h.

CAS-VSR-W1k [16], which contains 1000 Chinese words, is a large-scale open dataset containing word utterance scenes. The state-of-the-art (SOTA) recognition accuracies of LRW and CAS-VSR-W1k are 94.1% [18] and 55.7% [19], respectively. Academically, it is desirable to target CAS-VSR-W1k, which has a high degree of task difficulty. Since we could not obtain CAS-VSR-W1k, we target LRW.

Chung et al., who published LRW, proposed four convolutional neural network (CNN) models based on VGG-M [4]. Among the four models, the multiple towers model obtained the highest recognition rate of 66.1%. This model has a structure in which the convolutional layers for all frames are provided in the first layer. The outputs of all subsequent convolutional layers are connected to one and input to the second convolutional layer. The following year, Chung et al. proposed a new network with a watch, listen, attend, and spell (WLAS) structure, obtaining 76.2% [10]. In WLAS, the encoder consists of a CNN model that is an improved version of VGG-M, which extracts features from each input frame image, and a long short-term memory (LSTM) that summarizes the output of the features from the CNN model. The decoder consists of LSTM, attention, and softmax.

According to the paper with the code site (https://paperswithcode.com/sota/lipreading-on-lip-reading-in-the-wild, accessed on 26 April 2023), SOTA in LRW is currently the result of Ma et al. [18]. The model architecture consists of 3D-Conv + ResNet18 in the front stage and a Temporal model in the backstage with a mouth region of interest (ROI) as input. The temporal model is an ensemble of three different models; densely connected temporal convolutional networks (DC-TCNs), multi-scale temporal convolutional networks (MS-TCNs), and bidirectional gated recurrent units (BGRUs). In addition to the model architecture, they applied data augmentation, self-distillation, and word boundary indicators to improve the recognition accuracy. Many other papers have recently discussed the model training strategy, but the model architecture is often 3D-Conv + ResNet18 + MS-TCN [6,8,20]. The second highest accuracy in LRW is 3D Conv + EfficientNetV2 + Transformer + TCN structure, which obtained a recognition rate of 89.5%, proposed by Koumparoulis et al. [21]. The structure with the third highest accuracy 88.7% is Vosk + MediaPipe + LS + MixUp + SA + 3DResNet-18 + BiLSTM +

CosineWR [22], where Vosk is a voice activity detection model, which can detect speech regions even in heavy acoustically noisy conditions. MediaPipe is a machine-learning library provided by Google (https://developers.google.com/mediapipe, accessed on 26 April 2023). LS means label smoothing [23], SA means a squeeze-and-attention (SA) module, and CosineWR means cosine annealing warm restarts. This study uses two datasets, LRW and RUSAVIC [17], for evaluation. The Russian Audio-Visual Speech in Cars (RUSAVIC) is a multi-speaker and multi-modal corpus. The number of speakers is 20, and the number of phrases is 68.

3. Basic Model

In this section, we explain preprocessing and summarize the basic models of deep learning, data augmentation, distance learning, and fine-tuning discussed in this paper.

3.1. Preprocessing

To begin the process, we follow the same preprocessing steps as the existing method by our previous research [15]. The datasets we are working with, namely OuluVS and CUAVE, contain not only the speaker's face but also their upper body and the background. Hence, we first extract the face rectangle from the input image using face detection processing. Several face detectors have been proposed, including non-deep learning approaches that use Haar-like features and histograms of oriented gradients (HOG) [24,25] and deep learning approaches such as RetinaFace [26]. This paper uses the face detector implemented in the dlib library (http://dlib.net/, accessed on 26 April 2023).

Facial landmark detection helps determine the location of facial parts such as eyes, eyebrows, nose, and lips. This is an important process for stable ROI extraction. In this paper, we utilize the method proposed by Kazami and Sullivan [27], which is a typical facial landmark detection method implemented in the dlib library. A total of 68 facial landmarks are detected.

The size and rotation normalization process is applied based on the detected facial landmarks. At first, two variables of d_{eye}, the distance between two eyes, and θ, the angle between two eyes, are calculated. Then, an affine transformation is applied using d_{eye} and θ. Specifically, the scale is changed so that d_{eye} becomes 200 pixels, and the image is rotated so that θ becomes 0 degrees.

Then, the following equation extracts the upper left coordinate (L, T), lower right coordinate (R, B), and the size of $S \times S$ pixels of the lipROI.

$$L = (x_{llip} + x_{rlip})/2 - S/2,$$
$$T = (y_{llip} + y_{rlip})/2 - S/3,$$
$$R = L + S,$$
$$B = T + S.$$

Here, the two points (x_{llip}, y_{llip}) and (x_{rlip}, y_{rlip}) are the landmark coordinates of the left and right corners of the mouth, respectively. The extracted lipROI is fed to the deep learning model.

3.2. Three-Dimensional Convolution

Many deep learning models investigated in this paper will be described later in Section 4, which extract features using ResNet. The input data are time-series image data (lipROIs). For this reason, we first apply a 3D convolution. Specifically, the structure shown in Figure 1 is used.

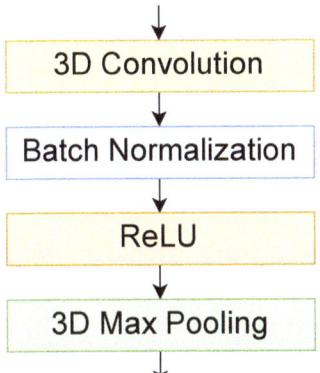

Figure 1. Structure of a 3D convolutional layer.

3.3. Resnet

Residual network (ResNet) [28] is a well-known model that introduces a residual block and a shortcut connection into an existing CNN model. The convolutional layer of CNN extracts features by combining them with the pooling layer, and it is thought that advanced and complex features can be extracted by stacking layers. However, when deep structures are used, there is a problem that training does not progress due to the gradient vanishing or exploding gradients. Therefore, ResNet solves the problem by training the residual function referenced from the layer's input instead of training the optimal output found in a layer. The residual block combines a convolutional layer and a skip connection, summing the outputs of the two passes. One of the residual blocks combines convolutional layers, and the other is the identity function. If this structure does not require transformations in additional layers, it can be handled by setting the weight to zero.

3.4. WideResNet

WideResNet [29] is an improved model of ResNet. The purpose of ResNet is to deepen the layers, but there is a problem stemming from the fact that deeper layers implies a lower computational efficiency in terms of performance. This is believed because many layer weights become meaningless, which is called the reduced feature reuse problem. WideResNet was proposed as a solution to this problem. WideResNet improves the computational efficiency and performance by increasing the number of channels for convolution in the residual block and introducing the dropout.

3.5. EfficientNet

In order to improve the accuracy of the image classification model, various measures, such as increasing the number of layers, widening the width (channel) of the model, and increasing the resolution of the input image, were implemented independently. On the other hand, EfficientNet [30] is a model that introduces a compound coefficient that simultaneously performs three changes in a well-balanced manner. EfficientNet proposed eight models, namely EfficientNet-B0–EfficientNet-B7, which are automatically designed using neural architecture search (NAS) [31]. NAS is a method that automatically optimizes a dedicated network structure to scale the composite coefficients in a balanced manner.

3.6. Transformer

Transformer is a model that uses only attention without recurrent neural network (RNN) or CNN [32]. Transformer is based on the encoder–decoder model and incorporates self-attention and a position-wise feed-forward network.

Self-attention calculates the similarity and importance among its own data. The input of the transformer is divided into Query Q, Key K, and Value V. Here, Q is the input data and represents what to search in the input data. K is used to measure how similar the object to be searched and the Q are. V is an element that outputs an appropriate V based on K. These features are transformed in the fully connected layer, and the inner product of K and V is taken. The inner product is then normalized by softmax so that the sum of the weights for a single Q is 1.0. Finally, the output is obtained by multiplying the obtained weight by V. A position-wise feed-forward network is an independent neural network for each data position that consists of two fully connected layers. Located after the attention layer, it linearly transforms the output of the attention layer.

3.7. ViT

Vision transformer (ViT) [33] is a model that divides an image into patches and treats each patch image as a word. Specifically, the image is divided into N patches x_N and passes through the transformer E to obtain one-dimensional E_{x_i}. Let x_{cls} be the classification token and $E_p \in R^{(N+1) \times D}$ be the location information, z_0, which is the input of the transformer, which is prepared from N E_{x_i}, where i is $i = 1, \ldots,$ and N and D is the number of dimensions of the latent vector. This feeds the input z into the transformer. The transformer consists of multi-head self-attention (MSA), layer normalization (LN), and multilayer perceptron (MLP).

3.8. ViViT

Since the video vision transformer (ViViT) [34] constructs the input token z from a video that does not handle 3D data, the method of obtaining the patch x is different from ViT. In ViT, an image is divided into patches and input tokens are obtained, whereas in ViViT, tablets are obtained by collecting patches on the spatio-temporal axis.

In addition, there are encoders with two different roles as a device to capture the time-series information. One is an encoder for capturing the spatial information. It extracts the tokens from the same time frames, interacts, and creates an average classified token x_{cls} through a transformer. The concatenated x_{cls} representing each time is input to the second time-series encoder. Classification is realized using x_{cls} output from the second encoder as a classifier.

3.9. MS-TCN

TCN [35] is a network that uses CNN for series data. It achieves higher accuracy than RNN, such as LSTM, in tasks for time-series data such as natural language and music. TCN consists of a combination of 1D fully convolutional networks and casual convolutions. Furthermore, Martinez et al. proposed a model using MS-TCN [6]. MS-TCN incorporates multiple timescales into the network to mix short-term and long-term information during feature coding.

3.10. Data Augmentation

In the research field of image recognition, data augmentation (DA) is widely used to increase the number of image data by applying operations such as slightly rotating the image or flipping it horizontally. This paper applies RandAugment (RA) [36], which randomly selects the DA method. Various transformations include identity, autocontrast adjustment, histogram equalization, rotation, solarization, color adjustment, posterization, contrast, brightness, sharpness, horizontal shearing, vertical shearing, horizontal translation, vertical translation, and generate N_{RA} images. This paper applies the MixUp [37] with a weight of 0.4. MixUp is a data augmentation technique that generates a weighted combination of random image pairs from the training data.

3.11. Distance Learning

Distance learning is a method of learning a function that maps data to a feature space so that similar data are brought closer to each other and dissimilar data are separated from each other. This paper applies ArcFace [38], which proposes distance learning using angles, one of the methods with high accuracy in face recognition.

In ArcFace, class classification can solve distance learning by replacing the softmax loss of class classification with angular margin Loss. Softmax Loss has the property of increasing the similarity between samples of the same class but not forcing the similarity of other classes to be low. In ArcFace, the input weights W and feature values x_i are normalized, and the bias b is set to 0. This gives $W_j^T x_i = \|W_j\| \|x_i\| \cos\theta_{j,i} = \cos\theta_{j,i}$. $\theta_{j,i}$ is the angular distance between the feature $x_{j,i}$ and the center position W_j of the j class. Therefore, $\cos\theta_{j,i}$ represents the cosine similarity between the feature x_i and the j class. In addition, the convergence of learning is stabilized by setting the scaling parameter s as a hyperparameter. Furthermore, a linear separation space is secured by directly adding the margin to the angle space.

3.12. Fine-Tuning

Fine-tuning (FT) is used to re-train the weights of the entire model using the weights of the trained network as the initial values to construct a highly accurate model. This paper uses four public datasets: LRW, OuluVS, CUAVE, and SSSD. Among them, LRW is larger than the other datasets. Therefore, FT using the model learned by LRW is applied to the three datasets excluding LRW.

4. Target Models

Referring to the SOTA model [18], this paper investigates six deep learning models shown in Figure 2. The numerical values in the figure indicate the layers that make up the model as one block and indicate the output size of each block. N_F is the number of input sequential image frames, and N_C is the number of classes, which is the number of units in the output layer.

4.1. 3D-Conv + ResNet18 + MS-TCN

An overview of the model diagram is shown in Figure 2a. Extract 512-dimensional features from input images using 3D-Conv + ResNet18. After that, it trains the temporal changes of the features obtained by MS-TCN. MS-TCN has three convolutional layers with kernel sizes of 3, 5, and 7, and obtains short-term and long-term information.

4.2. 3D-Conv + ResNet18 + ViT

As shown in Figure 2b, this model extracts 512-dimensional features from input images using 3D-Conv + ResNet18 and then trains temporal changes using ViT. Normally, the input of ViT is an image, but in this paper, the extracted features are regarded as image patches and input to ViT for training.

4.3. 3D-Conv + WiderResNet18 + MS-TCN

As shown in Figure 3a, this paper uses a model with permuted layers of ResNet. The activation function is changed from ReLU to swish (SiLU). Training is performed in the same way as ResNet18 + MS-TCN. In WideResNet, it is desirable to expand the number of dimensions of features to be extracted. However, this paper uses the same dimensions as ResNet18 due to resource constraints, as shown in Figure 3b.

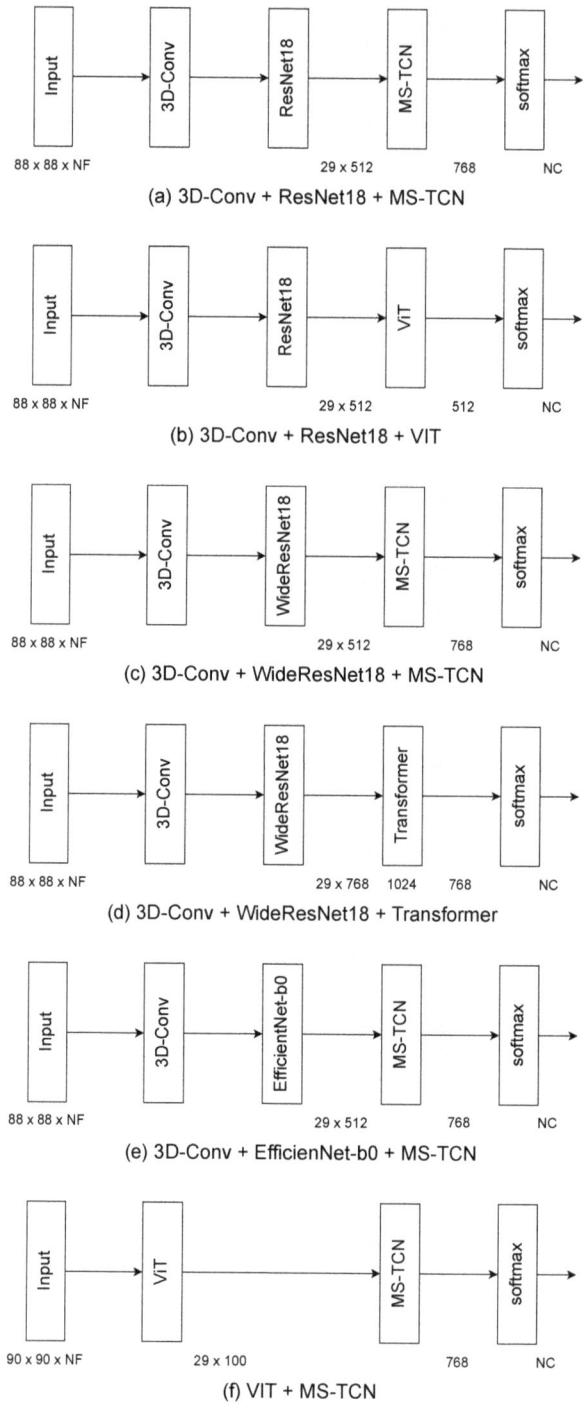

Figure 2. Target models.

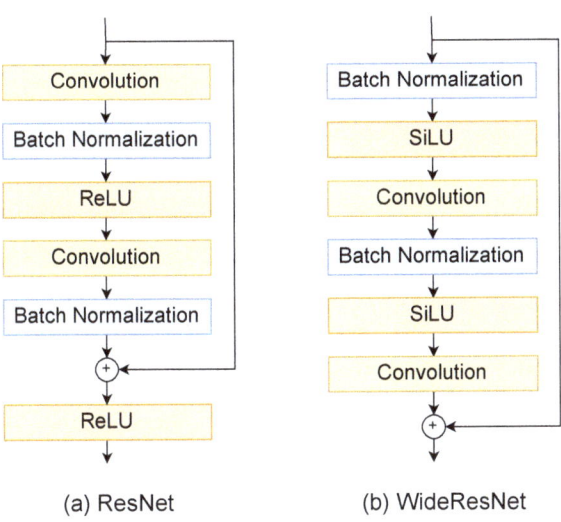

Figure 3. Structures of ResNet and WideResNet.

4.4. 3D-Conv + WiderResNet18 + Transformer

3D-Conv + WideResNet18 extracts 768-dimensional features from input images and then trains temporal changes using transformer. The transformer is originally multi-head self-attention, but single attention is used in this experiment.

4.5. 3D-Conv + EfficientNet-b0 + MS-TCN

3D-Conv + EfficientNet-b0 extracts 512-dimensional features from input images and MS-TCN trains temporal changes.

4.6. ViT + MS-TCN

It is possible to recognize by ViT alone, but in this paper, ViT is used as a feature extractor, as shown in Figure 2f. ViT extracts 100-dimensional features from each frame image, and MS-TCN trains temporal changes. In ViT, a feature vector called a class token inserted at the beginning of each frame image is extracted as a feature value of each frame image.

5. Evaluation Experiment

Several datasets have been published in the lip-reading field. Four public datasets shown in Table 1 are used in this experiment.

Table 1. Overview of the four datasets used in our experiments.

Name	Year	Language	# of Speakers	Content
LRW [4]	2016	English	1000+	500 words
OuluVS [13]	2009	English	20	10 greeting phrases
CUAVE [14]	2002	English	36	10 digits
SSSD [15]	2018	Japanese	72	25 words

We applied the preprocessing described in Section 3.1 to extract the grayscale lipROIs. The image size $S \times S$ of the lipROIs of LRW, OuluVS, CUAVE, and SSSD are 96×96 pixels, 64×64 pixels, 64×64 pixels, and 64×64 pixels, respectively. Figure 4 shows the lipROIs of OuluVS, CUAVE, and SSSD. Inputs to 3D-Conv, ViT, and ViViT are image data randomly extracted from 88×88 pixels, 90×90 pixels, and 87×87 pixels, respectively.

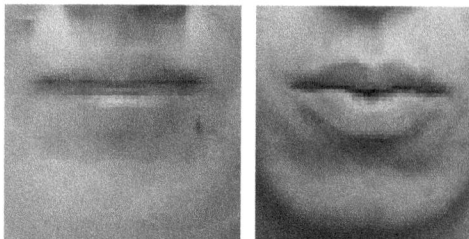

Figure 4. Extracted lipROIs (**left**: OuluVS, **center**: CUAVE, **right**: SSSD).

We used PyTorch to implement each model. To train our network, we utilized Adam with the decoupled weight decay (AdamW) [39] along with certain parameters such as $\beta_1 = 0.9$, $\beta_2 = 0.999$, $\epsilon = 10^{-8}$, and a weight decay of 0.01. The training was conducted for 80 epochs, using an initial learning rate of 0.0003 and a mini-batch size 32 for models using ResNet18 and WideResNet18 and 16 for other models. As for the transformer, single-head attention was used due to insufficient resources. We employed a cosine annealing schedule to decay the learning rate without warm-up steps. We gave $N_{RA} = 2$ as a parameter in RA.

5.1. LRW

Table 2 shows the top-1 accuracy and the number of parameters for each model described in Section 4. In the table, the upper row is the accuracy of other representative papers, and the lower row is the accuracy of this paper. "—" means that the reference does not show the number of parameters in this model. Although the recognition accuracy of this research did not reach the recognition accuracy of SOTA, the following things were clarified by examining various models.

Table 2. Recognition results (LRW).

Model	Top-1 Acc. (%)	Params $\times 10^6$
Multi-Tower 3D-CNN [4]	61.1	—
WLAS [10]	76.2	—
3D-Conv + ResNet34 + Bi-LSTM [40]	83.0	—
3D-Conv + ResNet34 + Bi-GRU [41]	83.39	—
3D-Conv + ResNet18 + MS-TCN [6]	85.3	—
3D-Conv + ResNet18 + MS-TCN + MVM [20]	88.5	—
3D-Conv + ResNet18 + MS-TCN + KD [8]	88.5	36.4
Alternating ALSOS + ResNet18 + MS-TCN [42]	87.0	41.2
Vosk + MediaPipe + LS + MixUp + SA + 3D-Conv + ResNet-18 + BiLSTM + Cosine WR [22]	88.7	—
3D-Conv + EfficientNetV2 + Transformer + TCN [21]	89.5	—
3D-Conv + ResNet18 + {DC-TCN, MS-TCN, BGRU} (ensemble) + KD + Word Boundary [18]	94.1	—
3D-Conv + ResNet18 + MS-TCN (ours)	87.4	36.0
3D-Conv + ResNet18 + MS-TCN + RA (ours)	85.3	36.0
3D-Conv + ResNet18 + MS-TCN + ArcFace (ours)	86.7	36.0
3D-Conv + ResNet18 + ViT (ours)	83.8	30.1
3D-Conv + WideResNet18 + MS-TCN (ours)	86.8	36.0
3D-Conv + WideResNet18 + Transformer (ours)	79.2	11.2
3D-Conv + EfficientNet-b0 + MS-TCN (ours)	80.6	32.3
ViT + MS-TCN (ours)	79.9	24.0
ViViT (ours)	72.4	3.9
ViViT + RA (ours)	75.6	3.9

As a feature extractor, it can be confirmed that ResNet18 is superior to other models. The analysis of mouth movements, which is the target of our experiment, has a smaller difference in movements than in other video classification tasks, such as human action recognition. In addition, other feature extractors did not obtain the expected accuracy because the image's resolution was small. The model used in this experiment is a model that achieves high accuracy in image recognition, but these are usually input with high-resolution images such as 224 × 224 pixels. This research is video recognition, and it is necessary to put many images in the memory at once, so the experiment was performed at a low resolution due to memory constraints. Therefore, effective feature values could not be obtained in deep layers.

It was found that MS-TCN tends to obtain recognition accuracy in the inference part of the latter stage. As for MS-TCN, we conducted experiments using RA and ArcFace, but the recognition accuracy decreased. We suspect that this is because the application of the RA-generated image data is unsuitable for the task or the model's generalization performance deteriorated due to the excessive application to the training data.

We consider the details of the recognition results for 3D-Conv + ResNet18 + MS-TCN, which had the highest accuracy among the models investigated in this paper. Among 500 words, 15 words with low recognition rates are listed in Table 3. In the table, the 4th column shows words with many mistakes, and the numbers in parentheses are the misrecognition rate. From the table, it can be confirmed that words with a low recognition rate are misrecognized as similar words. Among the 500 words, 252 had a recognition rate of 90.0% or higher.

Table 3. Fifteen words with low recognition rate (LRW).

Order	Word	Top-1 Acc. (%)	Most Misrecognized Word
486	ABOUT	64	AMONG (10)
	BECAUSE	64	ABUSE (10)
488	ACTUALLY	62	ACTION (6)
	COULD	62	EUROPEAN, SHOULD (4)
	MATTER	62	AMONG (6)
	NEEDS	62	YEARS (8)
	THINGS	62	YEARS (6)
493	THEIR	60	THERE (20)
	UNDER	60	DURING, LONDON (4)
	UNTIL	60	STILL (8)
496	SPEND	58	SPENT (18)
	THESE	58	THINGS (8)
498	THING	56	BEING, NOTHING, THESE (4)
499	THINK	50	THING (16)
500	THERE	44	THEIR (12)

5.2. OuluVS

OuluVS [13] contains ten sentences spoken by 20 speakers, comprising 17 males and 3 females. The contents of the 10 sentences are (1) "excuse me", (2) "good bye", (3) "have a good time", (4) "hello", (5) "how are you", (6) "I am sorry", (7) "nice to meet you", (8) "see you", (9) "thank you", and (10) "you are welcome". For each speaker, five utterance scenes are recorded for each sentence. The image size is 720 × 576 pixels, the frame rate is 25 fps, and the speaker speaks in front of a white background.

The leave-one-person-out cross-validation method was applied to 20 speakers in the evaluation experiment, and the average recognition rate was obtained. Here, the training and test data per speaker are 19 speakers × 10 sentences × 5 scenes = 950 scenes and 1 speaker × 10 sentences × 5 scenes = 50 scenes, respectively. Table 4 shows the recognition rate of the four training conditions and other methods. Here, $N_{RA} = 3$ was set for RA, and the training data were padded. AE, MF, and AU in [5,7,43] stand for feature names based on the auto-encoder, motion feature, and action unit, respectively. FOMM in [7] means a first-order motion model and generates utterance scenes. From the table, 3D-Conv + ResNet18 + MS-TCN + RA + FT obtained the highest recognition accuracy of 97.2%. It can be confirmed that the recognition accuracy is improved by fine-tuning with LRW.

Table 4. Recognition results (OuluVS).

Model	Top-1 Acc. (%)
Multi-Tower 3D-CNN [4]	91.4
AE + GRU [43]	81.2
FOMM → AE + GRU [7]	86.5
{MF + AE + AU} + GRU [5]	86.6
3D-Conv + ResNet18 + MS-TCN (ours)	90.1
3D-Conv + ResNet18 + MS-TCN + RA (ours)	93.1
3D-Conv + ResNet18 + MS-TCN + FT (ours)	95.1
3D-Conv + ResNet18 + MS-TCN + RA + FT (ours)	97.2

5.3. CUAVE

In CUAVE [14], utterance scenes are taken from 36 speakers, comprising 19 males and 17 females. The utterance contents are "zero", "one", "two", "three", "four", "five", "six", "seven", "eight", and "nine". A feature of CUAVE is that it includes frontal-face speech scenes and side-face speech scenes. Furthermore, not only utterances in which the speaker is standing still but scenes in which the speaker speaks while moving are recorded. It also includes scenes in which two speakers speak at the same time. The image size is 720×480 pixels, the frame rate is 29.97 fps, and the speaker speaks in front of a green background. In this experiment, we use a scene where the speaker stands still and speaks five samples per word.

The same leave-one-person-out cross-validation method as OuluVS was applied in the experiment, and the average recognition rate was obtained. The training and test data per speaker are 35 speakers $\times$ 10 sentences $\times$ 5 scenes = 1750 scenes and 1 speaker $\times$ 10 sentences $\times$ 5 scenes = 50 scenes, respectively. Table 5 shows the recognition rate of experimental conditions and other methods. However, $N_{RA} = 3$. From the table, 3D-Conv + ResNet18 + MS-TCN + RA + FT obtained the highest recognition accuracy as well as OuluVS.

Table 5. Recognition results (CUAVE).

Model	Top-1 Acc. (%)
AE + GRU [43]	72.8
FOMM → AE + GRU [7]	79.8
{MF + AE + AU} + GRU [5]	83.4
3D-CNN (ours)	84.4
3D-Conv + ResNet18 + MS-TCN (ours)	87.6
3D-Conv + ResNet18 + MS-TCN + RA (ours)	90.0
3D-Conv + ResNet18 + MS-TCN + FT (ours)	93.7
3D-Conv + ResNet18 + MS-TCN + RA + FT (ours)	94.1

5.4. SSSD

SSSD [15] consists of 25 utterances, comprising 10 Japanese numeric words and 15 greetings (https://www.saitoh-lab.com/SSSD/index_en.html, accessed on 26 April 2023). The 25 words are (1) /ze-ro/ (zero), (2) /i-chi/ (one), (3) /ni/ (two), (4) /sa-N/ (three), (5) /yo-N/ (four), (6) /go/ (five), (7) /ro-ku/ (six), (8) /na-na/ (seven), (9) /ha-chi/ (eight), (10) /kyu/ (nine), (11) /a-ri-ga-to-u/ (thank you), (12) /i-i-e/ (no), (13) /o-ha-yo-u/ (good morning), (14) /o-me-de-to-u/ (congratulation), (15) /o-ya-su-mi/ (good night), (16) /go-me-N-na-sa-i/ (I am sorry), (17) /ko-N-ni-chi-wa/ (good afternoon), (18) /ko-N-ba-N-wa/ (good evening), (19) /sa-yo-u-na-ra/ (goodbye), (20) /su-mi-ma-se-N/ (excuse me), (21) /do-u-i-ta-shi-ma-shi-te/ (you are welcome), (22) /ha-i/ (yes), (23) /ha-ji-me-ma-shi-te/ (nice to meet you), (24) /ma-ta-ne/ (see you), and (25) /mo-shi-mo-shi/ (hello).

Unlike OuluVS, CUAVE, and LRW, SSSD is filmed using a smart device. An image of the lower half of a face of 300×300 pixels extracted after normalization processing is applied for scale, and rotation is provided. The frame rate is 30 fps. The number of provided scenes is 72 speakers $\times$ 25 words $\times$ 10 samples = 18,000 scenes. As a competition using SSSD, the second machine lip-reading challenge was held in 2019, and 5000 scenes of test data were released.

For the accuracy evaluation, we used 18,000 scenes as training data and an extra 5000 scenes as test data, using the same task as the second machine lip-reading challenge to obtain recognition accuracy. The results are shown in Table 6. N_{RA} gave 3 like OuluVS and CUAVE. From the table, 3D-Conv + ResNet18 + MS-TCN + FT obtained the highest recognition accuracy of 95.14%. While OuluVS and CUAVE obtained high recognition accuracy when RA was applied, SSSD obtained the highest recognition rate when RA was not applied. We presume that SSSD has more training data than OuluVS and CUAVE and can train sufficiently without applying RA.

Table 6. Recognition results (SSSD).

Model	Top-1 Acc. (%)
LipNet	90.66
3D-Conv + ResNet18 + MS-TCN (ours)	93.08
3D-Conv + ResNet18 + MS-TCN + RA (ours)	93.68
3D-Conv + ResNet18 + MS-TCN + FT (ours)	95.14
3D-Conv + ResNet18 + MS-TCN + RA + FT (ours)	94.86

6. Conclusions

We conducted a study on word lip-reading using deep-learning models. Our goal was to find an effective model for this task. We explored different combinations of models such as ResNet, WideResNet, EfficientNet, Transformer, and ViT, referring to the SOTA model. While many papers use one or two datasets, recognition experiments were conducted using four public datasets, namely LRW, OuluVS, CUAVE, and SSSD, with different sizes and languages. As a result, we found that 3D-Conv + ResNet18 is a good model for feature extraction, and MS-TCN is a good model for inference. Although we did not propose a model that surpasses SOTA, our study confirmed the effectiveness of these models.

This paper investigates an effective word lip-reading model on four public datasets. There are other lip-reading datasets not used in this paper. In the future, we will work on experiments including other datasets. Since it has been clarified that the model structure is effective for lip-reading, we will also verify the training method of the model in the future. The recognition target of this paper is words, but sentence lip-reading has also been actively researched in recent years. Sentence lip-reading is also a target task for the future.

Author Contributions: Conceptualization, T.A. and T.S.; methodology, T.A. and T.S.; software, T.A.; validation, T.A.; formal analysis, T.A. and T.S.; investigation, T.A. and T.S.; resources, T.S.; data curation, T.A.; writing—original draft preparation, T.S.; writing—review and editing, T.S.; visualization, T.S.; supervision, T.S.; project administration, T.S.; funding acquisition, T.S. All authors have read and agreed to the published version of the manuscript.

Funding: This research was funded by the JSPS KAKENHI Grant No. 19KT0029.

Institutional Review Board Statement: Not applicable.

Informed Consent Statement: Not applicable.

Data Availability Statement: The data used in this article come from [4,13–15].

Conflicts of Interest: The authors declare no conflict of interest.

Abbreviations

The following abbreviations are used in this manuscript:

AE	Auto-Encoder
AU	Action Unit
BBC	British Broadcasting Corporation
BGRU	Bidirectional Gated Recurrent Unit
CNN	Convolutional Neural Network
DA	Data Augmentation
DC-TCN	Densely Connected Temporal Convolutional Network
FOMM	First-Order Motion Model
FT	Fine-Tuning
HOG	Histograms of Oriented Gradient
LN	Layer Normalization
LRW	Lip Reading in the Wild
LS	Label Smoothing
LSTM	Long Short-Term Memory
MF	Motion Feature
MLP	Multilayer Perceptron
MSA	Multi-head Self-Attention
MS-TCN	Multi-Scale Temporal Convolutional Network
NAS	Neural Architecture Search
RA	RandAugment
ResNet	Residual Network
RNN	Recurrent Neural Network
ROI	Region of Interest
RUSAVIC	Russian Audio-Visual Speech in Cars
SA	Squeeze-and-Attention
SiLU	Swish
SSSD	Speech Scene by Smart Device
SOTA	State of the Art
ViT	Vision Transformer
ViViT	Video Vision Transformer
WLAS	Watch, Listen, Attend, and Spell

References

1. Saitoh, T.; Konishi, R. Profile Lip Reading for Vowel and Word Recognition. In Proceedings of the 20th International Conference on Pattern Recognition (ICPR2010), Istanbul, Turkey, 23–26 August 2010. [CrossRef]
2. Nakamura, Y.; Saitoh, T.; Itoh, K. 3D CNN-based mouth shape recognition for patient with intractable neurological diseases. In Proceedings of the 13th International Conference on Graphics and Image Processing (ICGIP 2021), Kunming, China, 18–20 August 2022; Volume 12083, pp. 775–782. [CrossRef]
3. Kanamaru, T.; Arakane, T.; Saitoh, T. Isolated single sound lip-reading using a frame-based camera and event-based camera. Front. Artif. Intell. 2023, 5, 298. [CrossRef] [PubMed]
4. Chung, J.S.; Zisserman, A. Lip Reading in the Wild. In Proceedings of the Asian Conference on Computer Vision (ACCV), Taipei, Taiwan, 20–24 November 2016.
5. Shirakata, T.; Saitoh, T. Lip Reading using Facial Expression Features. Int. J. Comput. Vis. Signal Process. 2020, 10, 9–15.
6. Martinez, B.; Ma, P.; Petridis, S.; Pantic, M. Lipreading using temporal convolutional networks. In Proceedings of the IEEE International Conference on Acoustics, Speech and Signal Processing (ICASSP), Barcelona, Spain, 4–8 May 2020; pp. 6319–6323. [CrossRef]
7. Kodama, M.; Saitoh, T. Replacing speaker-independent recognition task with speaker-dependent task for lip-reading using First Order Motion Model. In Proceedings of the 13th International Conference on Graphics and Image Processing (ICGIP 2021), Kunming, China, 18–20 August 2022; Volume 12083, pp. 652–659. [CrossRef]
8. Ma, P.; Martinez, B.; Petridis, S.; Pantic, M. Towards Practical Lipreading with Distilled and Efficient Models. In Proceedings of the IEEE International Conference on Acoustics, Speech and Signal Processing (ICASSP), Toronto, ON, Canada, 6–11 June 2021; pp. 7608–7612. [CrossRef]
9. Fu, Y.; Lu, Y.; Ni, R. Chinese Lip-Reading Research Based on ShuffleNet and CBAM. Appl. Sci. 2023, 13, 1106. [CrossRef]
10. Chung, J.S.; Senior, A.; Vinyals, O.; Zisserman, A. Lip Reading Sentences in the Wild. In Proceedings of the IEEE Conference on Computer Vision and Pattern Recognition (CVPR), Honolulu, HI, USA, 21–26 July 2017; pp. 6447–6456. [CrossRef]

11. Arakane, T.; Saitoh, T.; Chiba, R.; Morise, M.; Oda, Y. Conformer-Based Lip-Reading for Japanese Sentence. In Proceedings of the 37th International Conference on Image and Vision Computing, Auckland, New Zealand, 24–25 November 2023; pp. 474–485. [CrossRef]
12. Jeon, S.; Elsharkawy, A.; Kim, M.S. Lipreading Architecture Based on Multiple Convolutional Neural Networks for Sentence-Level Visual Speech Recognition. *Sensors* **2022**, *22*, 72. [CrossRef]
13. Zhao, G.; Barnard, M.; Pietikainen, M. Lipreading with local spatiotemporal descriptors. *IEEE Trans. Multimed.* **2009**, *11*, 1254–1265. [CrossRef]
14. Patterson, E.K.; Gurbuz, S.; Tufekci, Z.; Gowdy, J.N. Moving-talker, speaker-independent feature study, and baseline results using the CUAVE multimodal speech corpus. *EURASIP J. Appl. Signal Process.* **2002**, *2002*, 1189–1201. [CrossRef]
15. Saitoh, T.; Kubokawa, M. SSSD: Speech Scene Database by Smart Device for Visual Speech Recognition. In Proceedings of the 24th International Conference on Pattern Recognition (ICPR2018), Beijing, China, 20–24 August 2018; pp. 3228–3232. [CrossRef]
16. Yang, S.; Zhang, Y.; Feng, D.; Yang, M.; Wang, C.; Xiao, J.; Long, K.; Shan, S.; Chen, X. LRW-1000: A Naturally-Distributed Large-Scale Benchmark for Lip Reading in the Wild. In Proceedings of the 14th IEEE International Conference on Automatic Face & Gesture Recognition (FG2019), Lille, France, 14–18 May 2019. [CrossRef]
17. Ivanko, D.; Ryumin, D.; Axyonov, A.; Kashevnik, A.; Karpov, A. Multi-Speaker Audio-Visual Corpus RUSAVIC: Russian Audio-Visual Speech in Cars. In Proceedings of the 13th Conference on Language Resources and Evaluation (LREC2022), Marseille, France, 21–23 June 2022; pp. 1555–1559.
18. Ma, P.; Wang, Y.; Petridis, S.; Shen, J.; Pantic, M. Training Strategies for Improved Lip-reading. In Proceedings of the IEEE International Conference on Acoustics, Speech and Signal Processing (ICASSP), Singapore, 23–27 May 2022. [CrossRef]
19. Feng, D.; Yang, S.; Shan, S.; Chen, X. Learn an Effective Lip Reading Model without Pains. *arXiv* **2020**, arXiv:2011.07557. [CrossRef]
20. Kim, M.; Yeo, J.H.; Ro, Y.M. Distinguishing Homophenes using Multi-head Visual-audio Memory for Lip Reading. In Proceedings of the 36th AAAI Conference on Artificial Intelligence (AAAI), Virtual, 22 February–1 March 2022.
21. Koumparoulis, A.; Potamianos, G. Accurate and Resource-Efficient Lipreading with Efficientnetv2 and Transformers. In Proceedings of the 2022 IEEE International Conference on Acoustics, Speech and Signal Processing (ICASSP), Singapore, 23–27 May 2022. [CrossRef]
22. Ivanko, D.; Ryumin, D.; Kashevnik, A.; Axyonov, A.; Karnov, A. Visual Speech Recognition in a Driver Assistance System. In Proceedings of the 30th European Signal Processing Conference (EUSIPCO), Belgrade, Serbia, 29 August–2 September 2022. [CrossRef]
23. Szegedy, C.; Vanhoucke, V.; Ioffe, S.; Shlens, J.; Wojna, Z. Rethinking the Inception Architecture for Computer Vision. In Proceedings of the IEEE Conference on Computer Vision and Pattern Recognition (CVPR), Las Vegas, NV, USA, 27–30 June 2016; pp. 2818–2826. [CrossRef]
24. Viola, P.; Jones, M. Rapid object detection using a boosted cascade of simple features. In Proceedings of the IEEE Computer Society Conference on Computer Vision and Pattern Recognition (CVPR), Kauai, HI, USA, 8–14 December 2001; Volume 1. [CrossRef]
25. Dalal, N.; Triggs, B. Histograms of oriented gradients for human detection. In Proceedings of the IEEE Conference on Computer Vision and Pattern Recognition (CVPR), San Diego, CA, USA, 20–25 June 2005; Volume 1, pp. 886–893. [CrossRef]
26. Deng, J.; Guo, J.; Ververas, E.; Kotsia, I.; Zafeiriou, S. RetinaFace: Single-Shot Multi-Level Face Localisation in the Wild. In Proceedings of the IEEE Conference on Computer Vision and Pattern Recognition (CVPR), Virtual, 14–19 June 2020; pp. 5203–5212.
27. Kazemi, V.; Sullivan, J. One millisecond face alignment with an ensemble of regression trees. In Proceedings of the IEEE Conference on Computer Vision and Pattern Recognition (CVPR), Columbus, OH, USA, 24–27 June 2014; pp. 1867–1874. [CrossRef]
28. He, K.; Zhang, X.; Ren, S.; Sun, J. Deep Residual Learning for Image Recognition. In Proceedings of the IEEE Conference on Computer Vision and Pattern Recognition (CVPR), Las Vegas, NV, USA, 26 June–1 July 2016; pp. 770–778. [CrossRef]
29. Zagoruyko, S.; Komodakis, N. Wide Residual Networks. In Proceedings of the British Machine Vision Conference (BMVC), York, UK, 19–22 September 2016; pp. 87.1–87.12. [CrossRef]
30. Tan, M.; Le, Q. EfficientNet: Rethinking Model Scaling for Convolutional Neural Networks. In Proceedings of the 36th International Conference on Machine Learning (ICMR), Long Beach, CA, USA, 9–15 June 2019.
31. Zoph, B.; Le, Q.V. Neural architecture search with reinforcement learning. In Proceedings of the 5th International Conference on Learning Representations (ICLR), Toulon, France, 24–26 April 2017. [CrossRef]
32. Vaswani, A.; Shazeer, N.; Parmar, N.; Uszkoreit, J.; Jones, L.; Gomez, A.N.; Kaiser, L.; Polosukhin, I. Attention Is All You Need. In Proceedings of the 31st Conference on Neural Information Processing Systems (NIPS2017), Long Beach, CA, USA, 4–9 December 2017.
33. Dosovitskiy, A.; Beyer, L.; Kolesnikov, A.; Weissenborn, D.; Zhai, X.; Unterthiner, T.; Dehghani, M.; Minderer, M.; Heigold, G.; Gelly, S.; et al. An Image is Worth 16x16 Words: Transformers for Image Recognition at Scale. In Proceedings of the International Conference on Learning Representations (ICLR), Virtual, 3–7 May 2021.
34. Arnab, A.; Dehghani, M.; Heigold, G.; Sun, C.; Lučić, M.; Schmid, C. ViViT: A Video Vision Transformer. In Proceedings of the IEEE/CVF International Conference on Computer Vision (ICCV), Virtual, 11–17 October 2021; pp. 6816–6826. [CrossRef]

35. Bai, S.; Kolter, J.Z.; Koltun, V. An Empirical Evaluation of Generic Convolutional and Recurrent Networks for Sequence Modeling. *arXiv* **2018**, arXiv:1803.01271. [CrossRef]
36. Cubuk, E.D.; Zoph, B.; Shlens, J.; Le, Q. RandAugment: Practical Automated Data Augmentation with a Reduced Search Space. In Proceedings of the Advances in Neural Information Processing Systems, Virtual, 14–19 June 2020; Volume 33, pp. 18613–18624.
37. Zhang, H.; Cisse, M.; Dauphin, Y.N.; Lopez-Paz, D. Mixup: Beyond empirical risk minimization. In Proceedings of the International Conference on Learning Representations (ICLR), Vancouver, BC, Canada, 30 April–3 May 2018. [CrossRef]
38. Deng, J.; Guo, J.; Xue, N.; Zafeiriou, S. ArcFace: Additive Angular Margin Loss for Deep Face Recognition. In Proceedings of the IEEE/CVF Conference on Computer Vision and Pattern Recognition (CVPR), Long Beach, CA, USA, 16–20 June 2019; pp. 4685–4694. [CrossRef]
39. Loshchilov, I.; Hutter, F. Decoupled weight decay regularization. In Proceedings of the International Conference on Learning Representations (ICLR), New Orleans, LA, USA, 6–9 May 2019. [CrossRef]
40. Stafylakis, T.; Tzimiropoulos, G. Combining Residual Networks with LSTMs for Lipreading. In Proceedings of the Conference of the International Speech Communication Association (Interspeech 2017), Stockholm, Sweden, 20–24 August 2018; pp. 3652–3656.
41. Petridis, S.; Stafylakis, T.; Ma, P.; Cai, F.; Tzimiropoulos, G.; Pantic, M. End-to-end Audiovisual Speech Recognition. In Proceedings of the IEEE International Conference on Acoustics, Speech, and Signal Processing (ICASSP), Calgary, AB, Canada, 15–20 April 2018; pp. 6548–6552.
42. Tsourounis, D.; Kastaniotis, D.; Fotopoulos, S. Lip Reading by Alternating between Spatiotemporal and Spatial Convolutions. *J. Imaging* **2021**, *7*, 91. [CrossRef] [PubMed]
43. Iwasaki, M.; Kubokawa, M.; Saitoh, T. Two Features Combination with Gated Recurrent Unit for Visual Speech Recognition. In Proceedings of the 14th IAPR Conference on Machine Vision Applications (MVA), Nagoya, Japan, 8–12 May 2017; pp. 300–303. [CrossRef]

Disclaimer/Publisher's Note: The statements, opinions and data contained in all publications are solely those of the individual author(s) and contributor(s) and not of MDPI and/or the editor(s). MDPI and/or the editor(s) disclaim responsibility for any injury to people or property resulting from any ideas, methods, instructions or products referred to in the content.

Article

A Pattern Recognition Analysis of Vessel Trajectories

Paolo Massimo Buscema [1,2,*], Giulia Massini [1], Giovanbattista Raimondi [3], Giuseppe Caporaso [4], Marco Breda [1] and Riccardo Petritoli [1]

1. Semeion Research Center of Sciences of Communication, Via Sersale 117, 00128 Rome, Italy; semeion@semeion.it (G.M.); m.breda@semeion.it (M.B.); r.petritoli@semeion.it (R.P.)
2. Department of Mathematical and Statistical Sciences, University of Colorado, 1201 Larimer St., Denver, CO 80204, USA
3. Comando per le Operazioni in Rete (COR), Italian Defense General Staff, Via Stresa, 31/b, 00135 Rome, Italy; giovanbattista.raimondi@marina.difesa.it
4. C4I and MSA Division and Development New System, Italian Navy, Via Stresa, 31/b, 00135 Rome, Italy; giuseppe.caporaso@marina.difesa.it
* Correspondence: m.buscema@semeion.it or mathstats-staff@ucdenver.edu; Tel.: +39-06-50652350

Abstract: The automatic identification system (AIS) facilitates the monitoring of ship movements and provides essential input parameters for traffic safety. Previous studies have employed AIS data to detect behavioral anomalies and classify vessel types using supervised and unsupervised algorithms, including deep learning techniques. The approach proposed in this work focuses on the recognition of vessel types through the "Take One Class at a Time" (TOCAT) classification strategy. This approach pivots on a collection of adaptive models rather than a single intricate algorithm. Using radar data, these models are trained by taking into account aspects such as identifiers, position, velocity, and heading. However, it purposefully excludes positional data to counteract the inconsistencies stemming from route variations and irregular sampling frequencies. Using the given data, we achieved a mean accuracy of 83% on a 6-class classification task.

Keywords: AIS; vessel classification; TOCAT

1. Introduction

The automatic identification system (AIS) serves as a vital monitoring apparatus for maritime vessel surveillance. It supplies essential input parameters that feed into naval traffic simulation models. These models are instrumental in conducting maritime risk analysis and devising strategies for incident prevention. The AIS enables the monitoring of maritime movements via the electronic exchange of navigational data between various entities. This system interlinks vessels, onboard transmitters, ground stations, and satellites, fostering a comprehensive network for efficient tracking. These data include information that is relevant to traffic safety. Although the exchange of AIS data is legally mandatory only for larger vessels, the usage is on the rise, enabling the deduction of various levels of contextual information, from the characterization of ports and offshore platforms to the spatial and temporal distribution of routes.

Numerous studies have harnessed AIS data to examine anomalies in ship behaviors, intending to pinpoint potential navigational threats. Unsupervised anomaly detection algorithms have been employed, using Ornstein–Uhlenbeck stochastic processes based on the analysis of historical routes [1], or identifying outliers derived from the clustering of behaviors and trajectories [2]. Other studies have concurrently used infrared images to discriminate noise, irrelevant objects, and suspicious vessels [3]. Recently, deep learning techniques have been applied, with models aimed at classifying suspicious trajectories using convolutional neural networks (CNNs) and generative-discriminative learning algorithms [4].

Within the specific context of vessel classification aimed at tracking fishing activities, and more broadly, in the pursuit of augmenting maritime situational awareness (MSA), supervised multiclass methodologies have been implemented. These methodologies employed several algorithms, frequently including random forest models [5–7] and light gradient-boosting machine (Light GBM) [8], to distinguish between different types of vessels. In the latter case, a classification model was effectively implemented using 60-dimensional feature vectors as input, although its application was confined to just three distinct categories of vessels. These feature vectors encapsulated various metrics derived from AIS trajectory data, such as the mean, first quartile, median, third quartile, standard deviation, and coefficient of dispersion associated with changes in speed, course, longitude, latitude, and displacement.

From the studies examined so far, it emerges that current approaches mainly rely on the creation of a single, albeit complex, neural network to which the entire task of recognition is delegated. From an innovative perspective, this paper proposes an alternative strategy rooted in the idea of building an ecosystem of neural networks, diverse in both topological and mathematical terms, and not based exclusively on gradient descent or decision trees. In this regard, each network is purposefully designed to focus on a particular statistical subset of ship trajectories, resulting in an enhancement in performance.

More specifically, this study has focused on the need to provide the Italian Navy with an accurate and efficient solution for monitoring and classifying ship trajectories, employing AIS signals in the context of the Mediterranean Sea. A significant problem arises when ships refuse to respond to AIS signals or provide potentially misleading responses, making it difficult for the Navy to correctly identify the type of ship and its intention. In this context, a solution was needed that allowed the Navy to filter and prioritize its interventions, enabling it to focus resources on cases that presented higher levels of risk or suspicion. This involved the development of a system capable of distinguishing, for example, between a fishing boat and a cruise ship based solely on trajectories, thereby improving the efficiency and effectiveness of the Navy's monitoring and response operations.

The approach presented here aims to recognize the type of vessel among N possible classes using adaptive algorithms specifically trained for this purpose. The data considered for the feature vector include the identifier, position, instantaneous velocity, and heading of each vessel. The adaptive algorithms have been trained using data acquired from radars on multiple vessels (after excluding the identifiers). Positional data are not included as vessels may follow new routes that are not present in the database, making the algorithm less reliable. Additionally, changes in vessel positions have been eliminated due to variations in sampling frequency, which could result in unreliable acceleration and deceleration data.

2. Materials and Methods

The experiments that were conducted during the search are summarized in three steps: (A) *data preparation*; (B) *experimentation*; (C) *results analysis*. The first step, data preparation, is structured in additional three phases: (A.I) *data cleaning*; (A.II) *data pre-processing*; (A.III) *model definition*;

2.1. A.I—Data Cleaning

To ensure a reliable dataset for the experiments, the following data-cleaning procedure was implemented (Figure 1):

Following this process, the analysis sample comprises 3669 vessels, each with a minimum of 100 consecutive radar detection points. Furthermore, in accordance with the Navy's approval, six distinct vessel classes have been identified (N = 6), serving as the targets for intelligent recognition by the adaptive models (Table 1). Note that it was not possible to measure the acceleration of naval vessels as their positions were detected at non-uniform time intervals.

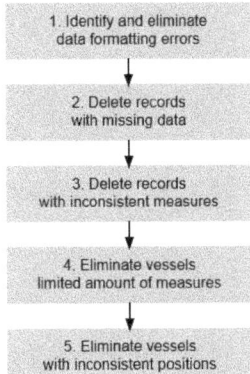

Figure 1. Flowchart of the data-cleaning procedure. (1) Identifying and eliminating any data-formatting errors; (2) removing records with missing data; (3) discarding records with incorrect data (such as velocity exceeding 100 knots, positional data on land, etc.); (4) excluding all vessels for which the number of sequential detections is below a certain threshold (100 points); (5) excluding any vessels that appear in two different and distant parts of the world within a short time span.

Table 1. Number of vessels belonging to each of the six classes, before and after data cleaning. The maximum, minimum, and average readings of each class and the standard deviation from the averages are also shown.

Data	PAX-TMP	TM	TMC	TMO	TU	TUG	TOT
No. Vessels	871	1074	576	2857	849	303	6522
No. Vessels (no. pts $\geq$ 100)	502	634	323	1717	304	194	3669
Min Route Distance	103	101	101	101	100	100	100
Max Route Distance	4774	5315	3550	4813	3610	4985	4985
No. Routes-avg	1240.08	1143.85	873.88	950.60	570.01	1280.52	~
No. Routes-std dev	956.32	971.77	602.44	650.53	495.49	937.40	~

2.2. A.II—Data Pre-Processing

The strategic objective of this experiment is to represent each vessel through the statistical profile of its route. In the dataset provided by the Navy, there are only two variables that characterize the route of a vessel: the velocity and direction of the bow at the time of radar detection. This is regardless of the stretch of sea crossed.

For example, by dividing the variable "velocity" into regular intervals (bins), you can measure how often the velocity of each vessel falls within each of them, during its journey. With an appropriate transformation, one can define the general probability with which a vessel can be found in each of these intervals. By establishing a defined number of intervals (bins) for each variable that characterizes the vessel's navigation, the statistical profile of each vessel's route can be established through the use of the probability density function.

Utilizing these intervals, we decided to characterize the only two variables available in the database for each vessel: Punctual velocity and direction at every radar detection point along the route. However, it is important to note that these measurements are taken in an unsystematic manner, rather than following a strict sampling plan.

A new variable is introduced to the statistical profile of the route of each vessel: delta velocity. Table 2 illustrates the variables that define the statistical profile of the route of each vessel, according to the pre-processing strategy adopted in experimentation no. 3.

Table 2. A total of 67 variables that will be used to rewrite the database of 3669 vessels.

ID	Variable Name	Code	Calculation	No. Intervals
1	Punctual velocity	$v(t)$	Available	22
2	Delta direction bow	$r(t, t-1)$	$|r(t) - r(t-1)|$	19
3	Delta velocity	$dv(t, t-1)$	$v(t) - v(t-1)$	23
4	Global average velocity			1 value
5	Global velocity variance			1 value
6	Global velocity variance with			1 value
	Total variables			67

For illustrative purposes, we provide the statistical profile of the velocity, $v(t)$, of the change of route, $r(t, t-1)$, and of the dynamic delta of the velocity, $dv(t, t-1)$, of a vessel randomly selected for each class. This is to give an idea of the type of input that the adaptive algorithms must manage to define an analysis model that allows the automatic classification of unknown vessels. See Figures 2–4.

Although Figures 2–4 represent examples of vessels randomly chosen in the database, it is evident that these new variables should provide a reliable portrait of the navigation style of each of the six types of vessels. In the following paragraphs, we will measure the accuracy and precision of this third pre-processing strategy.

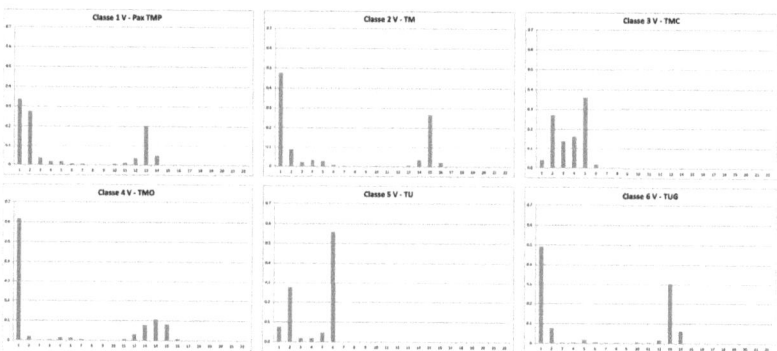

Figure 2. Statistical profile of the punctual velocity of 6 vessels randomly selected, each belonging to one of the 6 classes, with the goal of automatic classification.

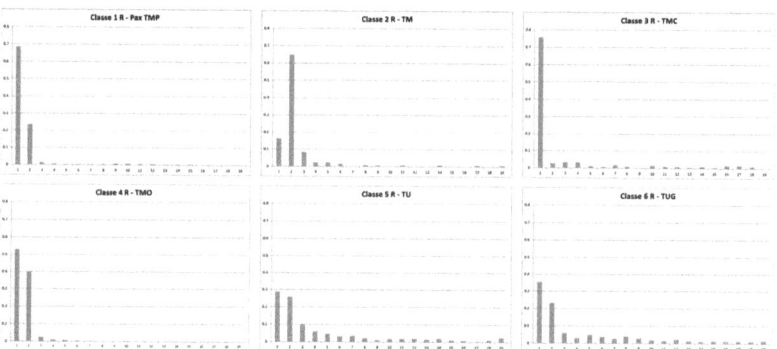

Figure 3. Statistical profile of the change of direction of route of 6 vessels randomly chosen, each belonging to 6 target classes used for automatic classification.

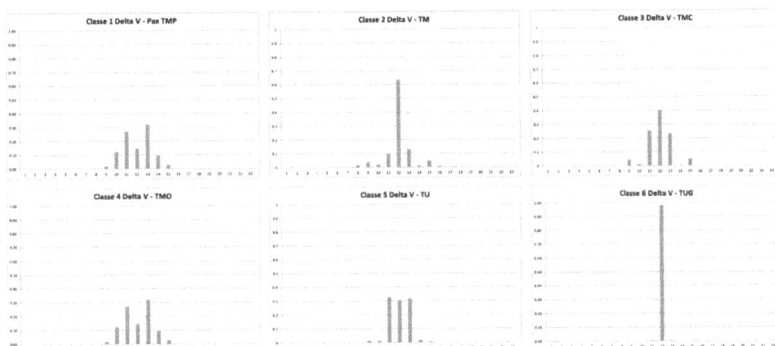

Figure 4. Statistical profile of the dynamic delta of the punctual velocity of 6 vessels randomly chosen, each belonging to the 6 target classes used for the automatic classification.

2.3. A.III—Model Definition

To address the complexities in this classification problem, we propose a new approach, denoted as 'Take One Class at a Time' (TOCAT). This classification strategy operates on two fundamental principles during the training/testing stage:

(a) *Breakdown* of a multinomial classification (1 of N) into N independent binary classifications (1 of 2). Each class is treated separately: all records of the focus class are assigned target 1 while all records belonging to other classes are assigned target 0. The advantage of this system is that even when faced with a high number of classes N, these are reduced to N binary classification processes. Therefore, instead of having a single process that must decide the class membership of a record to N possible classes, there will be N processes, each specialized in a single class, which must decide whether the record is of that class or not;

(b) *Free identification* of the best algorithm for each class. As each class is treated separately, each will be reviewed by multiple Machine Learning methods to select the type of algorithm and structure that obtains the best predictive result in testing. This procedure, which allows for the selection of a different algorithm for each class, can improve the overall performance of the system.

TOCAT, therefore, is not tied to a specific algorithm, but draws on a variety of algorithms to tackle individual bimodal classifications. At the end of the training/testing stage, each class will be associated with the algorithm that achieved the best result, as assessed by the values of Sensitivity and Specificity on the confusion matrix. Only the algorithm with the best performance will be used for the recall stage on new records.

The output of the recall stage of the TOCAT system, in which unclassified new input records are assigned a target by the N networks, is complex. Each network specialized in a single target is called to respond; thus, there may be conflicts in the assignments. This feature is an advantage when it is important to identify ambiguous records that need to be reported by the system. Therefore, using the TOCAT strategy, it is possible for each new pattern to be attributed a fuzzy membership (from 0 to 1) with each of the N Classes.

For the training phase, many adaptive algorithms are used (machine learning and artificial neural networks). The results of these algorithms were finally filtered by a neural meta-network [9,10], which significantly exceeded the results of the best basic algorithm. The different algorithms were implemented through two types of Software, accredited for scientific research: (a) *Supervised ANNs (version 27.5, Semeion, 1999–2017)*; (b) *Meta Net Multi Train (version 3.5, Semeion, 2010–2015)*.

For the experimental phase, a set of algorithms was utilized to explore and analyze the data. Some of the algorithms are commonly used in the literature (kNN [11,12], naive Bayes [13–17], majority vote [9,10,18]), while the remaining algorithms have been specifically developed by Semeion for pattern recognition. The objective was to compare the effectiveness and efficiency of the algorithms and determine which ones would yield the most accurate and reliable results: (a) *backpropagation (Bp)* [19–21]; (b) *deep learning (Deep)* [22–24]; (c) *adaptive vector quantization (AVQ)* [25–28]; (d) *kNN* [11,12]; (e) *meta Bayes (Mb)* [29]; (f) *Conic Net* [30]; (g) *Sine Net (Sn)* [31]; (h) *bimodal (Bm)* [32]; (i) *majority vote (Mv)* [9]; (j) *naive Bayes* [13–17]; (k) *supervised contractive map (SVCm)* [32]. The validation protocol used for all the algorithms is the training–testing protocol [33–35] (Figure 5: Validation protocol—5 × 2 CV (training–testing)).

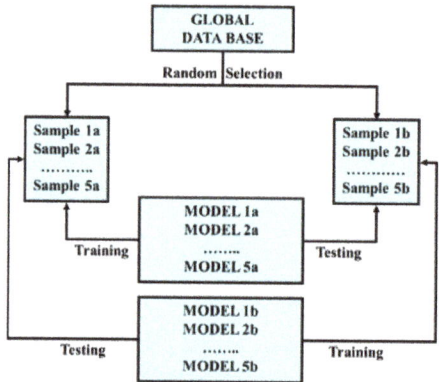

Figure 5. Validation protocol—5 × 2 CV (training–testing)).

In this experiment, three classification lines are used: (a) *Classification 1 of 5*: The attempt is to automatically classify each vessel in one of the five possible classes. In this specific case, class 2 (TM9) is excluded, as the results appear to be difficult to interpret and potentially confusing. This complexity arises from the indiscriminate inclusion of vastly different types of vessels within this class. (b) *Classification 1 of 6*: The attempt is to automatically classify each vessel in one of the six classes provided; (c) *Classification "Take One Class at a Time" (TOCAT)*: Adaptive algorithms are trained to recognize when each vessel belongs or not to a specification of the six possible classes, and the operation is repeated by placing one class at a time in relation to all the others. This procedure generates six different datasets, each of which is subjected to two-class validation (focused class versus other classes).

3. Results

3.1. Exploratory Analysis

At this juncture, an unsupervised neural network is deployed once more to discern the extent to which our pre-processing step can spontaneously segregate the database into the six distinct vessel classes. A self-organizing map (SOM) [28] is used for this purpose with a square grid of 15 × 15, capable of generating 225 codebooks, where each codebook represents a similar group of vessels; Figure 6 shows the results of the SOM software.

Figure 6 illustrates a more defined spontaneous separation of vessels into the six targeted classes for classification, compared to the outputs produced by previous pre-processing strategies. Class 2 (TM) remains the most difficult to characterize and therefore to be separated from the others. In all cases, even with this pre-processing strategy, the definition of the classification model is very complex due to the notable non-linear separability of some classes.

Figure 7 shows the projection grid of the SOM with the overlap percentage of vessels of different classes in each of the 225 codebook cells.

The results of the analysis using SOMs highlight the potential of the selected set of variables. Each class is distributed in specific areas of the map, even though the resulting overlaps between classes make it difficult to obtain an accurate classification system that can be used in operational mode.

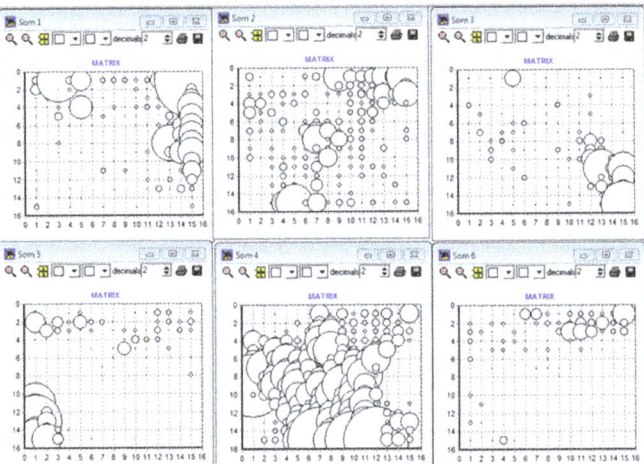

Figure 6. PAX-TMP = Som1; TM = Som2; TMC = Som3; TMO = Som4; TU = Som5; TUG = Som6.

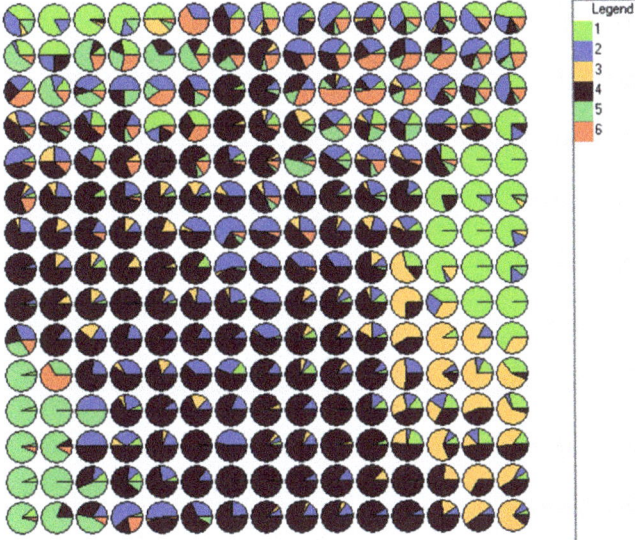

Figure 7. A 15x15 grid of the SOM with the overlap percentage in each cell of vessels belonging to different classes (PAX-TMP = 1; TM = 2; TMC = 3; TMO = 4; TU = 5; TUG = 6).

3.2. Results Analysis

3.2.1. Classification with Five Classes

Table 3 shows the results of classifications one of five.

3.2.2. Classification with Six Classes

Table 4 shows the results obtained by repeating the experiment with all six classes. In this case, the use of the meta-network is ineffective.

The outcomes of this experiment hold merit academically, but their practical applicability remains limited.

3.2.3. Class Classification of TOCAT

In this experiment, we present results from the six one-of-two classifications (class vs. other classes) to determine the sensitivity and specificity for each classification, as detailed in (Tables 5–10).

Table 3. Results of the classifications on 5 classes. Class 2 (TM) is excluded.

Adaptive Algorithms	PAX-TMP	TMC	TMO	TU	TUG	A.Mean	W.Mean	Error	SW
Mb	0.7602	0.6757	0.9468	0.8593	0.6591	0.7802	0.8622	193	Sem. no. 55
Mv	0.6878	0.6622	0.885	0.837	0.7614	0.7667	0.818	255	Sem. no. 55
DeepSn	0.6742	0.6554	0.8702	0.8148	0.7727	0.7575	0.8051	273	Sem. no. 12
kNN_N1	0.6471	0.7365	0.9345	0.8667	0.5682	0.7506	0.8387	226	Sem. no. 12
DeepBp	0.6109	0.6959	0.8307	0.7852	0.8295	0.7504	0.7773	312	Sem. no. 12
DeepBm	0.7376	0.6351	0.8863	0.8222	0.6591	0.7481	0.8158	258	Sem. no. 12
K-CM	0.6471	0.723	0.9333	0.8593	0.5682	0.7461	0.8358	230	Sem. no. 12
DeepConic	0.6968	0.6216	0.8764	0.837	0.6818	0.7427	0.8051	273	Sem. no. 12
Bm	0.6968	0.6892	0.8591	0.8222	0.6364	0.7407	0.798	283	Sem. no. 12
Sn	0.6516	0.7095	0.8739	0.8667	0.5909	0.7385	0.803	276	Sem. no. 12
Conic	0.6561	0.6081	0.8826	0.8519	0.6932	0.7384	0.803	276	Sem. no. 12
Bp	0.7104	0.7162	0.8467	0.8148	0.5909	0.7358	0.7923	291	Sem. no. 12

Table 4. Results of the classification on 6 classes.

Adaptive Algorithms	PAX-TMP	TM	TMC	TMO	TU	TUG	A.Mean	W.Mean	Error	SW
kNN_1	63.89%	50.95%	75.51%	86.79%	80.95%	42.16%	66.71%	73.45%	481	Sem. no. 12
Bm	71.43%	21.84%	67.35%	84.91%	78.91%	55.88%	63.39%	68.49%	571	Sem. no. 12
Conic	68.65%	38.61%	67.35%	81.72%	71.43%	45.10%	62.14%	68.32%	574	Sem. no. 12
DeepBp	71.83%	37.97%	61.90%	76.53%	79.59%	34.31%	60.36%	65.84%	619	Sem. no. 12
SVCm	71.83%	27.53%	63.95%	83.96%	72.79%	41.18%	60.21%	67.49%	589	Sem. no. 12
DeepConic	59.52%	37.34%	63.95%	77.48%	71.43%	42.16%	58.65%	64.40%	645	Sem. no. 12
AVQ	50.00%	33.86%	71.43%	81.72%	74.83%	30.39%	57.04%	64.68%	640	Sem. no. 12
Naive Bayes	11.11%	0.32%	41.50%	75.83%	0.68%	97.06%	37.75%	45.97%	979	Sem. no. 12

Table 5. Classification of Class 1: PAX-TMP.

ANN	Class1	Others	A.Mean	W.Mean	Errors
Conic(C1)	81.12%	91.03%	86.08%	89.65%	185
DeepConic(C1)	80.32%	91.61%	85.97%	90.04%	178
DeepBm(C1)	80.32%	91.61%	85.97%	90.04%	178
FFBp(C1)	79.92%	90.57%	85.25%	89.09%	195
DeepBp(C1)	78.71%	93.50%	86.11%	91.44%	153
DeepSn(C1)	76.31%	92.65%	84.48%	90.37%	172
kNN(C1)	74.30%	95.77%	85.04%	92.78%	129

Table 6. Classification of Class 2: PAX-TM.

ANN	Class1	Others	A.Mean	W.Mean	Errors
Conic(C2)	72.03%	70.01%	71.02%	70.35%	539
DeepConic(C2)	68.81%	70.74%	69.77%	70.41%	538
DeepBp(C2)	64.95%	75.71%	70.33%	73.87%	475
FFBm(C2)	62.70%	76.05%	69.37%	73.76%	477
DeepBm(C2)	62.06%	77.24%	69.65%	74.64%	461
kNN(C2)	58.52%	93.56%	76.04%	87.57%	226

Table 7. Classification of Class 3: TMC.

ANN	Class1	Others	A.Mean	W.Mean	Errors
Deep_Conic(C3)	76.36%	88.70%	82.53%	87.57%	223
FFBm(C3)	75.76%	90.67%	83.21%	89.30%	192
Conic(C3)	73.94%	93.43%	83.69%	91.64%	150
Conic(C3)	72.73%	93.25%	82.99%	91.36%	155
DeepBp(C3)	71.52%	92.69%	82.11%	90.75%	166
kNN(C3)	70.30%	98.16%	84.23%	95.60%	79

Table 8. Classification of Class 4: TMO.

ANN	Class1	Others	A.Mean	W.Mean	Errors
kNN(C4)	88.71%	87.34%	88.03%	87.99%	218
FFBp(C4)	84.52%	82.74%	83.63%	83.58%	298
DeepBm(C4)	85.10%	81.17%	83.14%	83.03%	308
DeepBp(C4)	84.52%	80.65%	82.58%	82.48%	318

Table 9. Classification of Class 5: TU.

ANN	Class1	Others	A.Mean	W.Mean	Errors
DeepBm(C5)	88.00%	88.11%	88.06%	88.10%	207
DeepBp(C5)	81.33%	93.90%	87.62%	92.82%	125
Conic(C5)	80.67%	96.42%	88.54%	95.06%	86
kNN(C5)	80.00%	98.93%	89.47%	97.30%	47
FFBp(C5)	80.00%	97.67%	88.84%	96.15%	67
DeepConic(C5)	79.33%	96.92%	88.13%	95.40%	80

Table 10. Classification of Class 6: TUG.

ANN	Class1	Others	A.Mean	W.Mean	Errors
Conic(C6)	88.24%	80.51%	84.37%	80.95%	346
DeepBm(C6)	84.31%	76.14%	80.23%	76.60%	425
DeepConic(C6)	63.73%	91.66%	77.69%	90.09%	180
DeepBp(C6)	65.69%	89.26%	77.48%	87.94%	219
FFBp(C6)	65.69%	88.62%	77.15%	87.33%	230
kNN(C6)	53.92%	98.72%	76.32%	96.20%	69

Table 11 summarizes, in an overview, the results of the best algorithms in each of the 6 classifications

Table 11. The best algorithms of the 6 classifications with the TOCAT procedure.

	Classes	Algorithm	Sensibility	Specificity	A.Mean	W.Mean	Errors
Class 1	PAX-TMP	Conic	81.12%	91.03%	86.08%	89.65%	185
Class 2	TM	Conic	72.03%	70.01%	71.02%	70.35%	539
Class 3	TMC	DeepConic	76.36%	88.70%	82.53%	87.57%	223
Class 4	TMO	kNN_1	88.71%	87.34%	88.03%	87.99%	218
Class 5	TU	DeepBm	88.00%	88.11%	88.06%	88.10%	207
Class 6	TUG	Conic	88.24%	80.51%	84.37%	80.95%	346
	Average × Class		82.41%	84.28%	83.35%	84.10%	286.33

4. Discussion

Significant findings have emerged. Firstly, the study highlighted the potential of an approach that involves transforming a sparse and incomplete dataset of vessel paths into a consistent set of features that could be universally applicable across all types of vessels, despite variations in the number of observations. While the individual techniques employed to convert vessel trajectories into fixed features were not new, the innovation lay in their combined application, resulting in the transformation of temporal flows into spatial features, derived from highly heterogeneous data.

The second significant result was the development of the TOCAT (take one class at a time) research design. While the TOCAT strategy itself is not entirely new, as it is already used in support vector machine algorithms [36,37] for multinomial classifications, our innovative application of this processing strategy was significant. Existing methods typically rely on the development of a singular, albeit sophisticated, convolutional or recurrent neural network, which shoulders the full responsibility of recognition. The TOCAT strategy, however, hinges on the concept of creating a diversified ecosystem of neural networks, varying both topologically and mathematically, which is not strictly reliant on gradient descent or decision trees. Each of these networks is adept at specializing in a statistical niche within the vessels' trajectories.

By utilizing different artificial neural networks (ANNs) and machine learning techniques for distinct "one of two" classification tasks, we achieved diverse mathematical and topological representations for each ANN used. This diversity enhanced classification accuracy, as each "one of two" classification tasks could leverage a specific ANN suitable for recognizing a particular class of vessels. The efficacy of this mathematical "biodiversity" in improving the final results represents an important milestone in this work. In summary, the collective integration of small, distinct artificial systems outperformed a single large ANN attempting to comprehend the entire scope independently.

5. Conclusions

This paper proposes a novel approach to vessel classification, using the normalization of sparse and incomplete vessel trajectory data into a universal set of features, which is applicable despite varying observation numbers. Our model utilizes the TOCAT (take one class at a time) design, a strategy typically used in support vector machine algorithms, but uniquely applied in this study to individual 'one versus rest' classification tasks using a variety of diverse artificial neural networks (ANNs). We derived distinct mathematical and topological representations from vessel trajectory data for each 'one versus rest' classification task, leveraging the proven capabilities of specific ANNs for recognizing certain vessel classes. The findings, which show a mean accuracy of 83% in a six-class classification task, suggest that the collaborative employment of these specialized ANNs could potentially outperform a single, larger ANN assigned to the entire classification task. During the experiments, we also pinpointed several critical aspects tied to data processing: primarily, the need for statistical sampling of the AIS signals from each vessel's trajectories. This method enables a robust estimation of vital parameters such as the vessel's speed, deceleration, and acceleration. Furthermore, as anticipated, the analysis highlights the importance of avoiding the use of overly complex algorithms, especially when the number of input variables is limited and the samples have not been collected through robust sampling procedures. Future research endeavors may explore the scalability and generalizability of the proposed approaches and extend their applications to other domains beyond vessel recognition and comparisons with the most recent techniques existing in the field.

Author Contributions: Conceptualization, P.M.B.; Methodology, P.M.B., G.M., M.B. and R.P.; Software, P.M.B. and G.M.; Validation, P.M.B. and G.M.; Formal Analysis, P.M.B.; Investigation, P.M.B., G.M., M.B. and R.P.; Resources, G.R. and G.C.; Data Curation, P.M.B., G.M., G.C., M.B. and R.P.; Writing—Original Draft Preparation, P.M.B., G.M., G.C., M.B. and R.P.; Writing—Review and Editing, P.M.B., M.B. and R.P.; Visualization, P.M.B., G.M., M.B. and R.P.; Supervision, P.M.B., G.M., M.B. and R.P.; Project Administration, P.M.B. and M.B.; Funding, P.M.B. All authors have read and agreed to the published version of the manuscript.

Funding: This research received no external funding.

Data Availability Statement: The data are not publicly available due to legal agreements between the research institution and collaborating parties.

Conflicts of Interest: The authors declare no conflict of interest.

References

1. Pallotta, G.; Horn, S.; Braca, P.; Bryan, K. Context-enhanced vessel prediction based on Ornstein-Uhlenbeck processes using historical AIS traffic patterns: Real-world experimental results. In Proceedings of the 17th International Conference on Information Fusion (FUSION), Salamanca, Spain, 7–10 July 2014; pp. 1–7.
2. Pallotta, G.; Vespe, M.; Bryan, K. Vessel Pattern Knowledge Discovery from AIS Data: A Framework for Anomaly Detection and Route Prediction. *Entropy* **2013**, *15*, 2218–2245. [CrossRef]
3. Teutsch, M.; Krüger, W. Classification of small boats in infrared images for maritime surveillance. In Proceedings of the 2010 International WaterSide Security Conference, Carrara, Italy, 3–5 November 2010; pp. 1–7, ISSN: 2166-1804. [CrossRef]
4. Duan, H.; Ma, F.; Miao, L.; Zhang, C. A semi-supervised deep learning approach for vessel trajectory classification based on AIS data. *Ocean. Coast. Manag.* **2022**, *218*, 106015. [CrossRef]
5. Breiman, L. Bagging predictors. *Mach. Learn.* **1996**, *24*, 123–140. [CrossRef]
6. Breiman, L. Random Forests. *Mach. Learn.* **2001**, *45*, 5–32. [CrossRef]
7. Zhong, H.; Song, X.; Yang, L. Vessel Classification from Space-based AIS Data Using Random Forest. In Proceedings of the 2019 5th International Conference on Big Data and Information Analytics (BigDIA), Kunming, China, 8–10 July 2019; pp. 9–12. [CrossRef]
8. Guan, Y.; Zhang, J.; Zhang, X.; Li, Z.; Meng, J.; Liu, G.; Bao, M.; Cao, C. Identification of Fishing Vessel Types and Analysis of Seasonal Activities in the Northern South China Sea Based on AIS Data: A Case Study of 2018. *Remote Sens.* **2021**, *13*, 1952. [CrossRef]
9. Kittler, J.; Hatef, M.; Duin, R.; Matas, J. On combining classifiers. *IEEE Trans. Pattern Anal. Mach. Intell.* **1998**, *20*, 226–239. [CrossRef]
10. Kuncheva, L. *Combining Pattern Classifiers: Methods and Algorithms: Second Edition*; Wiley: Hoboken, NJ, USA, 2004; Volume 47. [CrossRef]
11. Kowalski, B.R.; Bender, C.F. K-Nearest Neighbor Classification Rule (pattern recognition) applied to nuclear magnetic resonance spectral interpretation. *Anal. Chem.* **1972**, *44*, 1405–1411. [CrossRef]
12. Aha, D.W.; Kibler, D.; Albert, M.K. Instance-based learning algorithms. *Mach. Learn.* **1991**, *6*, 37–66. [CrossRef]
13. Friedman, N.; Geiger, D.; Goldszmidt, M. Bayesian Network Classifiers. *Mach. Learn.* **1997**, *29*, 131–163. [CrossRef]
14. Watson, T.J. An empirical study of the naive Bayes classifier. In Proceedings of the IJCAI 2001 Workshop on Empirical Methods in Artificial Intelligence, Seattle, WA, USA, 4–10 August 2001.
15. Zhang, H. The Optimality of Naive Bayes. In Proceedings of the Seventeenth International Florida Artificial Intelligence Research Society Conference, FLAIRS 2004, Miami Beach, FL, USA, 17–19 May 2004.
16. Nielsen, S.H.; Nielsen, T.D. Adapting Bayes network structures to non-stationary domains. *Int. J. Approx. Reason.* **2008**, *49*, 379–397. [CrossRef]
17. John, G.H.; Langley, P. Estimating continuous distributions in Bayesian classifiers. *arXiv* **2013**, arXiv:1302.4964.
18. Lam, L.; Suen, S. Application of majority voting to pattern recognition: An analysis of its behavior and performance. *Appl. Major. Voting Pattern Recognit. Anal. Its Behav. Perform.* **1997**, *27*, 553–568. [CrossRef]
19. Rumelhart, D.E.; McClelland, J.L.; Group, P.R. *Parallel Distributed Processing: Explorations in the Microstructure of Cognition: Foundations*; The MIT Press: Cambridge, MA, USA, 1986. [CrossRef]
20. Rumelhart, D.; Hinton, G.; Williams, R. Learning Internal Representations by Error Propagation. In *Readings in Cognitive Science*; Elsevier: Amsterdam, The Netherlands, 1988; pp. 399–421. . [CrossRef]
21. Lecun, Y. A Theoretical Framework for Back-Propagation. In *Proceedings of the 1988 Connectionist Models Summer School, CMU, Pittsburg, PA*; Morgan Kaufmann: Burlington, MA, USA, 1988; pp. 21–28.
22. Bengio, Y.; Simard, P.; Frasconi, P. Learning long-term dependencies with gradient descent is difficult. *IEEE Trans. Neural Netw.* **1994**, *5*, 157–166. [CrossRef]
23. Hinton, G.E.; Osindero, S.; Teh, Y.W. A fast learning algorithm for deep belief nets. *Neural Comput.* **2006**, *18*, 1527–1554. [CrossRef] [PubMed]
24. Bengio, Y. *Learning Deep Architectures for AI*; Now Publishers, Inc.: Norwell, MA, USA, 2009; Volume 2, pp. 1–127. [CrossRef]
25. Kohonen, T. Improved versions of learning vector quantization. In Proceedings of the 1990 IJCNN International Joint Conference on Neural Networks, San Diego, CA, USA, 17–21 June 1990; IEEE: Piscataway, NJ, USA; Volume 1, pp. 545–550. [CrossRef]
26. Kosko, B. *Neural Networks and Fuzzy Systems: A Dynamical Systems Approach to Machine Intelligence*; Prentice Hall: Hoboken, NJ, USA, 1991.
27. Kosko, B. *Neural Networks for Signal Processing*; Prentice Hall: Hoboken, NJ, USA, 1992.
28. Kohonen, T. Learning Vector Quantization. In *Self-Organizing Maps*; Kohonen, T., Ed.; Springer Series in Information Sciences; Springer: Berlin/Heidelberg, Germany, 1995 ; pp. 175–189. [CrossRef]

29. Buscema, M.; Tastle, W.J.; Terzi, S. Meta Net: A New Meta-Classifier Family. In *Data Mining Applications Using Artificial Adaptive Systems*; Springer: New York, NY, USA, 2013; pp. 141–182.
30. Buscema, P.M.; Massini, G.; Fabrizi, M.; Breda, M.; Della Torre, F. The ANNS approach to DEM reconstruction. *Comput. Intell.* **2018**, *34*, 310–344. [CrossRef]
31. Buscema, M.; Terzi, S.; Breda, M. Using sinusoidal modulated weights improve feed-forward neural network performances in classification and functional approximation problems. *WSEAS Trans. Inf. Sci. Appl.* **2006**, *3*, 885–893.
32. Buscema, M.; Benzi, R. Quakes Prediction Using Highly Non Linear Systems and A Minimal Dataset. In *Advanced Networks, Algorithms and Modeling for Earthquake Prediction*; River Publishers: Aalborg, Denmark, 2011.
33. Kohavi, R. A study of cross-validation and bootstrap for accuracy estimation and model selection. In Proceedings of the 14th International Joint Conference on Artificial Intelligence, Montreal, QC, Canada 20–25 August 1995; Morgan Kaufmann Publishers Inc.: Burlington, MA, USA; Volume 2, pp. 1137–1143.
34. Dietterich, T.G. Approximate statistical tests for comparing supervised classification learning algorithms. *Neural Comput.* **1998**, *10*, 1895–1923. [CrossRef]
35. Arlot, S.; Celisse, A. A survey of cross-validation procedures for model selection. *Statist. Surv.* **2010**, *4*, 40–79. [CrossRef]
36. Kecman, V. *Learning and Soft Computing, Support Vector Machines, Neural Networks, and Fuzzy Logic Models*; MIT Press: Cambridge, MA, USA, 2001.
37. Keerthi, S.S.; Shevade, S.K.; Bhattacharyya, C.; Murthy, K.R.K. Improvements to Platt's SMO Algorithm for SVM Classifier Design. *Neural Comput.* **2001**, *13*, 637–649. [CrossRef]

Disclaimer/Publisher's Note: The statements, opinions and data contained in all publications are solely those of the individual author(s) and contributor(s) and not of MDPI and/or the editor(s). MDPI and/or the editor(s) disclaim responsibility for any injury to people or property resulting from any ideas, methods, instructions or products referred to in the content.

Article

Automated Segmentation of Optical Coherence Tomography Images of the Human Tympanic Membrane Using Deep Learning

Thomas P. Oghalai [1], Ryan Long [2], Wihan Kim [2], Brian E. Applegate [2] and John S. Oghalai [2,*]

1 Department of Electrical and Computer Engineering, University of Wisconsin-Madison, Madison, WI 53706, USA; toghalai@wisc.edu
2 Caruso Department of Otolaryngology-Head and Neck Surgery, University of Southern California, Los Angeles, CA 90033, USA; ryanlong@usc.edu (R.L.); wihankim@usc.edu (W.K.); brianapp@usc.edu (B.E.A.)
* Correspondence: oghalai@usc.edu

Abstract: Optical Coherence Tomography (OCT) is a light-based imaging modality that is used widely in the diagnosis and management of eye disease, and it is starting to become used to evaluate for ear disease. However, manual image analysis to interpret the anatomical and pathological findings in the images it provides is complicated and time-consuming. To streamline data analysis and image processing, we applied a machine learning algorithm to identify and segment the key anatomical structure of interest for medical diagnostics, the tympanic membrane. Using 3D volumes of the human tympanic membrane, we used thresholding and contour finding to locate a series of objects. We then applied TensorFlow deep learning algorithms to identify the tympanic membrane within the objects using a convolutional neural network. Finally, we reconstructed the 3D volume to selectively display the tympanic membrane. The algorithm was able to correctly identify the tympanic membrane properly with an accuracy of ~98% while removing most of the artifacts within the images, caused by reflections and signal saturations. Thus, the algorithm significantly improved visualization of the tympanic membrane, which was our primary objective. Machine learning approaches, such as this one, will be critical to allowing OCT medical imaging to become a convenient and viable diagnostic tool within the field of otolaryngology.

Keywords: deep learning algorithm; tympanic membrane; Tensorflow; optical coherence tomography; convolutional neural network

1. Introduction

The tympanic membrane (TM), also known as the eardrum, is vital to hearing because it converts sound pressure waves in air into mechanical vibrations. These vibrations are then passed through the ossicular chain (the three middle ear bones) into the cochlea, where the vibrations are transduced into neural signals which our brain interprets as sound [1,2]. The structure of the TM is important for normal hearing. Many diseases cause hearing loss by altering the TM, such as perforations [3–7], retraction pockets [8–11], and cholesteatoma [11–13]. Other ear diseases cause hearing loss by affecting the middle ear space, which lies directly behind the TM, such as ear infections, middle ear fluid, or tumors of the middle ear. Thus, proper evaluation of the ear is critical for the diagnosis and treatment of these conditions [14]. Current medical management involves using an otoscope [15], which uses a small light with a magnifying glass to visualize the TM deep within the ear canal. However, this method of examining the ear is limited in the information it provides. It does allow the physician to see the surface of the TM. It can sometimes permit a limited ability to view the middle ear space and its underlying structures through the translucent nature of the TM. Sometimes, the illumination can be inadequate, making

otoscopy challenging. Furthermore, accurate otoscopic evaluations require extensive training and familiarity with the appearance of normal anatomy and pathology on examination, with conclusions often depending based on physician experience. Thus, making the clinical diagnosis of an ear infection is surprisingly fraught with difficulty, and it is common for this to be missed by routine clinical exam [16,17].

A potential adjunct to otoscopes that could provide additional anatomical and functional information on the TM and middle ear space to improve the diagnosis and management of ear disease is the use of Optical Coherence Tomography (OCT) [18]. OCT is an interferometry-based non-invasive imaging modality in which laser light is focused into a biological tissue of interest. The back-reflected light is analyzed to provide a depth profile of the tissue (Z-direction), which is comparable to how ultrasound works. By scanning the laser beam in the X and Y directions, 3D morphological images of the sample are obtained. OCT is already in wide clinical use in ophthalmology to help in the diagnosis and treatment of various eye pathologies [19,20]. It is now being explored for applications within the field of otolaryngology, commonly known as Ear, Nose, and Throat [21]. Our research group has been using it to study the ear [22–24]. The theoretical benefit of OCT for imaging the human ear is that it provides a 3D image of the TM, as opposed to only the 2D surface view that one gets from looking through an otoscope. Thus, it should improve the diagnostic sensitivity for ear disease and ultimately permit machine learning approaches to automate the diagnostic process.

We developed a handheld OCT device that provides 3D volumes of the ear [25] that allows one to view beyond the surface level of the TM and image deeper structures often affected by middle ear pathologies (Figure 1) [26]. However, the large, high-resolution image volume stacks are difficult to interpret. It is challenging to distinguish between the various anatomic structures within the ear (e.g., the TM, the ossicles, the cochlea), artifacts that stem from strong tissue reflections that saturate the detector, and fixed pattern artifacts that come from reflections within the optical device itself. Furthermore, the orientation of the image will vary depending on the way the physician holds the device relative to the patient and the patient's individual anatomy. Therefore, while it seems like it should be simple to collect a volume scan of a patient's ear and look at the results immediately, all these factors make it difficult to expeditiously interpret the imaging and make a clinical decision at the point of care. Here, we showed how an artificial intelligence approach based on a deep learning algorithm can be used to address this limitation. We developed an algorithm to quickly locate and segment normal TM anatomy. The main contribution of this algorithm is that it removes artifacts from the OCT images, leaving behind the key structure of interest, which is the TM. Deep learning algorithms for segmentation and diagnostics are starting to be used within the medical field [27–31]. This algorithm, thus, represents an early stage in AI-automated diagnosis of ear disease.

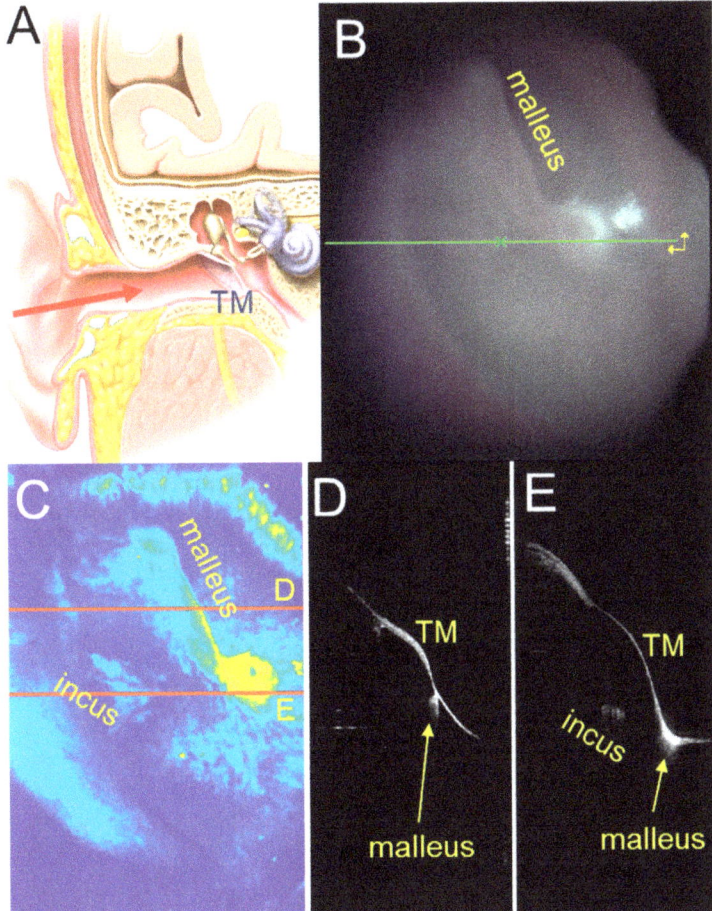

Figure 1. OCT images of the human ear. (**A**) Illustration of the ear. The tympanic membrane (*TM*) is visible by looking down the ear canal (red arrow). (**B**) A video image of the TM, looking through an otoscope. The malleus (the first of the three middle ear bones) is visible along its attachment to the TM. The green line helps the users know where the line scan is being performed. (**C**) A summed voxel projection created from an OCT image stack of a human TM. It has been pseudocolored so that yellow indicates higher reflectivity and dark blue indicates lower reflectivity. The malleus is visible. Also, the incus (the second middle ear bone) is visible because it can be detected by OCT. The red lines illustrate the X-Z slices shown in *D* and *E*. (**D**,**E**) X-Z slices from the image stack collected in the locations indicated by the two red lines in C. Note the thin curved appearance of the TM. The malleus and incus can also be seen under the TM.

2. Methods

2.1. Patient Dataset

This study was approved by the institutional review board at the University of Southern California (protocol HS-17-01014). We collected data from sixteen participants with no known history of ear pathology from the senior author's clinic. All participants had normal ear exams on the day of OCT imaging. Participants ages ranged from 26 to 45 years of age.

OCT volume scans of the posterior half of the tympanic membrane were collected using the same techniques as described in our previous studies [25,32]. Briefly, the participant was seated in an exam chair, and the speculum of the hand-held OCT device was inserted into the ear canal. The TM was brought into focus and the volume scan was collected. This took about 20 s per ear. We used a mixture of both right and left ears for this study.

Each OCT volume scan consisted of 417 X-Z slices (B-scans) that were 417 pixels wide by 669 pixels deep, with a total imaging range of 8 mm (X) by 8 mm (Y) by 12.83 mm (Z). The optical resolution of our OCT device was 35 μm in the X, Y, and Z dimensions, and so, the data were oversampled. Scans were anonymized and then used for this research. We used eight of the volume scans for training the models and the other eight volume scans for validating the finished algorithm (i.e., 8 volume scans * 417 B-scans per volume scan = 3336 images for training and another 3336 images for validating).

2.2. Overview of Our AI methodology

Creating and training our AI methodology involved multiple sequential steps (Figure 2). First, large objects within each X-Z slice of the volume stack were detected using the *large object detection algorithm*. We then manually looked at each object and manually classified it as either TM or non-TM. Next, these large objects were used to train the *large image recognition algorithm* using machine learning. Then, all large objects classified as TM were then re-evaluated using the *small object detection algorithm*, which detected smaller objects that were near the TM. Each new object was then manually classified as either TM or non-TM. Finally, these new smaller objects were used to train the *small image recognition algorithm* using machine learning.

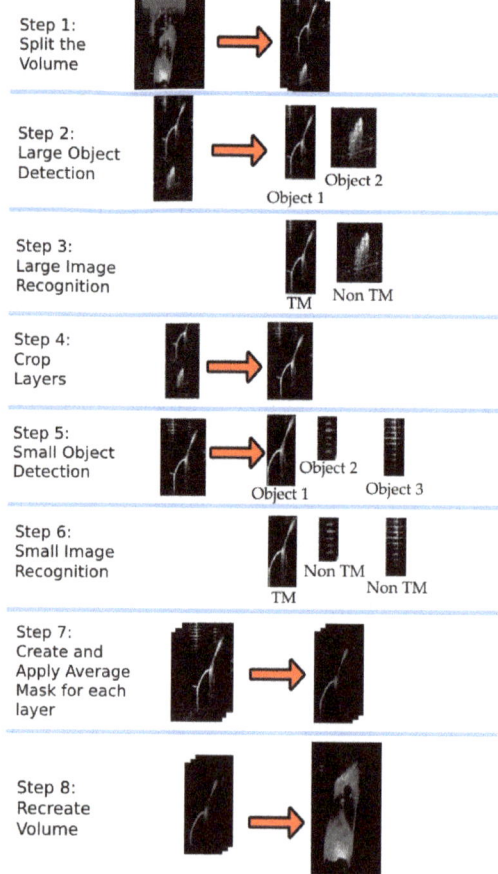

Figure 2. The sequence of steps in our deep-learning-based AI algorithm.

Once trained, using our two-stage AI methodology (first detecting large objects then detecting small objects) was completely automatic and did not require any user involvement. A 3D image stack was sent to the function and a 3D image stack containing only the segmented TM was returned. The entire algorithm was implemented on a PC-compatible desktop computer with an AMD Ryzen 5 3600X 6-Core Processor CPU running Python code. All code and datasets are available for download on our GitHub site [33]. Each step in our AI methodology is explained in detail below.

2.3. Large Object Detection Algorithm

The goal of the large object detection algorithm is to distinguish large structures within the collected images. Each 3D volume image collected by our OCT device was a tiff stack of 417 image slices (B-scans). Each slice represented a 2D cross-section of the volume with the X dimension on the horizontal axis and the Z dimension on the vertical axis. The Y coordinate of each image represents its order in the stack. Each image slice was analyzed sequentially.

First, the image was thresholded to convert each pixel to either black or white, to highlight structures. A higher threshold value resulted in fewer potential objects. This made the process faster, but also lowered accuracy and increased the chances of cropping out portions of the TM that were less intense. Then, the image was blurred to reduce the chance of one object being split up into multiple smaller objects. However, increasing the blur value too much increased the risk of artifacts or middle ear structures appearing to be part of the TM. Next, each large aggregation of white pixels was identified and contour detection was performed. Finally, the size of each aggregation was calculated. If it was higher than the minimum size we defined, the aggregation of white pixels was termed an "object" and the coordinates were saved. A higher minimum size requirement sped up the algorithm but could miss some of the TM objects. Thus, there were three parameters we considered when detecting objects: the threshold value, the blur value, and the minimum size. We iteratively varied these parameters until we achieved a decent balance between accuracy and speed.

From the images collected from the eight participants used for training, this algorithm identified roughly 45,000 large objects. We then went through these manually one-by-one and found that ~43,000 objects were non-TM and ~2000 objects were TM. The reason that there were fewer TM objects was that about half of the slices within each volume stack contained the TM and this would be identified as only one object. In contrast, each slice had multiple distinct artifacts, each identified as separate objects.

2.4. Small Object Detection Algorithm

Within the 2000 TM objects, we found that there were many artifacts that blurred together with the TM. To overcome this, we implemented a second small object detection algorithm, with the goal being to sort through only the TM objects and remove artifacts within them. The benefit of having this two-stage approach was that the more time-intensive process of separating TM from nearby non-TM objects could be focused on fewer and smaller images (i.e., just the 2000 TM objects).

First, the TM objects were selected from the original image data. These were rectangular-sized images with a corner position, length, and width that came from the TM objects detected with the large object detection algorithm. The same sequence of steps was then used, threshold, blurring, and aggregation size, to detect all objects within the image. However, the parameters were different than what was used for the large object detection algorithm, so as to detect most white pixels within the image. Thus, many more objects were found within this single TM object. We found that a good tradeoff in accuracy versus speed was to set the threshold value lower, the blur value lower, and the minimum size just slightly higher. This is due to the fact that more small objects will be detected. By increasing the maximum size threshold, the program will remove more objects before identifying them and complete its process more quickly.

Again, we went through each newly detected object and manually classified each one as either TM or non-TM. Thus, the 2000 TM objects could be broken down and re-evaluated into ~1000 TM objects and ~9000 non-TM objects.

2.5. Image Recognition Algorithms

We used TensorFlow to create two neural network-based image recognition algorithms [13]. Both algorithms used the same convolutional neural network approach (Figure 3). The only difference was that the large image recognition algorithm was trained on objects from the large object detection algorithm. Similarly, the small image recognition algorithm was trained on objects from the small object detection algorithm.

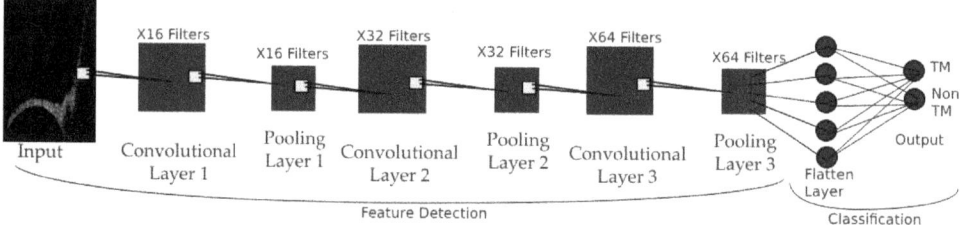

Figure 3. The structure of the convolutional neural network.

Data augmentation was performed by rotating each TM object in varying amounts, thus generating additional TM objects for training both neural networks. Data augmentation allowed the model to better identify the TM more accurately; however, it does carry the risk of overfitting the model to the training data [34–36].

2.6. 3D Reconstruction Algorithm

Once all the slices have TM and non-TM objects classified using the two-stage approach, the 3D TM object needs to be reconstructed. To carry this out, the TM object locations were used as a mask in order to remove all other objects from each slice. The algorithm then reconstructed the volume so that it could be viewed as a 3D object.

3. Results

3.1. Model Results

After training the two-stage algorithm on the data from the eight human participant, we tested it on eight additional volume sets collected from different human participant. This demonstrated that the TM was correctly identified in 95% of the objects. This was determined by going through each individual object identified from within each slice afterward by hand and assessing the accuracy of the detection algorithm. The algorithm only required the 3D volume, with no human intervention necessary. Using our PC-compatible desktop computer, analyzing all eight image stacks took about 66 min total, meaning that the average time to segment the TM from one image stack was 8.25 min.

The representative examples of single slices after segmentation revealed isolated TMs, with most artifacts removed (Figure 4). After every slice was parsed for the TM, the algorithm reconstructed the volume so that it could be viewed as a 3D object (Figure 5).

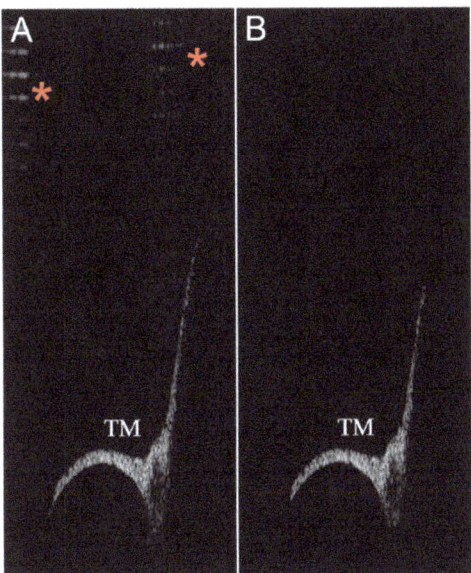

Figure 4. Representative example demonstrating how the algorithm segments out the TM and removes artifacts. (**A**) One slice from the original image. Artifacts are indicated (*red asterisks*). (**B**) After automatically segmenting the TM, the artifacts have been removed.

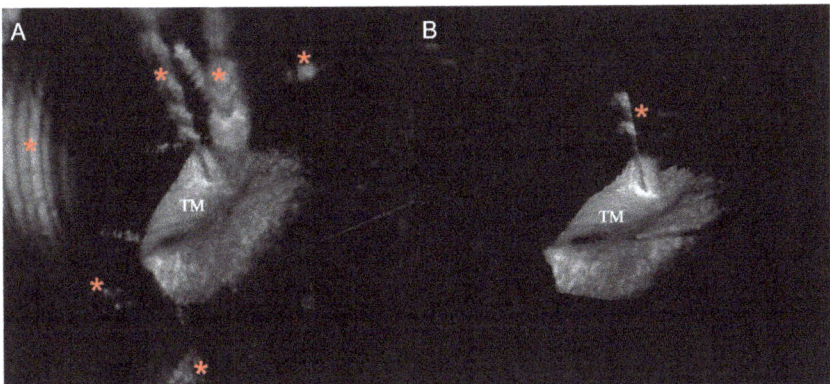

Figure 5. Representative example of 3D reconstructions of one OCT volume of the tympanic membrane. (**A**) Before and (**B**) after running the algorithm to segment out the tympanic membrane (*TM*). Artifacts (*red asterisks*) were either completely removed or greatly reduced in size after segmentation.

The two most common instances of artifacts across hundreds of images were a set of white lines at the top of the image (due to background fixed pattern noise) and vertical lines (due to strong tissue reflections). Our algorithm provided a reasonably good approach to removing both artifacts (Figures 5 and 6).

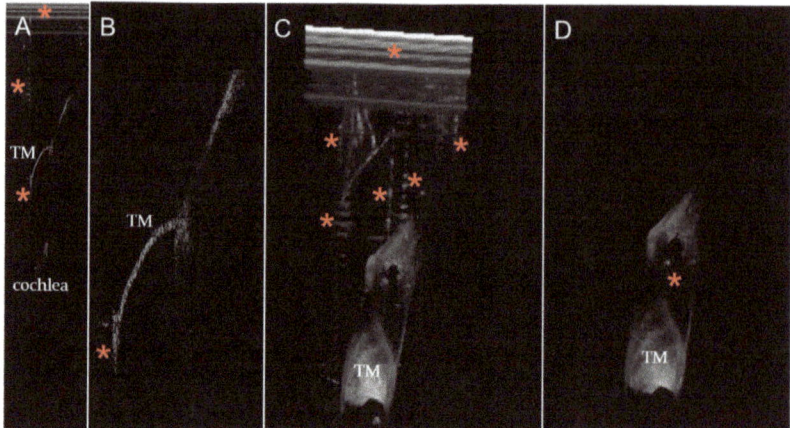

Figure 6. Examples of artifacts that could and could not be removed from the TM. (**A**) One slice from an original volume stack. Artifacts are identified (*red asterisks*). (**B**) The TM object from the same slice. Note that while the cochlea and most of the artifact was segmented out, some residual artifact remained at the bottom of the TM. (**C**) The original 3D volume projection contained many artifacts. (**D**) The segmented 3D volume projection of the TM was much cleaner. However, there was still some artifact within the TM that could not be segmented out.

3.2. Model Details and Rationale for Overfitting

As described earlier, the accuracy of the model on new data (i.e., the image stacks from the eight subjects used for validation) was 95%. To achieve these results, we evaluated the training accuracy and loss for both the large and small image recognition algorithms (Figure 7). Training accuracy and loss were computed by TensorFlow automatically during the training process, and so, only image stacks from the eight training participant were used to generate these plots. The training accuracy corresponded to the percent identified correctly within the subset of the volume stacks used to train the algorithm. The validation accuracy represents the percentage of images that the program correctly identified within the remaining volume stacks. Thus, training accuracy will always be higher than validation accuracy. The training loss represents the penalty involved in failing to identify an image correctly and goes up with overfitting the models.

The x-axis on each graph was measured in samples. Each sample represents one set of training data, and we used 45 samples per trained model. Each sample contained 182 images, and so, the program plotted its training accuracy and loss every 150 images. This was because 20% of the total data was saved randomly to be tested through the model every time the set of 150 images was processed. This means that every time a set of training images (150) was used, the program tested the validation data of all 45 image sets, leading each image set to validate using the same 1440 images for all of them. Thus, the orange line representing validation accuracy and validation loss used 1440 of the same images every time and corresponded to more training data being input over time.

For the large image recognition algorithm, the training loss of our models declined to near 0%, whereas the validation loss ended up at around 17% (Figure 7A). The output of this algorithm was then fed into the small image recognition algorithm. The training loss was 0% and the validation loss ended up at about 16% (Figure 7B). Both algorithms had accuracies of 100% on the training data and ~98% on the validation data.

These data mean that the final output of the two-stage algorithm, as shown in Figure 7B, was quite good. However, the validation loss indicated overfitting. We kept this because, on a practical basis, it was not so overfitted as to adversely affect the functional output of the algorithm. In fact, we purposefully kept the algorithms slightly overfit, as we recognized that the additional training cycles led to improved TM detection by eye, even though this was not represented in these plots. To demonstrate this, we re-created the small image recognition algorithm using only 25% of the data (Figure 7C). After completing all of the training, the accuracy and training loss were similar to what was found using the full training set (i.e., Figure 7B), but the validation loss was only ~10%. We then visually inspected the images that were output using new data and found that the overfitted model had slightly better artifact removal (Figure 8).

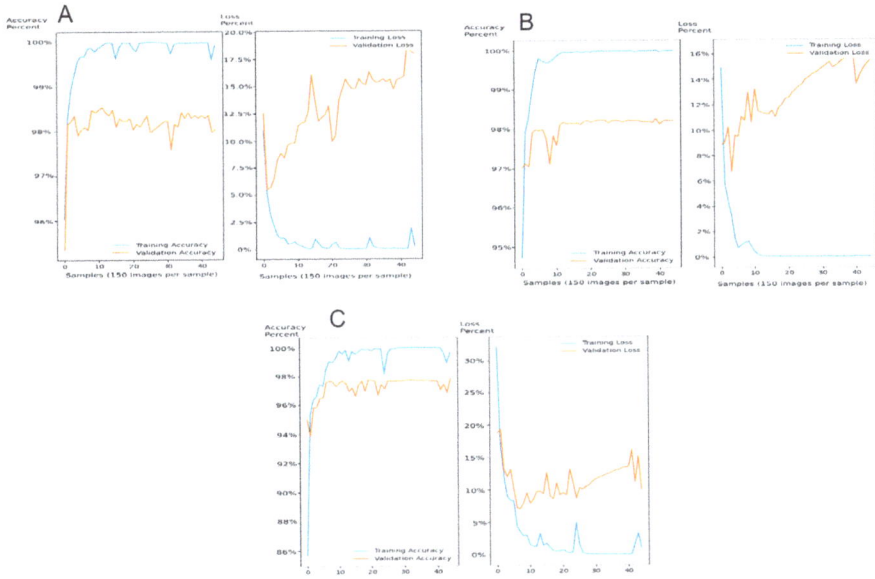

Figure 7. Accuracy and loss during training and validation. (**A**) The large image recognition model. (**B**) The small image recognition model. (**C**) The small image recognition model trained with only 25% of the data.

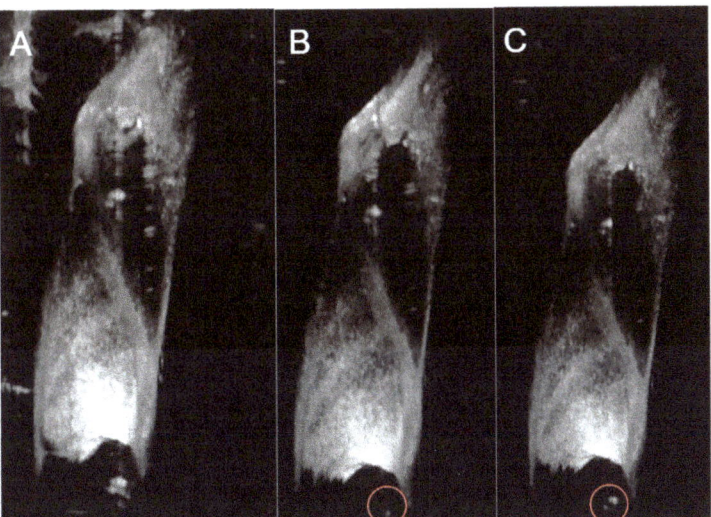

Figure 8. A representative 3D image of the TM before and after segmentation. (**A**) The original image. Note artifacts in the upper left. (**B**) The image after segmentation of the TM using the overfit algorithm from Figure 7B. (**C**) The image after segmentation of the TM using the algorithm from Figure 7C. The images in (**B**,**C**) are quite similar, but there is a little more artifact in (**C**) (red circle).

4. Discussion

OCT provides a unique non-invasive way to assess human tissues. It offers the benefits of ease of use within a routine clinical setting and there are no known safety concerns. However, because swept-source laser light is used for imaging, artifacts caused by saturation of the photodetectors by reflected light or electrical noise within the high-speed analog-to-digital sampling system can obscure the image. Furthermore, because the high resolution of OCT provides near-cellular resolution, only a section of the tissue being studied can be imaged at once. Finally, OCT identifies multiple structural layers, including surface and sub-surface structures. These three issues (the presence of artifacts, the small field of view, and the long depth of view) can make it difficult for physicians to clinically interpret an OCT image. Here, we presented a technique to automatically segment out the TM from OCT images of the human ear. It removes artifacts and subsurface structures that are not the TM and then labels the TM specifically. Thus, the physician is presented with a clear and obvious OCT view of just the TM, which will allow the identification of pathology that is not detectable by routine otoscopy.

While neural-network-based automated segmentation of CT images of the human ear has been carried out before [37–40], a similar approach for segmentation of OCT images of the human ear is novel. Automatic segmentation is particularly important for OCT since provides a small field of view compared to CT or MRI, which can image the entire body. To provide OCT coverage of an entire structure, multiple 3D images from multiple angles might be needed. Once the structure of interest has been segmented out from each image, the 3D images could then be stitched together to provide a comprehensive analysis of the area of interest. For example, each of our microscopic OCT images can image about 2/3rds of the TM. Now that the TM can be segmented, we could envision a future where the user slowly moves the imaging probe around in a circle, collecting many overlapping 3D volumes, and then the algorithm automatically cleans them and stitches them together. In this way, we expect to collect large, high-resolution 3D images of the ear in clinic.

4.1. Deep Learning and Model Limitations

We chose to use deep learning for our algorithm because of its potential to revolutionize medical imaging for disease diagnosis [35,41,42]. Furthermore, we needed our algorithm to be rigorous and dynamic, so that, regardless of the section of the TM that was being imaged, the TM could be automatically segmented. We used TensorFlow to create most of the neural network, but the key to obtaining accurate results was the use of two sequential classification steps, each with distinct object detection algorithms.

As shown, our algorithm does a good job of removing most artifacts within the image. However, sometimes, the artifacts overlap or are very close to the TM, creating a conjunction of pixels that appears as one single object. Thus, most but not all artifact is removed. In the future, we may consider adding a third neural network with a separate image recognition algorithm specifically to identify and remove these types of artifacts, which tend to be vertical lines that originate from strong tissue reflections. While this may improve the image somewhat, it would also increase the processing time. Furthermore, it is unlikely to improve the image substantially enough to add further value to the clinician.

Most of the time creating this algorithm was spent manually classifying the objects. The object-detection algorithms created thousands of images from each 3D image. However, the first step was to optimize the large-object-detection algorithm so that it best recognized the TM as one single object but did not include nearby artifacts. This tradeoff required lowering the threshold value of the object detection algorithm so that the entire TM was recognized as one object instead of being broken up into several smaller objects. However, doing this led to more artifacts within the TM image. As described in the methods section, we handled this by adding the second stage to the algorithm, using more accurate sorting methods for the smaller object area of the TM. Hardware improvements to reduce the number and intensity of the artifacts within the original image would improve the accuracy of this algorithm and speed training of future iterations of this algorithm.

4.2. The Future: Automated Diagnosis of Ear Pathology

This algorithm to segment out the normal TM is the first step in the automatic detection of pathologic TMs. We anticipate that training the algorithm on images collected from patients with diseases, such as a tympanic membrane perforation, a retraction pocket, a cholesteatoma, or a tumor, will permit artificial intelligence to enhance, and perhaps even automate, the diagnosis of ear disease. Ultimately, we also plan to expand this algorithm to not only segment the TM but also other structures visible within the volume stack, including the ossicles and the cochlea. This will expand the number of diseases that AI-enhanced OCT could detect. Our goal is for an inexperienced clinician to be able to image their patient's ear and for the software to automatically provide a tentative diagnosis with high accuracy.

5. Conclusions

Automatic segmentation is a core component needed to use AI to analyze 3D medical images. This algorithm demonstrates that deep learning can segment out the TM from within an artifact-filled OCT volume stack of images. This will be critical to the successful clinical implementation of OCT technology to detect ear disease. It allows a physician to have their computer create a 3D model of the TM quickly and ultimately, which will permit disease detection. Furthermore, it allows for immediate dialogue with the patient about the status of their ear; thus, it provides a point-of-care diagnostic. This provides an additional example of how deep learning is a powerful tool for medical imaging.

Author Contributions: Conceptualization, T.P.O. and J.S.O.; methodology, T.P.O.; software, T.P.O.; writing—original draft preparation, T.P.O.; writing—review and editing, T.P.O., R.L., W.K., B.E.A. and J.S.O.; visualization, R.L. and W.K.; funding acquisition, J.S.O. and B.E.A. All authors have read and agreed to the published version of the manuscript.

Funding: This work was supported by NIH grants R01 DC014450, R01 DC013774, R01 DC017741, and R25 DC019700 (to JSO) and R01 EB027113 (to BEA).

Data Availability Statement: Data and code supporting the findings of this study are available online in our lab's GitHub account (https://github.com/jso111/) Accessed on 16 September 2023.

Acknowledgments: We thank James Kingsley, Lucille Finnerty, Rafael Choudhoury, and Edward Han for contributing to the sorting of data to train the models.

Conflicts of Interest: The authors declare no conflict of interest.

References

1. Luers, J.C.; Hüttenbrink, K.B. Surgical anatomy and pathology of the middle ear. *J. Anat.* **2016**, *228*, 338–353. [CrossRef] [PubMed]
2. Geisler, C.D. *From Sound to Synapse: Physiology of the Mammalian Ear*; Oxford University Press: New York, NY, USA, 1998; ISBN 0195100255.
3. Saliba, I.; Abela, A.; Arcand, P. Tympanic membrane perforation: Size, site and hearing evaluation. *Int. J. Pediatr. Otorhinolaryngol.* **2011**, *75*, 527–531. [CrossRef] [PubMed]
4. Lerut, B.; Pfammatter, A.; Moons, J.; Linder, T. Functional Correlations of Tympanic Membrane Perforation Size. *Otol. Neurotol.* **2012**, *33*, 379–386. [CrossRef]
5. Lou, Z.C.; Lou, Z.H.; Zhang, Q.P. Traumatic Tympanic Membrane Perforations: A Study of Etiology and Factors Affecting Outcome. *Am. J. Otolaryngol.* **2012**, *33*, 549–555. [CrossRef]
6. Ibekwe, T.S.; Adeosun, A.A.; Nwaorgu, O.G. Quantitative analysis of tympanic membrane perforation: A simple and reliable method. *J. Laryngol. Otol.* **2009**, *123*, e2. [CrossRef] [PubMed]
7. Xydakis, M.S.; Bebarta, V.S.; Harrison, C.D.; Conner, J.C.; Grant, G.A.; Robbins, A.S. Tympanic-Membrane Perforation as a Marker of Concussive Brain Injury in Iraq. *N. Engl. J. Med.* **2007**, *357*, 830–831. [CrossRef]
8. Bateman, L.; Borsetto, D.; Boscolo-Rizzo, P.; Mochloulis, G.; Vijendren, A. A narrative review of the management of pars flaccida tympanic membrane retractions without cholesteatoma. *Clin. Otolaryngol.* **2023**. [CrossRef]
9. Maddineni, S.; Ahmad, I. Updates in Eustachian Tube Dysfunction. *Otolaryngol. Clin. N. Am.* **2022**, *55*, 1151–1164. [CrossRef]
10. Spinos, D.; Mallick, S.; Judd, O. Management of retraction pockets: Historic and novel approaches. *J. Laryngol. Otol.* **2022**, *136*, 582–587. [CrossRef]
11. Urík, M.; Tedla, M.; Hurník, P. Pathogenesis of Retraction Pocket of the Tympanic Membrane—A Narrative Review. *Medicina* **2021**, *57*, 425. [CrossRef]

12. Gutierrez, J.A.; Cabrera, C.I.; Stout, A.; Mowry, S.E. Tympanoplasty in the Setting of Complex Middle Ear Pathology: A Systematic Review. *Ann. Otol. Rhinol. Laryngol.* **2023**, *132*, 1453–1466. [CrossRef]
13. Piras, G.; Sykopetrites, V.; Taibah, A.; Russo, A.; Caruso, A.; Grinblat, G.; Sanna, M. Long term outcomes of canal wall up and canal wall down tympanomastoidectomies in pediatric cholesteatoma. *Int. J. Pediatr. Otorhinolaryngol.* **2021**, *150*, 110887. [CrossRef]
14. Nicholas Jungbauer, W.; Jeong, S.; Nguyen, S.A.; Lambert, P.R. Comparing Myringoplasty to Type I Tympanoplasty in Tympanic Membrane Repair: A Systematic Review and Meta-analysis. *Otolaryngol. Head Neck Surg.* **2023**, *168*, 922–934. [CrossRef] [PubMed]
15. Mankowski, N.; Raggio, B. Otoscope Exam. Available online: https://www.statpearls.com/point-of-care/27339 (accessed on 25 July 2023).
16. Schilder, A.G.M.; Chonmaitree, T.; Cripps, A.W.; Rosenfeld, R.M.; Casselbrant, M.L.; Haggard, M.P.; Venekamp, R.P. Otitis media. *Nat. Rev. Dis. Primers* **2016**, *2*, 16063. [CrossRef] [PubMed]
17. Carr, J.A.; Valdez, T.A.; Bruns, O.T.; Bawendi, M.G. Using the shortwave infrared to image middle ear pathologies. *Proc. Natl. Acad. Sci. USA* **2016**, *113*, 9989–9994. [CrossRef]
18. Aumann, S.; Donner, S.; Fischer, J.; Müller, F. Optical Coherence Tomography (OCT): Principle and Technical Realization. In *High Resolution Imaging in Microscopy and Ophthalmology*; Springer: Berlin/Heidelberg, Germany, 2019; pp. 59–85.
19. Kudsieh, B.; Fernández-Vigo, J.I.; Flores-Moreno, I.; Ruiz-Medrano, J.; Garcia-Zamora, M.; Samaan, M.; Ruiz-Moreno, J.M. Update on the Utility of Optical Coherence Tomography in the Analysis of the Optic Nerve Head in Highly Myopic Eyes with and without Glaucoma. *J. Clin. Med.* **2023**, *12*, 2592. [CrossRef] [PubMed]
20. Mahendradas, P.; Acharya, I.; Rana, V.; Bansal, R.; Ben Amor, H.; Khairallah, M. Optical Coherence Tomography and Optical Coherence Tomography Angiography in Neglected Diseases. *Ocul. Immunol. Inflamm.* **2023**, 1–8. [CrossRef] [PubMed]
21. Tan, H.E.I.; Santa Maria, P.L.; Wijesinghe, P.; Francis Kennedy, B.; Allardyce, B.J.; Eikelboom, R.H.; Atlas, M.D.; Dilley, R.J. Optical Coherence Tomography of the Tympanic Membrane and Middle Ear: A Review. *Otolaryngol. Head Neck Surg.* **2018**, *159*, 424–438. [CrossRef]
22. Badash, I.; Quiñones, P.M.; Oghalai, K.J.; Wang, J.; Lui, C.G.; Macias-Escriva, F.; Applegate, B.E.; Oghalai, J.S. Endolymphatic Hydrops is a Marker of Synaptopathy Following Traumatic Noise Exposure. *Front. Cell Dev. Biol.* **2021**, *9*, 3163. [CrossRef] [PubMed]
23. Dewey, J.B.; Altoè, A.; Shera, C.A.; Applegate, B.E.; Oghalai, J.S. Cochlear outer hair cell electromotility enhances organ of Corti motion on a cycle-by-cycle basis at high frequencies in vivo. *Proc. Natl. Acad. Sci. USA* **2021**, *118*, e2025206118. [CrossRef]
24. Kim, J.; Xia, A.; Grillet, N.; Applegate, B.E.; Oghalai, J.S. Osmotic stabilization prevents cochlear synaptopathy after blast trauma. *Proc. Natl. Acad. Sci. USA* **2018**, *115*, E4853–E4860. [CrossRef]
25. Lui, C.G.; Kim, W.; Dewey, J.B.; Macías-Escrivá, F.D.; Ratnayake, K.; Oghalai, J.S.; Applegate, B.E. In vivo functional imaging of the human middle ear with a hand-held optical coherence tomography device. *Biomed. Opt. Express* **2021**, *12*, 5196–5213. [CrossRef] [PubMed]
26. Merchant, G.R.; Siegel, J.H.; Neely, S.T.; Rosowski, J.J.; Nakajima, H.H. Effect of Middle-Ear Pathology on High-Frequency Ear Canal Reflectance Measurements in the Frequency and Time Domains. *J. Assoc. Res. Otolaryngol.* **2019**, *20*, 529–552. [CrossRef]
27. Deliwala, S.S.; Hamid, K.; Barbarawi, M.; Lakshman, H.; Zayed, Y.; Kandel, P.; Malladi, S.; Singh, A.; Bachuwa, G.; Gurvits, G.E.; et al. Artificial intelligence (AI) real-time detection vs. routine colonoscopy for colorectal neoplasia: A meta-analysis and trial sequential analysis. *Int. J. Color. Dis.* **2021**, *36*, 2291–2303. [CrossRef]
28. Suzuki, H.; Yoshitaka, T.; Yoshio, T.; Tada, T. Artificial intelligence for cancer detection of the upper gastrointestinal tract. *Dig. Endosc.* **2021**, *33*, 254–262. [CrossRef]
29. Mahmood, H.; Shaban, M.; Rajpoot, N.; Khurram, S.A. Artificial Intelligence-based methods in head and neck cancer diagnosis: An overview. *Br. J. Cancer* **2021**, *124*, 1934–1940. [CrossRef] [PubMed]
30. Liu, Y.; Han, G.; Liu, X. Lightweight Compound Scaling Network for Nasopharyngeal Carcinoma Segmentation from MR Images. *Sensors* **2022**, *22*, 5875. [CrossRef] [PubMed]
31. Zhi, Y.; Hau, W.K.; Zhang, H.; Gao, Z. Vessel Contour Detection in Intracoronary Images via Bilateral Cross-Domain Adaptation. *IEEE J. Biomed. Health Inform.* **2023**, *27*, 3314–3325. [CrossRef] [PubMed]
32. Kim, W.; Kim, S.; Oghalai, J.S.; Applegate, B.E. Stereo Microscope Based OCT System Capable of Subnanometer Vibrometry in the Middle Ear. In Proceedings of the Progress in Biomedical Optics and Imaging—Proceedings of SPIE, Prague, Czech Republic, 23 June 2019; Volume 11078.
33. Oghalai, J. GitHub Repository. Available online: https://github.com/jso111/linear-mixed-effect-modeling (accessed on 16 September 2023).
34. Snider, E.J.; Hernandez-Torres, S.I.; Hennessey, R. Using Ultrasound Image Augmentation and Ensemble Predictions to Prevent Machine-Learning Model Overfitting. *Diagnostics* **2023**, *13*, 417. [CrossRef]
35. Shorten, C.; Khoshgoftaar, T.M. A survey on Image Data Augmentation for Deep Learning. *J. Big Data* **2019**, *6*, 60. [CrossRef]
36. Khosla, C.; Saini, B.S. Enhancing Performance of Deep Learning Models with different Data Augmentation Techniques: A Survey. In Proceedings of the 2020 International Conference on Intelligent Engineering and Management (ICIEM), London, UK, 17–19 June 2020; pp. 79–85. [CrossRef]
37. Neves, C.A.; Tran, E.D.; Kessler, I.M.; Blevins, N.H. Fully automated preoperative segmentation of temporal bone structures from clinical CT scans. *Sci. Rep.* **2021**, *11*, 116. [CrossRef]

38. Lv, Y.; Ke, J.; Xu, Y.; Shen, Y.; Wang, J.; Wang, J. Automatic segmentation of temporal bone structures from clinical conventional CT using a CNN approach. *Int. J. Med. Robot. Comput. Assist. Surg.* **2021**, *17*, e2229. [CrossRef] [PubMed]
39. Wang, J.; Lv, Y.; Wang, J.; Ma, F.; Du, Y.; Fan, X.; Wang, M.; Ke, J. Fully automated segmentation in temporal bone CT with neural network: A preliminary assessment study. *BMC Med. Imaging* **2021**, *21*, 166. [CrossRef] [PubMed]
40. Ding, A.S.; Lu, A.; Li, Z.; Sahu, M.; Galaiya, D.; Siewerdsen, J.H.; Unberath, M.; Taylor, R.H.; Creighton, F.X. A Self-Configuring Deep Learning Network for Segmentation of Temporal Bone Anatomy in Cone-Beam CT Imaging. *Otolaryngol. Head Neck Surg.* **2023**. [CrossRef] [PubMed]
41. Windsor, G.O.; Bai, H.; Lourenco, A.P.; Jiao, Z. Application of artificial intelligence in predicting lymph node metastasis in breast cancer. *Front. Radiol.* **2023**, *3*, 928639. [CrossRef]
42. Koseoglu, N.D.; Grzybowski, A.; Liu, T.Y.A. Deep Learning Applications to Classification and Detection of Age-Related Macular Degeneration on Optical Coherence Tomography Imaging: A Review. *Ophthalmol. Ther.* **2023**, *12*, 2347–2359. [CrossRef] [PubMed]

Disclaimer/Publisher's Note: The statements, opinions and data contained in all publications are solely those of the individual author(s) and contributor(s) and not of MDPI and/or the editor(s). MDPI and/or the editor(s) disclaim responsibility for any injury to people or property resulting from any ideas, methods, instructions or products referred to in the content.

Article

Neural Network Based Approach to Recognition of Meteor Tracks in the Mini-EUSO Telescope Data

Mikhail Zotov [1,*], Dmitry Anzhiganov [1,2], Aleksandr Kryazhenkov [1,2], Dario Barghini [3,4,5], Matteo Battisti [3], Alexander Belov [1,6], Mario Bertaina [3,4], Marta Bianciotto [4], Francesca Bisconti [3,7], Carl Blaksley [8], Sylvie Blin [9], Giorgio Cambiè [7,10], Francesca Capel [11,12], Marco Casolino [7,8,10], Toshikazu Ebisuzaki [8], Johannes Eser [13], Francesco Fenu [4,†], Massimo Alberto Franceschi [14], Alessio Golzio [3,4], Philippe Gorodetzky [9], Fumiyoshi Kajino [15], Hiroshi Kasuga [8], Pavel Klimov [1,6], Massimiliano Manfrin [3,4], Laura Marcelli [7], Hiroko Miyamoto [3], Alexey Murashov [1,6], Tommaso Napolitano [14], Hiroshi Ohmori [8], Angela Olinto [13], Etienne Parizot [9,16], Piergiorgio Picozza [7,10], Lech Wiktor Piotrowski [17], Zbigniew Plebaniak [3,4,18], Guillaume Prévôt [9], Enzo Reali [7,10], Marco Ricci [14], Giulia Romoli [7,10], Naoto Sakaki [8], Kenji Shinozaki [18], Christophe De La Taille [19], Yoshiyuki Takizawa [8], Michal Vrábel [18] and Lawrence Wiencke [20]

1. Skobeltsyn Institute of Nuclear Physics, Lomonosov Moscow State University, Moscow 119991, Russia
2. Faculty of Computational Mathematics and Cybernetics, Lomonosov Moscow State University, Moscow 119991, Russia
3. INFN, Sezione di Torino, Via Pietro Giuria, 1, 10125 Torino, Italy
4. Dipartimento di Fisica, Università di Torino, Via Pietro Giuria, 1, 10125 Torino, Italy
5. INAF, Osservatorio Astrofisico di Torino, Via Osservatorio 20, Pino Torinese, 10025 Torino, Italy
6. Faculty of Physics, M.V. Lomonosov Moscow State University, Moscow 119991, Russia
7. INFN, Sezione di Roma Tor Vergata, Via della Ricerca Scientifica 1, 00133 Roma, Italy
8. RIKEN, 2-1 Hirosawa, Wako, Saitama 351-0198, Japan
9. AstroParticule et Cosmologie, CNRS, Université Paris Cité, F-75013 Paris, France
10. Dipartimento di Fisica, Universita degli Studi di Roma Tor Vergata, Via della Ricerca Scientifica 1, 00133 Roma, Italy
11. Max Planck Institute for Physics, Föhringer Ring 6, D-80805 Munich, Germany; capel.francesca@gmail.com
12. Department of Particle and Astroparticle Physics, KTH Royal Institute of Technology, SE-100 44 Stockholm, Sweden
13. Department of Astronomy and Astrophysics, The University of Chicago, Chicago, IL 60637, USA
14. INFN—Laboratori Nazionali di Frascati, 00044 Frascati, Italy
15. Department of Physics, Konan University, Kobe 658-8501, Japan
16. AstroParticule et Cosmologie, Institut Universitaire de France (IUF), CEDEX 05, 75231 Paris, France
17. Faculty of Physics, University of Warsaw, 02-093 Warsaw, Poland
18. National Centre for Nuclear Research, Ul. Pasteura 7, PL-02-093 Warsaw, Poland
19. Omega, Ecole Polytechnique, CNRS/IN2P3, 91128 Palaiseau, France
20. Department of Physics, Colorado School of Mines, Golden, CO 80401, USA
* Correspondence: zotov@eas.sinp.msu.ru
† Current address: Agenzia Spaziale Italiana, Via del Politecnico, Roma, Italy.

Citation: Zotov, M.; Anzhiganov, D.; Kryazhenkov, A.; Barghini, D.; Battisti, M.; Belov, A.; Bertaina, M.; Bianciotto, M.; Bisconti, F.; Blaksley, C.; et al. Neural Network Based Approach to Recognition of Meteor Tracks in the Mini-EUSO Telescope Data. *Algorithms* **2023**, *16*, 448. https://doi.org/10.3390/a16090448

Academic Editors: Frank Werner, Chih-Lung Lin, Bor-Jiunn Hwang, Shaou-Gang Miaou and Yuan-Kai Wang

Received: 20 June 2023
Revised: 28 July 2023
Accepted: 22 August 2023
Published: 19 September 2023

Copyright: © 2023 by the authors. Licensee MDPI, Basel, Switzerland. This article is an open access article distributed under the terms and conditions of the Creative Commons Attribution (CC BY) license (https:// creativecommons.org/licenses/by/ 4.0/).

Abstract: Mini-EUSO is a wide-angle fluorescence telescope that registers ultraviolet (UV) radiation in the nocturnal atmosphere of Earth from the International Space Station. Meteors are among multiple phenomena that manifest themselves not only in the visible range but also in the UV. We present two simple artificial neural networks that allow for recognizing meteor signals in the Mini-EUSO data with high accuracy in terms of a binary classification problem. We expect that similar architectures can be effectively used for signal recognition in other fluorescence telescopes, regardless of the nature of the signal. Due to their simplicity, the networks can be implemented in onboard electronics of future orbital or balloon experiments.

Keywords: machine learning; neural network; pattern recognition; meteor; fluorescence telescope; orbital experiment; UV illumination; atmosphere

1. Introduction

The JEM-EUSO (Joint Exploratory Missions for Extreme Universe Space Observatory) collaboration is developing a program of studying ultra-high energy cosmic rays (UHECRs) with a wide angle telescope from a low Earth orbit [1–3]. The idea is based on the possibility to register fluorescence and Cherenkov radiation in the ultraviolet (UV) range that is emitted during development of extensive air showers generated by primary particles hitting the atmosphere [4]. There are several benefits of this technique in comparison with ground-based experiments: (i) it can provide a huge exposure necessary for collecting sufficient statistics of these extremely rare events; (ii) the celestial sphere can be observed almost uniformly, which is important for anisotropy studies; and (iii) the whole sky can be observed with one instrument.

It became clear at early stages of the development of the JEM-EUSO program that an orbital telescope aimed at studying UHECRs can serve as a tool for exploring other phenomena that manifest themselves in the UV range in the nocturnal atmosphere of Earth [5]. It was demonstrated by TUS, the world's first orbital fluorescence telescope aimed for testing the technique of studying UHECRs from space, that such an instrument can provide data on transient luminous events, thunderstorm activity, meteors, anthropogenic illumination of different kinds, and other types of signals [6,7]. In particular, observations of meteors are considered as an important branch of studies in the JEM-EUSO program [8,9].

The JEM-EUSO program is being implemented in a number of steps aimed at development and testing of different aspects of a full-blown orbital experiment. In particular, laser shots were successfully registered by a fluorescence telescope looking down on the atmosphere within the EUSO-Balloon mission [10]. A wide program of studies is being performed with the EUSO-TA experiment [11]. In 2018–2019, the Mini-EUSO (Multiwavelength Imaging New Instrument for the EUSO) telescope was built by the JEM-EUSO collaboration. It was brought to the International Space Station (ISS) on 22 August 2019, by the Soyuz MS-14 vehicle and has been operated since then as a part of an agreement between the Italian Space Agency (Agenzia Spaziale Italiana; ASI) and Roscosmos (Russia) [12–15]. The EUSO-SPB2 stratospheric balloon equipped with a fluorescence and Cherenkov telescopes made a short flight from Wanaka, New Zealand, in May 2023 [16–18]. All these instruments are aimed to be pathfinders and test beds for full-size orbital experiments like K-EUSO [19] and POEMMA [20].

Similar to the other projects of the JEM-EUSO collaboration, the Mini-EUSO telescope is registering multiple types of UV emission taking place in the nocturnal atmosphere of Earth, among them signals of meteors. A series of studies is dedicated to their search and analysis [21,22]. In the present paper, we continue our earlier research aimed at developing a method of recognizing meteor tracks in the Mini-EUSO data with neural networks [23]. A motivation for the study is the following. A conventional approach to finding signals of meteors in the Mini-EUSO data is time consuming and prone to numerous false positives. Thus, it is interesting to figure out if an approach based on machine learning (ML) and artificial neural networks (ANNs) can demonstrate higher efficacy than the conventional one so that both approaches complement each other. If so, it is interesting to test if results can be achieved with simple neural networks that can be implemented in the forthcoming orbital experiments, which are unlikely to have powerful onboard processors. These results can also be useful for recognizing tracks of extensive air showers in the future experiments since such signals resemble shapes and kinematics similar to those produced be meteors, though at completely different time scales. Finally, in case of the successful development of an ANN-based pipeline for recognizing meteor signals in the Mini-EUSO data, it can be applied for a search of track-like signals of different nature, including those that mimic extensive air showers. In what follows, we present a pipeline consisting of two basic neural networks that demonstrate high performance and can be trained on an ordinary PC. The work continues a series of studies fulfilled within the JEM-EUSO collaboration on application of machine learning and neural networks to analysis of data of fluorescence telescopes [24–27]. We do not present any results of applying the suggested method to data analysis since this will be covered in detail in a dedicated paper.

2. Mini-EUSO Experiment

The main components of the Mini-EUSO telescope include two Fresnel lenses and a focal surface (FS). The lenses have a diameter of 25 cm with the focal distance of the optical system equal to 300 mm. The FS has a square shape with 2304 pixels. It is built of 36 multi-anode photomultiplier tubes (MAPMTs) Hamamatsu R11265-M64 each consisting of 8×8 pixels. All MAPMTs are grouped into nine so-called elementary cell (EC) units. Every EC unit has its own high-voltage system, which operates independently of the others providing necessary control of sensitivity of the respective MAPMTs. A 2-mm thick UV filter manufactured of the BG3 glass is located in front of each MAPMT. The size of one pixel equals 2.88 mm $\times$ 2.88 mm. The point spread function (PSF) has a size of $\sim$1.2 pixels. Mini-EUSO has a wide field of view (FoV) of $44° \times 44°$ with spatial resolution (FoV of one pixel) equal to 6.3 km $\times$ 6.3 km. From the orbit of the ISS, an area observed by the telescope exceeds 300 km $\times$ 300 km. A detailed description of the instrument can be found in [12].

Mini-EUSO collects data in three modes. The D1 mode has a time resolution of 2.5 µs. This is called a D1 gate time unit (GTU). The D2 mode records data integrated over 128 D1 GTUs. Finally, the D3 mode operates with data integrated over 128×128 D1 GTUs resulting in time resolution of 40.96 ms. To the contrary to the D1 and D2 modes, the D3 mode does not have a trigger, and its data can be considered as a series of videos with "seasons" corresponding to sessions of observations and "episodes" corresponding to night segments of the ISS orbit during a session. Each session takes around 12 h. With the orbital period of the ISS equal to 92.9 min, a typical session includes eight subsets of data taken during nocturnal segments of an orbit, with each of them taking slightly longer than 1/3 of the period. Every "video" has a resolution of 48×48 pixels and consists of $T/40.96$ µs frames, where T is the duration of one nocturnal segment. Observations are performed approximately twice per month through the UV-transparent window at the Zvezda module with the schedule coordinated with other experiments. Due to this, background illumination varies strongly from one session to another depending on the phase of the Moon and the season. The D3 mode allows for registering meteors and other slow phenomena taking place in the night atmosphere of Earth. In what follows, we use only data recorded during sessions 5–8 and 11–14 taken from 19 November 2019 to 1 April 2020. All artificial neural networks discussed below were trained using data of seven sessions and tested on the remaining session. This way, we checked all possible combinations of the eight sessions.

3. Meteor and Background Signals

Signals of meteors registered with Mini-EUSO have some features important for the presented analysis:

- A signal produced by a meteor in a pixel has a shape resembling the bell-like curve similar to the probability density function of the normal distribution.
- Meteor signals produce quasi-linear tracks in the focal surface.
- The number of hit ("active") pixels in more than 75% of meteor tracks is $\leq$5, so that their footprints on the focal surface are small.
- Peaks of a meteor signal shift from one pixel to another (except for arrival directions close to nadir).
- There are multiple signals in the data with the shape similar to that of meteors but simultaneously illuminating large areas of the FS.
- Meteors are often registered on strong and quickly varying background illumination.
- The amplitude of a meteor signal is typically lower than amplitudes of some other signals in the FoV of Mini-EUSO registered simultaneously with the meteor.
- In some cases, it is impossible to judge unequivocally if a signal originated from a meteor or another source.

Let us discuss the most important of these features taking as an example signals demonstrated in Figures 1 and 2.

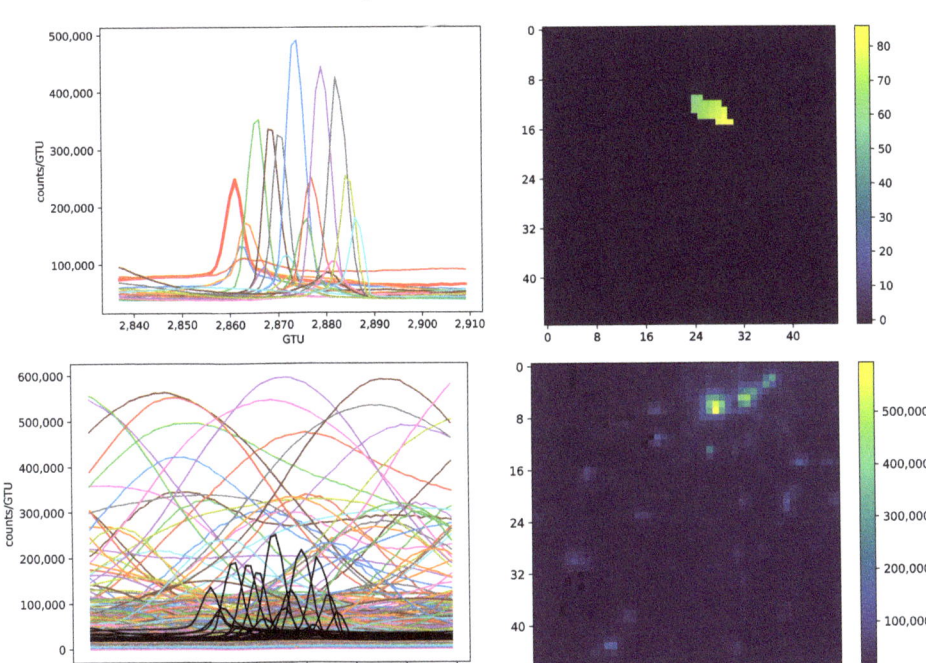

Figure 1. An example of a clearly pronounced meteor signal registered by Mini-EUSO on 20 October 2019. (**Top left**): signals in pixels that constitute the meteor signal. Signals in different pixels are shown in different colors. (**Top right**): location of meteor pixels in the focal surface. Colors denote time shift of the peaks with respect to the first one (in units of D3 GTUs). (**Bottom left**): all signals registered by Mini-EUSO simultaneously with the meteor. The black curves show the meteor signal. (**Bottom right**): a snapshot of the focal surface made at the moment of maximum of the brightest meteor pixel (GTU 2874).

Figure 1 provides an example of a bright and clearly pronounced meteor signal with numerous active pixels. The top row shows only the meteor signal, with the background illumination omitted. It can be seen that signals in every pixel have a typical shape resembling the bell curve (see the left panel). The peaks are shifted in time with respect to each other due to the meteor moving in the FoV of the instrument, resulting in a quasi-linear track on the focal surface (see the right panel). The task of recognizing meteor signals might look trivial after looking at these "pure" signals. However, the FoV of Mini-EUSO covers a huge area resulting in numerous different signals being registered simultaneously, with many of them being much brighter than those of meteors. This is demonstrated in the second row of Figure 1. The left panel shows shapes of all other signals recorded simultaneously with the meteor with the meteor signal shown in black. The right panel presents a snapshot of the FS made at the moment of the maximum of the meteor signal. The brightest pixel of the meteor has coordinates (row, column) = (13, 27) and can be seen as a small spot below a much brighter and extended area that appeared due to anthropogenic illumination (sine-like curves in the left panel). It is important to remark that the bottom rows of Figures 1 and 2 demonstrate signals that were flat-fielded during an offline analysis for the sake of clarity.

However, all results presented below were obtained using raw data as they are recorded by the instrument. This was performed in order to understand how effectively is our method if implemented in onboard electronics.

An example of a typical meteor is shown in Figure 2. It has only four active pixels, and it is so dim in comparison with other illumination registered simultaneously that it is hardly possible to find it by eye in the bottom right panel presenting a snapshot of the focal surface at the moment of the maximum brightness of the meteor.

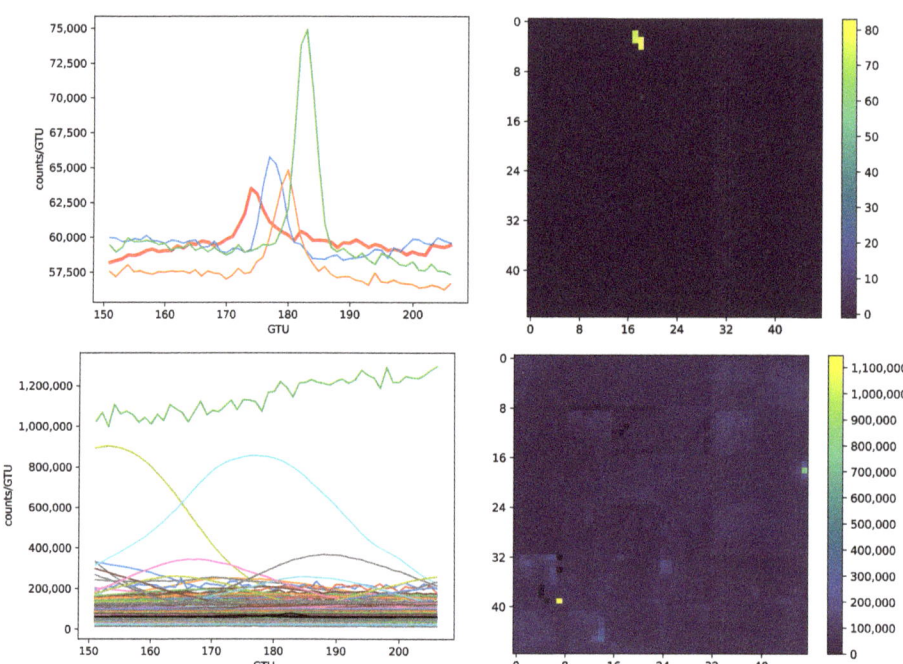

Figure 2. A typical meteor signal registered by Mini-EUSO. (**Top left**): signals in pixels that constitute the meteor signal. (**Top right**): location of meteor pixels in the focal surface. Colors denote a time shift of the peaks with respect to the first one (in units of D3 GTUs). (**Bottom left**): all signals registered by Mini-EUSO simultaneously with the meteor. The black curves show the meteor signal. (**Bottom right**): a snapshot of the focal surface made at the moment of the brightest meteor signal (GTU 184).

A conventional way to find meteor signals in the Mini-EUSO or TUS data would be to look for signals that can be fitted with the probability density function of a Gaussian distribution or its sum with a polynomial in case of non-stationary background illumination [21,22,28]. The known range of possible speeds of meteors (11–72 km s^{-1}) together with information on the orbital speed of the ISS ($\sim$7.7 km s^{-1}), the FoV of one pixel, and the PSF size allows one to estimate the variance of a Gaussian fit. This also allows for verifying the kinematics of a signal moving across the FS. The latter step is of crucial importance since there are multiple signals in the Mini-EUSO data that can be fitted with a Gaussian distribution but take place simultaneously in big pixel groups without forming a track on the focal surface.

The task of recognizing meteor patterns in the Mini-EUSO dataset with machine learning methods can be considered as a binary classification problem since one basically needs to separate meteor signals from all the rest. Seemingly the most obvious way to tackle the task with artificial neural networks is to employ supervised learning. Within this approach, one can train an ANN using a labeled dataset. The dataset can be either prepared by simulations or extracted from real experimental data. It is tempting to choose the first way since meteor signals mostly have a bell curve shape similar to the density of the normal distribution. However, realistic simulations are not trivial since the background illumination is diverse, and sensitivity of different pixels on the FS is not known. This made us adopt the second approach.

We took two meteor datasets obtained by the JEM-EUSO collaboration and complemented them by our own analysis to prepare a dataset suitable for training and testing an ANN. It is necessary to stress once again that the source of a considerable number of bell curve-like signals registered with Mini-EUSO cannot be identified with confidence. Signals like those shown in Figures 1 and 2 do not pose a problem in this respect. However, the shape and kinematics of tracks produced by meteors consisting of ≤ 4 pixels are often confusing. Another difficulty arises from dim signals on a strong and varying background. As a result, it is impossible to obtain ground-truth labels basing exclusively on the existing data set. After several tests, we confined the labeled dataset to signals the nature of which causes little doubt. In particular, we excluded almost all signals occupying two pixels. The resulting dataset used for training and testing ANNs discussed below consisted of 1068 meteor signals. Every record in the dataset included a timestamp of a meteor, coordinates of active pixels on the focal surface, and positions of the respective signal peaks.

Since data obtained in the D3 mode do not have a trigger but are similar to a series of videos each consisting of thousands of "frames" representing "snapshots" of the FS, there is a question how to extract samples containing meteor and non-meteor signals for training and testing datasets. For example, one can create data chunks of size $48 \times 48 \times T$, where T is the number of time frames (D3 GTUs) large enough to fit all meteors in the dataset and either center them on meteor peaks or put them in a fixed position with respect to the beginning of a meteor signal. This will allow one to obtain a "unified" representation of meteor signals to an ANN thus simplifying the task of their recognition. This way, non-meteor samples can be extracted from the rest of the data in a random fashion. However, this is not the way the data flow can be analyzed onboard. Besides this, the above approach will leave us with mere 1068 meteor samples, which is not sufficient to effectively train an ANN. This made us use a sliding window producing chunks that overlap by dt GTUs. In what follows, we present results obtained for $dt = 8$ GTUs that allowed us to avoid losing short meteor tracks and provided reasonable accuracy of their recognition. The procedure of labeling data chunks extracted this way will be explained in detail in Section 4.1.

We split the task of meteor signal recognition into two steps. First, we trained an ANN to recognize three-dimensional data chunks that contained meteor signals. After this, we employed another ANN to select pixels containing respective signals. Each ANN thus solved a task of binary classification. This allowed us to obtain lists of meteors registered with Mini-EUSO together with their active pixels thus providing information necessary for their subsequent analysis (reconstruction of brightness, arrival directions etc.).

4. Results

An important question to discuss before presenting ANNs is how to evaluate their performance. It is usually advised to use balanced datasets whenever possible, both for training and testing. In this case, the Receiver Operating Characteristic Area Under the Curve (ROC AUC) is one of the popular performance metrics [29]. Recall that the ROC curve is a plot of the true positive rate against the false positive rate at various threshold settings. Given one randomly selected positive instance and one randomly selected negative instance, AUC is the probability that the classifier will be able to tell which one is which. Due to its definition, ROC AUC does not depend on the sample size.

However, the number of meteor signals in the Mini-EUSO data is negligibly small in comparison with the number of non-meteor signals, so that using balanced datasets for testing would provide unrealistic results, while using them for training would not represent the full diversity of non-meteor signals thus resulting in lower performance and numerous false positives during tests. Thus, we unavoidably arrive at the necessity to use strongly imbalanced datasets both for training and testing. It is argued in the literature that ROC AUC does not act as a fully adequate performance metric in this case, and other metrics should be used instead, see, e.g., [30–32]. In what follows, we provide results obtained in terms of three more metrics besides ROC AUC. These are the Precision–Recall (PR) AUC, the Matthews correlation coefficient (MCC), and the F_1 score. One more metric will be introduced below.

Recall that the PR AUC equals an area under the plot of precision vs. recall with

$$\text{Precision} = \frac{\text{TP}}{\text{TP} + \text{FP}}, \qquad \text{Recall} = \text{TPR} = \frac{\text{TP}}{\text{TP} + \text{FN}},$$

where TP, FP, and FN are the number of true positives, false positives, and false negatives as classified by the model, respectively. The Matthews correlation coefficient can be calculated from the confusion matrix as

$$\text{MCC} = \frac{\text{TP} \cdot \text{TN} - \text{FP} \cdot \text{FN}}{\sqrt{(\text{TP} + \text{FP})(\text{TP} + \text{FN})(\text{TN} + \text{FP})(\text{TN} + \text{FN})}},$$

where TN is the number of true negatives. Finally, the F_1 score is the harmonic mean of the precision and recall. It can be presented as

$$F_1 = \frac{2\text{TP}}{2\text{TP} + \text{FP} + \text{FN}}.$$

Notice that these metrics will be applied to three-dimensional data chunks that partially overlap due to the employment of a sliding window for preparing input datasets. This can lead to a situation when a part of chunks containing a meteor signal are classified as non-meteor ones while others are classified properly, so that the value of a performance metric might be misleading. Since we are interested in maximizing the number of recognized meteors but not meteor chunks, it can be beneficial to also introduce a metric expressed in terms of the original 1068 meteors. Probably the most straightforward way is to use $1 - \text{FNR (met)}$, where FNR (met) is the false negative rate of meteor signals represented as the number of meteors lost by the classifier divided by the total number of meteors in a session used for testing.

All these metrics are equal to 1 for a perfect model. The MCC equals -1 for the worst possible model; other metrics give 0 in this case.

4.1. Recognition of Meteor Data Samples

One of the crucial questions to be solved before training an ANN is how to prepare and organize input data. In our case, the question is twofold. We need to decide how to label three-dimensional chunks into those that contain a signal of a meteor and those that do not. Besides this, we need to choose the size of data chunks $P \times P \times T$, where $P \times P$ defines the size of a square on the focal surface (measured in pixels), and T is the number of time frames ("snapshots" of the FS).

A data chunk was labeled as containing a meteor signal in case there were at least two meteor pixels inside a $P \times P$ area with peaks of the signals located within T GTUs. The reason is that we do not have a straightforward way to decide if a signal originates from a meteor with just a single active pixel. On the other hand, putting a more strict cut on the number of active pixels inside a chunk (≥ 3) results in a loss of short meteor tracks that have only two active pixels.

As for the number of time frames in data chunks, we tested $T = 8, 16, 32, 48, 64$, and 128, which covers the range of duration of meteor signals in the dataset. Values $T = 32, 48,$ and 64 demonstrated the best results in our tests in terms of all metrics mentioned above with different combinations of training and testing sessions regardless of the value of P, with $T = 48$ showing in average marginally better performance than the other two values. This value is used in all figures and tables presented below.

In [33], a simple convolutional neural network (CNN) was employed to perform binary classification of two types of signals registered with the TUS telescope. The instrument had a focal surface of 16×16 pixels, and data arranged in $16 \times 16 \times T$ chunks worked well. Thus, we first tried training a CNN for Mini-EUSO with data chunks of the size $48 \times 48 \times T$. The input data was standardized according to the formula $(X_i - \langle X_i \rangle)/\sigma(X_i)$, where X_i is the signal in pixel i, $\langle X_i \rangle$ and $\sigma(X_i)$ are estimations of the mean and the standard deviation during T time frames. However, as it was briefly reported in [23], this approach did not allow us to obtain acceptable results. We tested numerous configurations of CNNs and long short-term memory networks but failed to obtain ROC AUC > 0.75 on testing datasets. A simple solution was found by splitting the FS into smaller squares. We tested splitting with $P = 24, 16, 12, 8,$ and 6. In order to avoid losing signals around the boundaries of these small areas, we used overlapping by $P/2$ pixels in both directions. Figure 3 shows the behavior of mean values of different performance metrics for varying P with models trained on all possible combinations of seven sessions and tested on the remaining one. In this case, the same architecture of a CNN was used for all tests. It can be seen that performance expressed in terms of any metric increases quickly for $P < 24$. The PR AUC, the MCC, and the F_1 score change in a very similar fashion while the values of the ROC AUC are close to those of $1-$ FNR (met) for small P. The best performance is reached for $P = 8$ with the MCC and F_1 metrics slightly decreasing for $P = 6$. Thus, data chunks of the size $8 \times 8 \times 48$ are used in what follows.

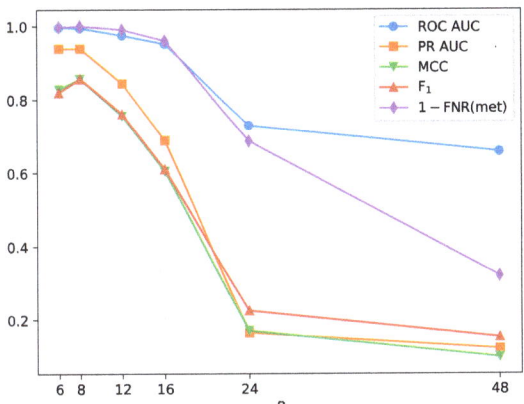

Figure 3. Mean values of different performance metrics as a function of the data chunk size P for models trained on all possible combinations of seven sessions of observations and tested on the remaining session. See the text for details.

A CNN that we adopted for classifying meteor chunks consists of a convolutional layer with 24 filters and a kernel of size 3. It utilizes ReLU as an activation function and the L2 kernel regularizer with factor 0.1. The convolutional layer was followed by a maxpooling and dropout layers and two fully connected layers with 256 and then 64 neurons. Adam was used as an optimization algorithm. Sigmoid was employed as an activation function in the output layer. The architecture is shown in Figure 4.

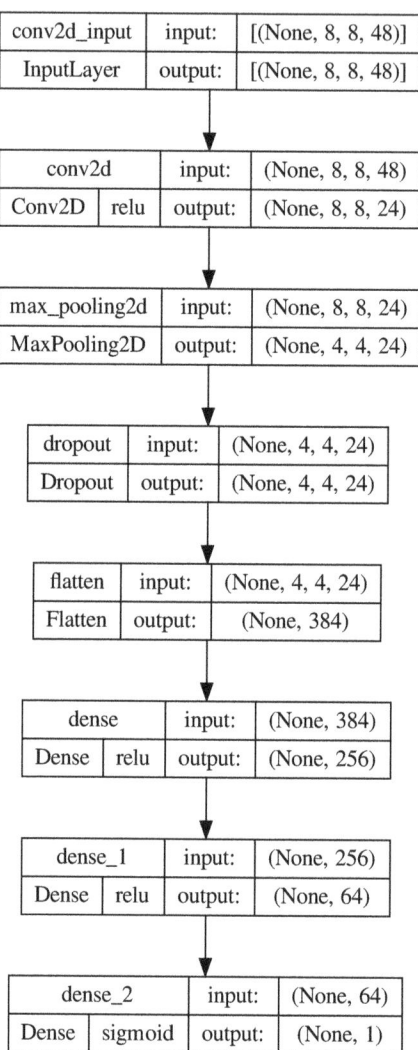

Figure 4. Architecture of the CNN used for binary classification in meteor and non-meteor data chunks. The total number of trainable parameters equals 125,465.

The way we employed to prepare input data allowed us to obtain ≈18,000–24,000 thousand meteor chunks of the size 8 × 8 × 48 depending on the set of sessions used for training, see details in Table 1. These chunks were augmented then by the standard procedure of image rotation thus providing four times more samples. Non-meteor data chunks were selected in a random fashion, with their number being six times larger than the total number of meteor chunks (after augmentation). Twenty percent of the training dataset were used for validation during training. PR AUC was utilized as a performance metric during the training process. The loss function was defined as binary crossentropy. Validation loss was employed to adjust the learning rate and to avoid overfitting. Testing datasets included 100,000 non-meteor samples and all meteor data chunks available for the particular session varying from 474 chunks for session 7 up to 6446 chunks for session 6, see Table 1.

Table 1. The top row: sessions of observations used for testing CNNs trained on data of all other sessions. The second row: the total number of $8 \times 8 \times 48$ chunks (after augmentation) with signals of meteors used for training the respective CNNs. The number of original chunks is four times less. The last two rows: the number of chunks with meteor signals and the real number of meteors, respectively, in test sessions.

Test Session	5	6	7	8	11	12	13	14
Training meteor chunks	91,060	72,124	95,924	81,188	80,724	88,456	89,228	86,820
Testing meteor chunks	1712	6446	474	4180	4274	2341	2148	2772
Testing meteors	65	280	18	186	193	106	90	130

Table 2 contains values of five performance metrics described above for different combinations of testing and training datasets in the task of classification of $8 \times 8 \times 48$ chunks into meteor and non-meteor groups. For example, the column labeled as "5" presents results obtained with sessions 6–14 employed for training, and session 5 used for testing the CNN. It can be seen that zero out of 1068 meteors were lost in all sessions. Notice however that values of the PR AUC, the MCC, and the F_1 score vary considerably from one session to another.

Table 2. Performance of the CNN on different sessions of observations. See the text for details.

Test Session	5	6	7	8	11	12	13	14
ROC AUC	0.992	0.994	0.999	0.993	0.994	0.998	0.993	0.996
PR AUC	0.937	0.955	0.876	0.933	0.946	0.973	0.931	0.956
MCC	0.872	0.894	0.732	0.857	0.888	0.921	0.782	0.901
F1	0.873	0.901	0.718	0.863	0.892	0.922	0.776	0.904
FNR (met)	0	0	0	0	0	0	0	0

4.2. Active Pixel Selection

At the second step, we want to separate pixels of 3-dimensional meteor chunks selected by the CNN into two groups: those containing the signal of a meteor (active pixels) and all the rest. Since meteor signals have a typical shape resembling the bell curve, as shown in the top left panels of Figures 1 and 2, it is straightforward to train a multilayer perceptron (MLP) to solve the task. The input dataset consists of vectors of length $T = 48$ now.

We followed the same way of training and testing the neural network as was used at the first stage. Namely, we trained an MLP on data extracted from seven sessions and tested it on the remaining one with all possible combinations of sessions. In order to avoid duplicate entries in the training dataset, we extracted data vectors from chunks of the size $48 \times 48 \times 48$. Due to a comparatively small shift $dt = 8$ GTUs, we obtained samples with meteor signal peaks located at almost all possible positions along the time axis. The number of vectors (pixels) containing meteor signals in training data sets was up to 30 thousand, with non-meteor samples ten times greater. The input data was standardized similar to the CNN case. Twenty percent of training samples were utilized for validation. Binary crossentropy was used as the loss function and validation loss was employed to adjust the learning rate and to avoid overfitting. Testing was performed on all vectors extracted from meteor chunks of the size $8 \times 8 \times 48$ selected by the CNN. The number of chunks used for testing of each particular session data can be found in Table 1.

We compared a number of possible configurations of simple MLPs with one, two, and three hidden layers and different number of neurons. An optimizer, activation functions and a performance metric used for training were the same as for the CNN described above. Table 3 presents results obtained with a two-layer MLP with 96 and 64 neurons in the two layers, respectively.

Table 3. Performance of the MLP on different sessions of observations. See the text for details.

Test Session	5	6	7	8	11	12	13	14
ROC AUC	0.992	0.995	0.993	0.994	0.996	0.993	0.995	0.995
PR AUC	0.899	0.932	0.877	0.916	0.932	0.887	0.936	0.928
MCC	0.790	0.841	0.744	0.826	0.847	0.812	0.809	0.835
F1	0.794	0.845	0.737	0.832	0.852	0.814	0.810	0.840
Mean IoU	0.808	0.853	0.773	0.843	0.861	0.829	0.825	0.850
FNR (pxl)	2/422	2/1428	0/80	7/928	5/958	0/492	0/457	2/630

Similar to Table 2, Table 3 presents results illustrating performance of the MLP trained and tested on different sessions of data collection. Values of one more performance metric are shown here, namely, the mean values of the intersection-over-union (IoU) score. This function and its versions are often used in tasks of labeling pixels of images. It can be written as

$$\text{IoU} = \frac{\text{TP}}{\text{TP} + \text{FP} + \text{FN}}.$$

It can be seen that values of the mean IoU metric are slightly above those of the F_1 score. The last line of the table shows the false negative rate calculated for pixels containing meteor signals. Two things can be easily noticed. First, the MLP did not properly recognize 18 out of 5395 active pixels, i.e., FNR (pxl) $\approx$ 0.33%. In this sense the accuracy can be estimated as 99.67% in average. On the other hand, the worst result was obtained for the test run on data from session 8 with FNR (pxl) $\approx$ 0.75% so that the accuracy for this particular session equals 99.25%.

We have tried to address the same task with a few other machine learning methods, among them logistic regression, K nearest neighbors, the random forest, and XGBoost. We have failed to outperform results demonstrated with the MLP with any of them.

5. Discussion

We have demonstrated that a pipeline made of two simple neural networks, a CNN and an MLP, can be used to effectively recognize meteor tracks in the data of the Mini-EUSO fluorescence telescope that is observing the nocturnal atmosphere of Earth in the UV band from the International Space Station. The CNN used to select three-dimensional data chunks containing meteor signals managed to recognize properly all 1068 meteors in the dataset. The MLP employed to recognize pixels with meteor signals in data chunks picked up by the CNN, reached an accuracy beyond 99%. Neither of the ANNs puts high demands on computer resources necessary for their training. Besides this, they perform surprisingly fast during the classification stage with a major part of time being spent on reading data from a data storage, thus strongly outperforming the conventional algorithm.

We have seen in [33] that an ANN trained on data with clearly pronounced signals is able to identify patterns with low signal-to-noise ratio. Such events are classified as false positives during tests but their closer analysis reveals that their considerable part contains "positive" signals that were not found by the conventional algorithm used to prepare training and testing datasets. A preliminary analysis of signals marked as false positives in our tests has demonstrated that the same situation takes place with meteor tracks, so that the list of meteors can be extended with these newly found signals. The same applies to the list of active pixels. This is an important advantage of ML-based methods over conventional approaches.

One can anticipate that new experimental data might present new patterns of non-meteor signals since observational conditions vary strongly from one session to another. In this case, it might be necessary to extend the training dataset and retrain the ANNs. However, we do not expect that the architecture of the CNN and the MLP will need to be modified considerably. On the other hand, it is clear that the presented ANNs are not the only way of recognizing meteor tracks in the Mini-EUSO data using methods of machine learning. Still, it might be not easy to exceed the accuracy of the suggested pipeline with simple models. We plan to analyze other possible approaches, especially for the segmentation part. We are also going to address the task of solving the same problem in one step, without splitting it into two parts.

Finally, it is worth mentioning that the presented method is not confined to recognizing meteor tracks. Preliminary results of applying the same approach and even the same trained models for recognizing signals that mimic illumination expected from extensive air showers born by ultra-high energy cosmic rays, are quite promising and will be reported elsewhere. We also expect that this pipeline or a similar one can be implemented in onboard electronics of future orbital missions to act as a trigger for track-like signals of different nature manifesting themselves at various time scales.

Author Contributions: Conceptualization, M.B. (Mario Bertaina), M.C., T.E., P.G., P.K., T.N., A.O., E.P., P.P., M.R. and L.W.; methodology, M.Z.; formal analysis, D.A. and A.K.; investigation, M.Z.; resources, M.B. (Matteo Battisti), A.B., M.B. (Marta Bianciotto), F.B., C.B., S.B., K.S., G.C., F.C., J.E., F.F., M.A.F., A.G., F.K., H.K., M.M., L.M., H.O., Z.P., G.P., E.R., G.R., N.S., C.D.L.T., Y.T. and M.V.; data curation, D.A., D.B., A.K., H.M., L.M., A.M. and L.W.P.; writing—original draft preparation, M.Z.; writing—review and editing, M.B. (Mario Bertaina) and M.V.; project administration, M.C. and P.K.; funding acquisition, M.C., P.K. and M.Z. All authors have read and agreed to the published version of the manuscript.

Funding: The research of M.Z., D.A. and A.K. was funded by grant number 22-22-00367 of the Russian Science Foundation (https://rscf.ru/project/22-22-00367/) (accessed on 20 June 2023).

Data Availability Statement: The data used in the study are not publicly available due to the current JEM-EUSO collaboration policy.

Acknowledgments: All neural networks discussed in the paper were implemented in Python with TensorFlow [34] and scikit-learn [35] software libraries.

Conflicts of Interest: The authors declare no conflict of interest.

Abbreviations

The following abbreviations are used in this manuscript:

ANN	Artificial neural network
AUC	Area under the curve
CNN	Convolutional neural network
EC	Elementary cell
FoV	Field of view
FS	Focal surface
ISS	International space station
MAPMT	Multi-anode photomultiplier
MCC	Matthews correlation coefficient
MLP	Multi-layer perceptron
PR	Precision-recall
PSF	Point spread function
ROC	Receiver operating characteristic
UHECR	Ultra-high energy cosmic ray
UV	ultraviolet

References

1. Adams, J.H., Jr.; Ahmad, S.; Albert, J.N.; Allard, D.; Anchordoqui, L.; Andreev, V.; Anzalone, A.; Arai, Y.; Asano, K.; Ave Pernas, M.; et al. The JEM-EUSO instrument. *Exp. Astron.* **2015**, *40*, 19–44. [CrossRef]
2. Adams, J.H., Jr.; Ahmad, S.; Albert, J.N.; Allard, D.; Anchordoqui, L.; Andreev, V.; Anzalone, A.; Arai, Y.; Asano, K.; Ave Pernas, M.; et al. The JEM-EUSO mission: An introduction. *Exp. Astron.* **2015**, *40*, 3–17. [CrossRef]
3. Bertaina, M.E. An overview of the JEM-EUSO program and results. In Proceedings of the 37th International Cosmic Ray Conference—PoS(ICRC2021), Berlin, Germany, 15–22 July 2021; Volume 395, p. 406. [CrossRef]

4. Benson, R.; Linsley, J. Satellite observation of cosmic ray air showers. In Proceedings of the 17th International Cosmic Ray Conference, Paris, France, 13–25 July 1981; Volume 8, pp. 145–148.
5. Adams, J.H., Jr.; Ahmad, S.; Albert, J.N.; Allard, D.; Anchordoqui, L.; Andreev, V.; Anzalone, A.; Arai, Y.; Asano, K.; Ave Pernas, M.; et al. Science of atmospheric phenomena with JEM-EUSO. *Exp. Astron.* **2015**, *40*, 239–251. [CrossRef]
6. Klimov, P.A.; Panasyuk, M.I.; Khrenov, B.A.; Garipov, G.K.; Kalmykov, N.N.; Petrov, V.L.; Sharakin, S.A.; Shirokov, A.V.; Yashin, I.V.; Zotov, M.Y.; et al. The TUS Detector of Extreme Energy Cosmic Rays on Board the Lomonosov Satellite. *Space Sci. Rev.* **2017**, *212*, 1687–1703.
7. Khrenov, B.A.; Klimov, P.A.; Panasyuk, M.I.; Sharakin, S.A.; Tkachev, L.G.; Zotov, M.Y.; Biktemerova, S.V.; Botvinko, A.A.; Chirskaya, N.P.; Eremeev, V.E.; et al. First results from the TUS orbital detector in the extensive air shower mode. *J. Cosmol. Astropart. Phys.* **2017**, *9*, 6.
8. Adams, J.H., Jr.; Ahmad, S.; Albert, J.N.; Allard, D.; Anchordoqui, L.; Andreev, V.; Anzalone, A.; Arai, Y.; Asano, K.; Ave Pernas, M.; et al. JEM-EUSO: Meteor and nuclearite observations. *Exp. Astron.* **2015**, *40*, 253–279. [CrossRef]
9. Abdellaoui, G.; Abe, S.; Acheli, A.; Adams, J.; Ahmad, S.; Ahriche, A.; Albert, J.N.; Allard, D.; Alonso, G.; Anchordoqui, L.; et al. Meteor studies in the framework of the JEM-EUSO program. *Planet. Space Sci.* **2017**, *143*, 245–255. . [CrossRef]
10. Adams, J.H.; Ahmad, S.; Allard, D.; Anzalone, A.; Bacholle, S.; Barrillon, P.; Bayer, J.; Bertaina, M.; Bisconti, F.; Blaksley, C.; et al. A Review of the EUSO-Balloon Pathfinder for the JEM-EUSO Program. *Space Sci. Rev.* **2022**, *218*, 3. [CrossRef]
11. Abdellaoui, G.; Abe, S.; Adams, J.; Ahriche, A.; Allard, D.; Allen, L.; Alonso, G.; Anchordoqui, L.; Anzalone, A.; Arai, Y.; et al. EUSO-TA—First results from a ground-based EUSO telescope. *Astropart. Phys.* **2018**, *102*, 98–111. . [CrossRef]
12. Bacholle, S.; Barrillon, P.; Battisti, M.; Belov, A.; Bertaina, M.; Bisconti, F.; Blaksley, C.; Blin-Bondil, S.; Cafagna, F.; Cambiè, G.; et al. Mini-EUSO Mission to Study Earth UV Emissions on board the ISS. *Astrophys. J. Suppl. Ser.* **2021**, *253*, 36.
13. Casolino, M.; Adams, J., Jr.; Anzalone, A.; Arnone, E.; Arnone, D.; Barghini, S.; Bartocci, M.; Battisti, R.; Bellotti, M.; Bertaina, F.; et al. The Mini-EUSO telescope on board the International Space Station: Launch and first results. In Proceedings of the 37th International Cosmic Ray Conference—PoS(ICRC2021), Berlin, Germany, 15–22 July 2021; Volume 395, p. 354. [CrossRef]
14. Marcelli, L.; Barghini, D.; Battisti, M.; Blaksley, C.; Blin, S.; Belov, A.; Bertaina, M.; Bianciotto, M.; Bisconti, F.; Bolmgren, K.; et al. Integration, qualification, and launch of the Mini-EUSO telescope on board the ISS. *Rend. Lincei. Sci. Fis. Nat.* **2023**, *34*, 23–35. [CrossRef]
15. Casolino, M.; Barghini, D.; Battisti, M.; Blaksley, C.; Belov, A.; Bertaina, M.; Bianciotto, M.; Bisconti, F.; Blin, S.; Bolmgren, K.; et al. Observation of night-time emissions of the Earth in the near UV range from the International Space Station with the Mini-EUSO detector. *Remote Sens. Environ.* **2023**, *284*, 113336.
16. Scotti, V.; Osteria, G.; JEM-EUSO Collaboration. The EUSO-SPB2 mission. *Nucl. Instrum. Methods Phys. Res. A* **2020**, *958*, 162164. [CrossRef]
17. Cummings, A.; Eser, J.; Filippatos, G.; Olinto, A.V.; Venters, T.M.; Wiencke, L. EUSO-SPB2: A sub-orbital cosmic ray and neutrino multi-messenger pathfinder observatory. *arXiv* **2022**, arXiv:2208.07466.
18. Eser, J.; Olinto, A.V.; Wiencke, L. Overview and First Results of EUSO-SPB2. In Proceedings of the 38th International Cosmic Ray Conference—PoS(ICRC2023), Nagoya, Japan, 26 July–3 August 2023; Volume 444, p. 397. [CrossRef]
19. Klimov, P.; Battisti, M.; Belov, A.; Bertaina, M.; Bianciotto, M.; Blin-Bondil, S.; Casolino, M.; Ebisuzaki, T.; Fenu, F.; Fuglesang, C.; et al. Status of the K-EUSO Orbital Detector of Ultra-High Energy Cosmic Rays. *Universe* **2022**, *8*, 88.
20. POEMMA Collaboration; Olinto, A.V.; Krizmanic, J.; Adams, J.H.; Aloisio, R.; Anchordoqui, L.A.; Anzalone, A.; Bagheri, M.; Barghini, D.; Battisti, M.; et al. The POEMMA (Probe of Extreme Multi-Messenger Astrophysics) observatory. *J. Cosmol. Astropart. Phys.* **2021**, *2021*, 7.
21. Barghini, D.; Battisti, M.; Belov, A.; Edoardo Bertaina, M.; Bisconti, F.; Capel, F.; Casolino, M.; Ebisuzaki, T.; Gardiol, D.; Klimov, P.; et al. Meteor detection from space with Mini-EUSO telescope. In Proceedings of the European Planetary Science Congress, EPSC2020-800. Online, 21 September–9 October 2020. [CrossRef]
22. Barghini, D.; Battisti, M.; Belov, A.; Bertaina, M.E.; Bertone, S.; Bisconti, F.; Capel, F.; Casolino, M.; Cellino, A.; Ebisuzaki, T.; et al. Analysis of meteors observed in the UV by the Mini-EUSO telescope onboard the International Space Station. In Proceedings of the European Planetary Science Congress, EPSC2021-243. Online, 13–24 September 2021. [CrossRef]
23. Zotov, M.; Sokolinskii, D. A Neural Network Approach for Selecting Track-like Events in Fluorescence Telescope Data. *Bull. Rus. Acad. Sci. Phys.* **2023**, *87*, 1054–1057. [CrossRef]
24. Vrábel, M.; Genci, J.; Bobik, P.; Bisconti, F. Machine Learning Approach for Air Shower Recognition in EUSO-SPB Data. In Proceedings of the 36th International Cosmic Ray Conference (ICRC2019), Madison, WI, USA, 24 July–1 August 2019; Volume 36, p. 456.
25. Szakács, P.; Vrábel, M.; Genči, J. Classification of EUSO-SPB data using convolutional neural networks (CNNs). In *Electrical Engineering and Informatics X, Proceedings of the Faculty of Electrical Engineering and Informatics of the Technical University of Košice*; Technical University of Košice: Staré Mesto, Slovakia, 2019; pp. 262–267.
26. Filippatos, G.; Battisti, M.; Bertaina, M.E.; Bisconti, F.; Eser, J.; Osteria, G.; Sarazin, F.; Wiencke, L.; JEM-EUSO Collaboration. Expected Performance of the EUSO-SPB2 Fluorescence Telescope. In Proceedings of the 37th International Cosmic Ray Conference, Berlin, Germany, 15–22 July 2022; p. 405.
27. Montanaro, A.; Ebisuzaki, T.; Bertaina, M. Stack-CNN algorithm: A new approach for the detection of space objects. *J. Space Saf. Eng.* **2022**, *9*, 72–82. [CrossRef]

28. Ruiz-Hernandez, O.I.; Sharakin, S.; Klimov, P.; Martínez-Bravo, O.M. Meteors observations by the orbital telescope TUS. *Planet. Space Sci.* **2022**, *218*, 105507. [CrossRef]
29. Fawcett, T. An introduction to ROC analysis. *Pattern Recognit. Lett.* **2006**, *27*, 861–874. [CrossRef]
30. Ferri, C.; Hernández-Orallo, J.; Modroiu, R. An experimental comparison of performance measures for classification. *Pattern Recognit. Lett.* **2009**, *30*, 27–38. [CrossRef]
31. Saito, T.; Rehmsmeier, M. The Precision-Recall Plot Is More Informative than the ROC Plot When Evaluating Binary Classifiers on Imbalanced Datasets. *PLoS ONE* **2015**, *10*, e0118432. [CrossRef] [PubMed]
32. Zhu, Q. On the performance of Matthews correlation coefficient (MCC) for imbalanced dataset. *Pattern Recognit. Lett.* **2020**, *136*, 71–80. [CrossRef]
33. Zotov, M. Application of Neural Networks to Classification of Data of the TUS Orbital Telescope. *Universe* **2021**, *7*, 221. [CrossRef]
34. Abadi, M.; Agarwal, A.; Barham, P.; Brevdo, E.; Chen, Z.; Citro, C.; Corrado, G.S.; Davis, A.; Dean, J.; Devin, M.; et al. TensorFlow: Large-Scale Machine Learning on Heterogeneous Systems. 2015. Available online: tensorflow.org (accessed on 30 April 2023).
35. Pedregosa, F.; Varoquaux, G.; Gramfort, A.; Michel, V.; Thirion, B.; Grisel, O.; Blondel, M.; Prettenhofer, P.; Weiss, R.; Dubourg, V.; et al. Scikit-learn: Machine Learning in Python. *J. Mach. Learn. Res.* **2011**, *12*, 2825–2830.

Disclaimer/Publisher's Note: The statements, opinions and data contained in all publications are solely those of the individual author(s) and contributor(s) and not of MDPI and/or the editor(s). MDPI and/or the editor(s) disclaim responsibility for any injury to people or property resulting from any ideas, methods, instructions or products referred to in the content.

Article

FenceTalk: Exploring False Negatives in Moving Object Detection

Yun-Wei Lin [1,*], Yuh-Hwan Liu [2], Yi-Bing Lin [3,4,5,6,*] and Jian-Chang Hong [1]

1. The College of Artificial Intelligence, National Yang Ming Chiao Tung University, Tainan 711, Taiwan; jim93073@gmail.com
2. Faculty of Technology Management, China Medical University, Taichung 404, Taiwan; qlyh@mail.cmu.edu.tw
3. College of Computer Science, National Yang Ming Chiao Tung University, Hsinchu 300, Taiwan
4. College of Biomedical Engineering, China Medical University, Taichung 404, Taiwan
5. Miin Wu School of Computing, National Cheng Kung University, Tainan 701, Taiwan
6. Research Center for Information Technology Innovation, Academia Sinica, Taipei 115, Taiwan
* Correspondence: jyneda@nycu.edu.tw (Y.-W.L.); liny@nctu.edu.tw (Y.-B.L.); Tel.: +886-6-3032121 (Y.-W.L.)

Abstract: Deep learning models are often trained with a large amount of labeled data to improve the accuracy for moving object detection in new fields. However, the model may not be robust enough due to insufficient training data in the new field, resulting in some moving objects not being successfully detected. Training with data that is not successfully detected by the pre-trained deep learning model can effectively improve the accuracy for the new field, but it is costly to retrieve the image data containing the moving objects from millions of images per day to train the model. Therefore, we propose FenceTalk, a moving object detection system, which compares the difference between the current frame and the background image based on the structural similarity index measure (SSIM). FenceTalk automatically selects suspicious images with moving objects that are not successfully detected by the Yolo model, so that the training data can be selected at a lower labor cost. FenceTalk can effectively define and update the background image in the field, reducing the misjudgment caused by changes in light and shadow, and selecting images containing moving objects with an optimal threshold. Our study has demonstrated its performance and generality using real data from different fields. For example, compared with the pre-trained Yolo model using the MS COCO dataset, the overall recall of FenceTalk increased from 72.36% to 98.39% for the model trained with the data picked out by SSIM. The recall of FenceTalk, combined with Yolo and SSIM, can reach more than 99%.

Keywords: deep learning; image object detection; internet of things; structural similarity index measure (SSIM)

1. Introduction

In the past, traditional techniques often employed simple cameras for security surveillance in specific areas. Applications included home security, farm monitoring, factory security, and more, aiming to prevent personal property or crops from being damaged. However, security personnel had to monitor the images at all times, incurring high costs. With the rise of internet of things (IoT) technology, security monitoring can be carried out using physical sensors such as infrared sensors, vibration sensors, or microwave motion sensors [1]. Nevertheless, these methods have varying accuracy due to different physical technologies or installation methods, and sensors can struggle to differentiate whether an object is a target, leading to false alarms.

Object detection is a task within the realm of computer vision, encompassing the identification and precise localization of objects of interest within an image. Its primary objective is not merely recognizing the objects present but also delineating their exact positions through the creation of bounding boxes. In the modern landscape of object detection, deep learning models such as convolutional neural networks (CNNs) [2,3] and pioneering architectures like Yolo (you only look once) [4] have come to the forefront. These technologies have significantly advanced the accuracy and efficiency of object detection tasks. Yolo, in particular, has played a pivotal role by enabling both efficient and accurate real-time object detection, cementing its status as a foundational tool across diverse computer vision applications.

What sets Yolo apart is its groundbreaking concept of single-shot detection, signifying its ability to detect and classify objects in an image during a single pass through a neural network. This innovative approach culminates in the prediction of bounding boxes, exemplified by the pink square box in Figure 1, which encompasses the identified objects. Similar objectives can be identified using the structural similarity index measure (SSIM) [5] and vision transformer (ViT) [6], both of which are efficient object detection algorithms.

Figure 1. FenceTalk user GUI. (Red polygon: fenced region; Pink square: the moving object).

Hence, object detection using Yolo finds wide-ranging practical applications, including security surveillance, where it excels at identifying and tracking individuals or moving objects within security camera footage. This paper presents FenceTalk, a user-friendly moving object detection system. FenceTalk adopts a no-code approach, allowing non-AI expert users to effortlessly track the objects they define. In Section 5, we will carry out a performance comparison between FenceTalk and the previously proposed approaches.

Figure 1 illustrates the window-based FenceTalk graphical user interface (GUI), where a red polygon can be easily created by dragging an area of interest. This area is designated as a fenced region within a fixed location. Through deep learning models, image recognition is performed to detect the presence of moving objects (depicted by the pink square in Figure 1) within the fenced area.

Combining with IoTtalk, an IoT application development platform [7], FenceTalk sends real-time alert messages to pre-bound devices through IoTtalk when a moving object is detected within the fenced area. This allows users to swiftly take appropriate measures to ensure the security of the area. A reliable image recognition model is crucial for FenceTalk. However, even after capturing images with targets in the desired new area for labeling and using this annotated data for training the deep learning model, the model often fails to recognize certain moving objects. Identifying these images where recognition was unsuccessful and using them as training data for the model can further enhance its accuracy.

Extracting image frames containing moving objects from a vast amount of image data as training data for the model incurs significant labor and time costs. For instance, at the common camera recording rate of 30 frames per second (FPS), a camera can generate 2,592,000 image frames in a single day. Given that the moving objects to be recognized in a fixed area exhibit movement characteristics, FenceTalk designs an algorithm to define a background image, and compares the current image frame with the background image using SSIM to analyze differences in the images. Subsequently, based on the brightness of the current image, FenceTalk employs an optimal threshold to separate background noise from the moving objects of interest. This process determines whether the current image frame contains a suspicious image with moving objects. By filtering these selected suspicious images, FenceTalk enables users to choose the model's training data with reduced human resource costs. To minimize hardware costs, we deployed FenceTalk on an embedded device, the Nvidia Jetson Nano, equipped with a graphics processing unit. The Jetson Nano GPU features 128 CUDA cores and 4 GB LPDDR4 memory, making it suitable for running CNN models that require extensive matrix operations.

This paper is organized as follows: Section 2 reviews the previous work; Section 3 describes the FenceTalk architecture; Section 4 describes how SSIM is used in FenceTalk; Section 5 evaluates the object detection performance of FenceTalk and the time and space complexities of Nvidia Jetson Nano.

2. Related Works

Deep learning models for moving object detection based on CNNs [8–11] can utilize camera images from specific areas as training data to learn the features of moving objects. This enables accurate identification of moving objects within the frame. In a previous work [4], the authors chose the Yolo model for recognizing humans and vehicles, applying it to a real-time recognition system in an advanced driver assistance system (ADAS). Compared to two-stage recognition models like Faster-RCNN, the one-stage Yolo model strikes a balance between accuracy and speed. Yolo and Yolo-tiny were employed for pedestrian detection, achieving a recall rate of more than 80%. Yolo-tiny, in particular, is well-suited to scenarios requiring real-time detection—aligning with the application context of FenceTalk in our study.

We evaluated different versions of YOLO [12,13]. Despite YOLOv6 [14] having 33% fewer parameters than YOLOv7-tiny and YOLOv4-tiny at the same scale, its inference speed improved by less than 10%. However, this reduction in parameters led to an accuracy drop of nearly 10% compared to YOLOv7-tiny. YOLOv7-tiny's optimization primarily emphasizes training efficiency and inference speed. On our hardware, YOLOv7-tiny demonstrated object detection approximately 10% faster than that of YOLOv4-tiny. Nevertheless, YOLOv7-tiny only achieved an accuracy of 90.86% on our dataset, falling short of YOLOv4-tiny's performance. As a result, we ultimately opted for YOLOv4-tiny as our choice.

Now we review the past methods for evaluating image similarity and the applications related to SSIM. Mean square error (MSE) is a common and straightforward approach to measure the similarity between two images. It calculates the mean squared error value of image pixels as an indicator of image similarity. A higher MSE value indicates greater disparity between the two images. The calculation method of MSE considers only the corresponding pixel values of the two images, making its results less reliable. In a study [15], using Einstein's image as an example, JPEG image compression and blurring were applied, resulting in significant differences between the two images. However, their MSE results were similar to the original image. MSE can also exhibit substantial variations in values due to minor pixel rearrangements. The authors of the study slightly shifted and scaled the Einstein image, maintaining a similar appearance to the original. Nevertheless, the MSE values for these modified images were significantly high, reaching 871 and 873, respectively, compared to the original image.

Peak signal-to-noise ratio (PSNR) [16] is defined based on MSE and is also widely used for measuring image similarity [17]. PSNR resolves this issue by incorporating peak intensity considerations. In particular, the logarithmic transformation enables us to express scores more concisely. Consequently, the advantages of PSNR over MSE can be summarized as follows: (1) facilitating the comparison of results obtained from images encoded with varying bit depths and (2) providing a more concise representation. However, it is important to note that, by definition, PSNR remains essentially a normalized variant of MSE. PSNR finds widespread utility in tasks related to image and video compression, restoration, and enhancement, effectively quantifying the extent of information loss during compression or processing. A higher PSNR value indicates greater similarity between the two images. Nevertheless, when it comes to background detection, it does not exhibit the same effectiveness as mixture of Gaussians (MOG) [18] and absolute difference (AD) [19].

MOG [18] is purpose-built for modeling the background within images by employing a mixture of Gaussian distributions. MOG demonstrates adaptability to changing lighting conditions and exhibits robust performance across various scenarios, making it suitable for real-time applications. AD [19] represents another background subtraction technique. While specific implementation details may vary, the core concept is to calculate the absolute pixel-wise difference between corresponding pixels in two input images. The result is a new image, known as the absolute difference image. Specifically, MOG garners favor for its robustness in accommodating evolving environments, whereas the efficacy of AD hinges on pixel-wise absolute difference.

In [20], it is argued that natural images possess highly structured characteristics with closely related neighboring pixels. As a result, the structural similarity index measure (SSIM) was introduced. SSIM is designed based on the human visual system (HVS) and evaluates image similarity by considering brightness, contrast, and structural factors. Unlike traditional methods like MSE or PSNR, which use a sum of errors, SSIM does not drastically change its results due to minor changes in image brightness or noise, better aligning with the human perception of image quality. The calculation method of SSIM will be further introduced in Section 4.

In [21], the authors advocate using SSIM to measure the distance between tracked objects and candidate bounding boxes in object tracking. Unlike the traditional method of object tracking based on object color classification, this paper demonstrates that even in challenging conditions such as temporary occlusion or changes in image brightness, using SSIM still yields stable and reliable object tracking results.

In [22], the authors compare detection methods for moving objects using a single Gaussian model, an adaptive Gaussian mixture model 2 (MOG2), and SSIM-based approaches. The traditional single Gaussian model tends to capture a significant amount of image noise, while MOG2 struggles to detect targets with similar colors to the background. However, the SSIM-based detection method can accurately detect moving objects. The study only utilized the first frame of the video as the SSIM background image and did not consider subsequent changes in the background. In contrast, FenceTalk incorporates an algorithm to update the SSIM background image and validates its stability using a large dataset.

In [23], a method for detecting geometric defects in digital printing images is proposed based on SSIM. The study inspects defects such as stains, scratches, and ink in simulated and real images. Compared to the AD-based inspection method, the SSIM-based approach effectively detects subtle defects and reduces misjudgments caused by environmental lighting factors.

ViT [6] is the state-of-the-art model for image classification, which is a revolutionary deep learning architecture that has gained significant attention and popularity in the field of computer vision. ViT has powerful image recognition capabilities, and when used for image classification, it can achieve excellent results. Based on practical experience, ViT's performance significantly surpasses that of earlier algorithms MSE, PSNR, MOG2, and AD. While running ViT does consume a significant amount of computational resources, its performance can serve as an upper-bound reference point for accuracy comparison.

In a home security automatic recording system described in [24], SSIM is used to compare the current and the next image frames to decide whether to trigger the recording process. This system achieves a high-accuracy and stable event detection, saving storage space and simplifying subsequent image analysis time. While the study used an average similarity index and standard deviation of 100 consecutive pairs of frames as the threshold for motion detection, FenceTalk differentiates images based on brightness to find the optimal threshold. It then separates background noise and moving objects using this threshold, automatically selecting images that Yolo failed to recognize. This reduces the time cost of manual labeling. The study reported a motion detection accuracy ranging from 0.985 to 0.995 in most experiments, while FenceTalk achieves a slightly better performance of 0.9994. In [25], SSIM-NET is introduced for defect detection in printed circuit boards (PCBs). SSIM-NET employs a two-stage approach where SSIM identifies regions of interest in the image, followed by classification using MobileNet-V3. Compared to Faster-RCNN, SSIM-NET offers more than 12 times faster recognition speed, 0.62% higher accuracy, and faster training time. However, unlike FenceTalk, PCB defect detection does not need to consider the impact of abrupt lighting changes on SSIM, and the paper does not mention how SSIM background images are updated.

3. FenceTalk Architecture

FenceTalk utilizes a network camera to transmit real-time images for image recognition. Figure 2 illustrates the FenceTalk architecture, comprising six key components. At its core is the FenceTalk server, which includes the FenceTalk engine (Figure 2 (1)), derived from IoTtalk [6], and a developer GUI (Figure 2 (2)). Following the IoTtalk philosophy, the FenceTalk server operates as an IoT server and employs an image database known as FenceTalk DataBase (FTDB) to store images extracted from the video streams (Figure 2 (3)). Additionally, the remaining four components within FenceTalk serve as IoT devices, encompassing the camera software module (Figure 2 (4)), the Yolo module (Figure 2 (5)), the SSIM module (Figure 2 (6)), and the user GUI module (Figure 2 (7)).

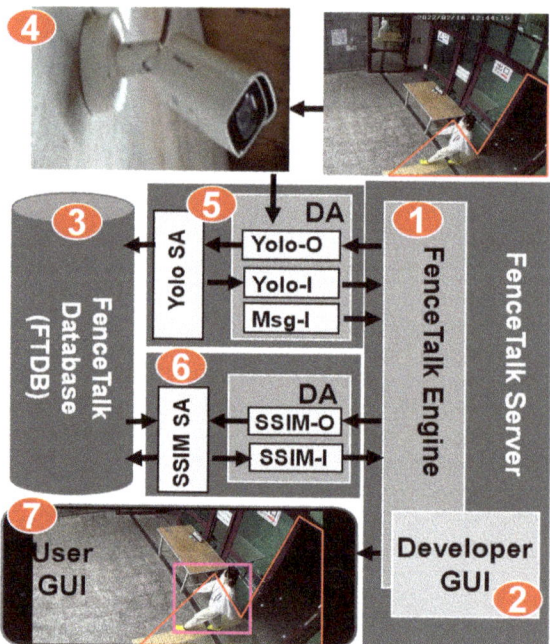

Figure 2. The FenceTalk architecture.

Each of these software modules consists of two parts: device application (DA) and sensor and actuator application (SA). The DA is responsible for communication with the FenceTalk server using HTTP communication ((5)↔(1), (6)↔(1), (7)↔(1) in Figure 2), which can be achieved through wired means (e.g., Ethernet) or wireless options (e.g., 5G or Wi-Fi) for communication. The SA part implements functionalities related to IoT devices, such as Yolo SA for object detection based on the Yolo model and SSIM SA for detecting moving objects based on SSIM.

The network camera (Figure 2 (4)) uses the real-time streaming protocol (RTSP) to stream the image frames. If the user draws a red polygon to define a fenced area within the desired field, FenceTalk will exclusively perform object detection within this specified area. The Yolo SA (Figure 2 (5)) continuously receives the latest streaming image frames and conducts object detection within the designated region using the Yolov4-tiny model. The results of the detection process, including object names, positions, and image paths, are stored in FTDB (Figure 2 (3)).

The ongoing Yolo recognition result is transmitted to the FenceTalk Engine (Figure 2 (1)) through the DA interface Yolo-I. The engine receives this data and has the capability to preprocess it using custom functions before transmitting it to connected IoT devices. Specifically, the engine forwards the Yolo recognition results to SSIM (Figure 2 (6)) through the DA interface SSIM-O for further evaluation. The outcomes of this evaluation, which include information about the presence of moving objects and corresponding image paths, are then stored in FTDB. Simultaneously, the SSIM assessment result is sent to the FenceTalk server via the DA interface SSIM-I, which is sent back to the Yolo module for further processing (See Section 4). The results of object recognition are subsequently displayed on user-defined devices (Figure 2 (7)).

In FenceTalk, IoT devices can be effortlessly connected and configured using the developer GUI (Figure 2 (2)) accessible through a web browser. Figure 3 depicts this GUI, wherein the involved IoT devices can be chosen from the model pulldown list (Figure 3 (1)). When we opt for the Yolo IoT device (i.e., the Yolo module), two icons are displayed within the GUI window. The icon on the left side of the window indicates the DA interfaces Yolo-I and Msg-I (Figure 3 (2)). The icon on the right side represents the DA interface Yolo-O (Figure 3 (3)). Similarly, we can choose the SSIM and the display devices. To establish connections between the IoT devices, all that is required is dragging the "join links." For instance, Join 1 connects Yolo-I to SSIM-O. This link creates an automatic communication pathway between the Yolo module and the SSIM module. Consequently, the configuration shown in Figure 3 yields the FenceTalk architecture displayed in Figure 2.

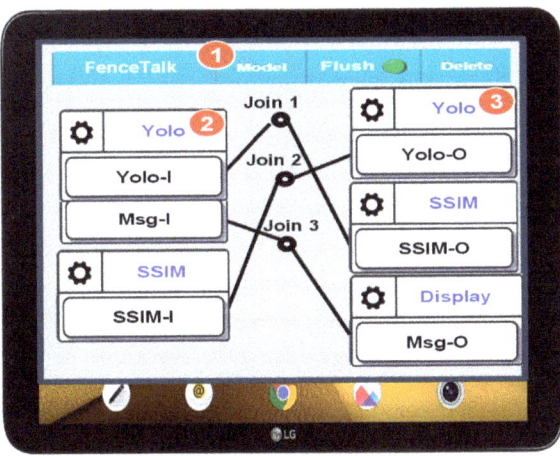

Figure 3. The FenceTalk developer GUI.

4. The SSIM Module

When the Yolo module processes an image and detects the moving objects, it places the image in the moving object database (Figure 4 (1)). If no moving objects are detected, the image is placed in the non-moving object database (Figure 4 (2)). Both databases are parts of FTDB (Figure 2 (3)). The detection accuracy of the Yolo module can be improved through re-training using false positive images from the moving object database and false negative images from the non-moving object database. The identification of false positive/negative images is typically performed manually. Unfortunately, our experience indicates that the size of the non-moving object database is usually substantial, making the identification of false positive images a highly tedious task. To resolve this issue, we developed the SSIM module.

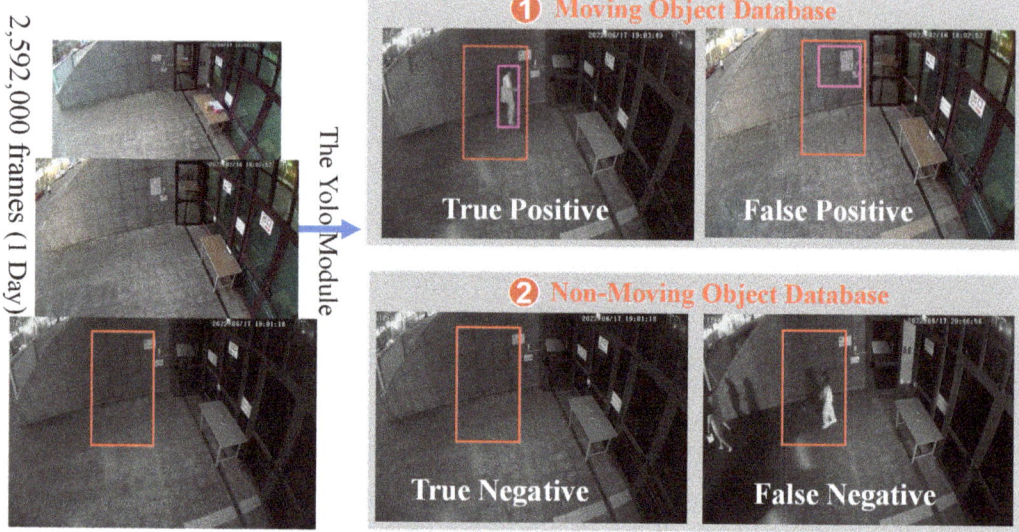

Figure 4. The operation of the Yolo module.

The structural similarity index (SSIM) is a metric used to measure the similarity between two images. It assesses the images based on their brightness, contrast, and structural similarity. It is commonly employed to determine the degree of similarity or distortion between two images. Given a background image x and a test photo y, both of size $m \times n$, SSIM is defined as:

$$\text{SSIM}(x,y) = [\Psi(x,y)]^\alpha [c(x,y)]^\beta [s(x,y)]^\gamma \tag{1}$$

where

$$\Psi(x,y) = \frac{2\mu_x \mu_y + C_1}{\mu_x^2 + \mu_y^2 + C_1} \tag{2}$$

$$c(x,y) = \frac{2\sigma_x \sigma_y + C_2}{\sigma_x^2 + \sigma_y^2 + C_2} \tag{3}$$

$$s(x,y) = \frac{\sigma_{xy} + C_3}{\sigma_x \sigma_x + C_3} \tag{4}$$

In $\Psi(x, y)$, grayscale values of the images are used to compare the similarity in average luminance between the two images. In the function $c(x, y)$, contrast similarity between the images is assessed by calculating the standard deviation of image pixels. The function $s(x, y)$ measures the similarity in structural content between the two photographs. In Equation (2), μ_x and μ_y represent the average values of the pixel intensities of the two images, while σ_x and σ_y in Equations (3) and (4) denote the covariance of the two images. Constants C_1, C_2, and C_3 are used to stabilize the function, where $C_1 = K_1 L^2$, $C_2 = K_2 L^2$, and $C_3 = \frac{C_2}{2}$. The values of K_1 and K_2 are set as 0.01 and 0.03, respectively, and L is the number of possible intensity levels in the image. For an 8-bit grayscale image, L would be 255. In [20], the values of α, β, and γ are set to 1, leading to a simplified SSIM formula of Equation (1):

$$\text{SSIM}(x,y) = \frac{(2\mu_x\mu_y + C_1)(2\sigma_{xy} + C_2)}{(\mu_x^2 + \mu_y^2 + C_1)(\sigma_x^2 + \sigma_y^2 + C_2)} \quad (5)$$

The structural similarity index (SSIM) yields larger values for more similar images and possesses three key properties. The symmetry property is

$$\text{SSIM}(x,y) = \text{SSIM}(y,x)$$

The bound property is

$$\text{SSIM}(x,y) \leq 1$$

The uniqueness property states that when two images are identical, i.e., $\mu_x = \mu_y$ and $\sigma_x = \sigma_y$, we have

$$\text{SSIM}(x,y) = 1$$

FenceTalk utilizes a predefined background image (Figure 5 (1)) and an image frame (Figure 5 (2)) to determine the presence of moving objects. Since SSIM requires two single-channel image frames as input, we convert the input images from RGB with three channels to single-channel grayscale images. Then, we use an $N \times N$ sliding window with a moving stride of 1. For each window, the SSIM is calculated. The output range of SSIM is [0, 1], while the pixel values of an 8-bit image are in the range of [0, 255]. To represent the SSIM values obtained from each sliding window in the full range of grayscale values, we linearly scale the SSIM values to the range of [0, 255]. This produces a complete single-channel grayscale (binarized) SSIM image (Figure 5 (3)).

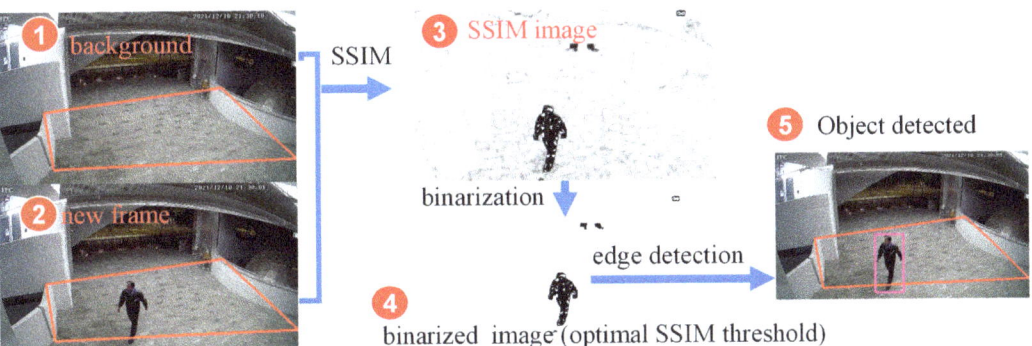

Figure 5. The operation of the SSIM module.

In the SSIM image, different degrees of differences are represented by varying shades of color. When the difference between the two images within a sliding window is larger, the SSIM value is lower, and the corresponding area on the SSIM output image is represented by darker shades. Conversely, when the difference between the two images within a sliding window is smaller, the SSIM value is higher, and the corresponding area on the SSIM output image is represented by lighter shades.

To determine the presence of moving objects in the SSIM image, a threshold value needs to be set to separate the background noise from the moving objects. Using Figure 5 (3) as an example, a threshold value of 125 is applied to this SSIM image to obtain the image in Figure 5 (4). This process involves image binarization and edge detection, filtering out small noise components, and determining whether the area within the fence boundary contains moving objects. The corresponding detection positions are then highlighted on the RGB image (Figure 5 (5)). The selection of an appropriate threshold to effectively distinguish background noise from moving objects will be further discussed and experimentally demonstrated in the next section.

The flowchart of the Yolo and the SSIM modules is illustrated in Figure 6. FenceTalk determines the presence of moving objects by comparing the current frame with a background image. Before processing the image frame detection, FenceTalk selects frames with varying brightness from the video as candidate background images, denoted as BG[l] ($0 \leq l \leq L = 255$). If the moving objects to be recognized are within the 80 classes of the MS COCO dataset [26], these candidate backgrounds are initially processed by a pre-trained Yolo model trained on the MS COCO dataset to identify images that do not contain the moving object (e.g., people). However, due to limitations in the pre-trained Yolo model's recognition capabilities or when the moving object is not in the MS COCO dataset, users need to examine and remove background images that FenceTalk incorrectly identified as not containing any moving objects. In FenceTalk, moving objects are defined as objects that change their position relative to the background image over time, i.e., they are not fixed in the background.

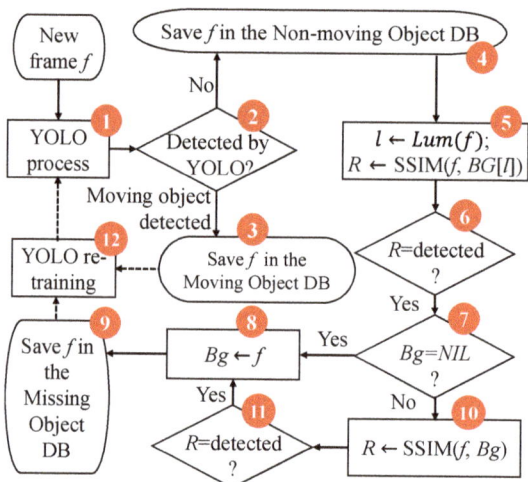

Figure 6. Operation of the Yolo and the SSIM modules.

FenceTalk reads a predefined background image BG[l] at the brightness level l, and from the video, it reads the next frame f to be processed. Frame f first undergoes the Yolo module for object recognition (Figure 6 (1)), where the Yolo model can be a pre-trained model on the MS COCO dataset or a user-trained model. If the Yolo module detects any objects (Figure 6 (2)), the images containing the moving objects are stored in the moving object database (Figure 6 (3)), and those without the moving object are stored in the non-moving object database (Figure 6 (4)). The process then moves on to the next frame. If the Yolo module does not detect any objects, FenceTalk enters the SSIM module to detect the false negatives images in the non-moving object database. The non-moving object database serves as a buffer due to the differing processing speeds of the Yolo module, which utilizes GPU, and the SSIM module, which operate on CPU. Consequently, the non-moving object database guarantees that all images from the Yolo module can be processed by the SSIM module without any loss of images.

FenceTalk calculates the brightness of frame f as l and selects the background image BG[l] with the same or closest brightness to f. The SSIM module then detects missing moving objects by comparing f and BG[l] (Figure 6 (5)). If the detection result (Figure 6 (6)) does not contain moving objects, the process continues to the next frame.

To reduce false positives caused by changes in background lighting or the addition of new objects to the background, if FenceTalk determines that f contains moving objects

(Figure 6 (6)), it performs a second check using a secondary background image, Bg. Initially, FenceTalk checks if background image Bg exists (Figure 6 (7)). If Bg does not exist (Initially, Bg = NIL), f is set as Bg (Figure 6 (8)) and stored (Figure 6 (9)) for subsequent model training. If Bg exists, SSIM is again applied to detect moving objects by comparing f and Bg (Figure 6 (10)). If the detection result (Figure 6 (11)) does not contain moving objects, the process continues to the next frame. If moving objects are detected, Bg is replace by f (Figure 6 (8)), and f is stored in the missing object database (Figure 6 (9)). The purpose of Bg is to prevent the repetitive detection and storage of the same moving object across a sequence of consecutive images. The utilization of SSIM(f, Bg) (Figure 6 (10)) ensures that each moving object is saved in the missing object database only once. After processing all desired image frames, users can retrieve all identified moving objects (Figure 6 (3)) and missing objects (Figure 6 (9)) from the database. These false negative images can be automatically annotated and used to retrain the Yolo model (Figure 6 (12)) to improve its accuracy. The retrained Yolo model can then be used to repeat the recognition process (Figure 6 (1)) for improved recognition accuracy.

Figure 7 illustrates the process of automatic background update. The SSIM image (Figure 7 (3)) is generated by comparing the background image (Figure 7 (1)) with the current image frame f_t at time t (Figure 7 (2)). Noticeable changes in shadows can lead to misjudgment of moving objects by FenceTalk (Figure 7 (4)). Upon detecting moving objects, FenceTalk updates the background image (Figure 7 (5)). The SSIM image (Figure 7 (7)) is generated by comparing the updated background image with the subsequent image frame f_{t+1} at time $t + 1$ (Figure 7 (6)). The result is a correct judgment that the image frame f_{t+1} does not contain any moving objects (Figure 7 (8)).

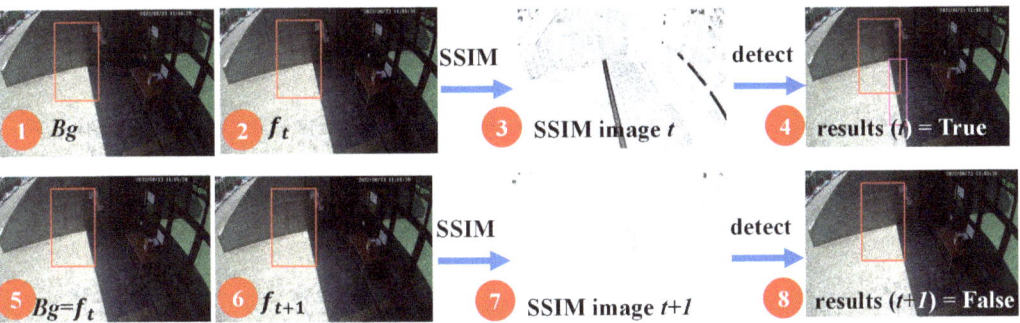

Figure 7. Background update.

5. FenceTalk Experiments

This section describes the datasets we collected and explains how we utilized these data to experimentally demonstrate the universality of the optimal threshold for SSIM. We will also discuss the accuracy of SSIM and Yolo in different subsets of the dataset. Finally, we will showcase the processing speed and resource usage of FenceTalk on the embedded device Jetson Nano.

We collected continuous camera footage from two outdoor locations, National Yang Ming Chiao Tung University and China Medical University, for a duration of six days each, using a recording frame rate of 30 FPS. Compared to indoor stable lighting conditions, the use of outdoor camera footage from these two locations provided a more robust evaluation of FenceTalk's performance in complex lighting environments. Dataset 1 was obtained from the entrance of the Electronic Information Building at No. 1001 University Road, National Yang Ming Chiao Tung University, Hsinchu City (as shown in Figure 8a). The data collection period was from 17 June 2022 to 23 June 2022, covering the entire day's camera footage. Dataset 2 was gathered from the Innovation and Research Building at No. 100, Section 1, Jingmao Road, Beitun District, China Medical University, Taichung

City (as shown in Figure 8b). The data collection period spanned from 10 December 2021 to 15 December 2022, capturing the full day's camera footage. In Dataset 1, images were collected at a rate of 10 FPS, resulting in a total of 4,832,579 images. Among these, there were 226,518 images containing moving objects (people). Dataset 2 comprised images collected at a rate of 15 FPS, with a total of 6,910,580 images. Within this dataset, 144,761 images were of moving objects. All images had a resolution of 1920 × 1080 pixels. These datasets were chosen to encompass a diverse range of lighting conditions and scenarios, enabling us to validate FenceTalk's performance robustness and reliability in real-world outdoor environments. It is noteworthy that due to this high collection frequency, the contents of any two consecutive images exhibited striking similarity. When it came to utilizing these images in training our model, a straightforward approach proved to be counterproductive, as it substantially consumed computational resources without yielding significant benefits to the model's performance. Therefore, we resampled the images at a rate of 5 FPS, which captured one image every 0.2 s. This adjustment allowed us to maintain a sufficiently fast capture rate to effectively track moving objects (people), and train a highly effective model while conserving computational power. Table 1 shows the total number of images utilized in Datasets 1 and 2 after the resampling process.

(**a**) Dataset 1   (**b**) Dataset 2

Figure 8. The locations for data collection.

Table 1. The images for training, validation, and testing.

Dataset 1	No. of Images	No. of Images with People
Training	862,258	23,098
Validation	808,153	35,358
Testing	786,443	40,700
Dataset 2		
Training	496,936	22,805
Validation	479,452	20,025
Testing	402,265	19,003

We utilized standard output measures for AI predictions, distinguishing them for the Yolo module, the SSIM module, and FenceTalk (Yolo + SSIM). All images processed by the Yolo module were classified into the following categories: TP (true positives), TN (true negatives), FN (false negatives), and FP (false positives). Therefore, we have

$$\text{Yolo : Precision} = \frac{TP}{TP + FP} \text{ and Recall} = \frac{TP}{TP + FN} \qquad (6)$$

In the SSIM module, the images in the non-moving object database were classified into TP^* true positives, TN^* true negatives, FN^* false negatives, and FP^* false positives. Therefore, we have

$$\text{SSIM}: \text{Precision} = \frac{TP^*}{TP^* + FP^*} \quad \text{andRecall} = \frac{TP^*}{TP^* + FP^*} \tag{7}$$

and finally, the output measures for FenceTalk (Yolo + SSIM) are

$$\text{FenceTalk}: \text{Precision} = \frac{TP + TP^*}{TP + FP + TP^* + FP^*} \tag{8}$$

and

$$\text{FenceTalk}: \text{Recall} = \frac{TP + TP^*}{TP + FN} \tag{9}$$

Specifically, the total count of all moving objects within a dataset is derived by $TP + FN$. The count of correctly predicted cases by Yolo is represented by TP. TP^* represents the count of accurately predicted cases by SSIM among the FN cases. Consequently, the total count of correctly predicted cases by FenceTalk (Yolo + SSIM) is $TP + TP^*$. Therefore, the recall is calculated as $TP + TP^*/TP + FN$. The F1 score is expressed as

$$\text{F1score} = 2 \left(\frac{\text{Precision} \times \text{Recall}}{\text{Precision} + \text{Recall}} \right) \tag{10}$$

In Equation (10), the F1 score takes into account both precision and recall.

In FenceTalk, to find and verify the universality of the optimal SSIM threshold in each dataset, we divided each dataset into three subsets: the training dataset, the validation dataset, and the testing dataset. In the FenceTalk experiments, we calculated the precision and recall for both Yolo and SSIM (Equations (6) and (7)).

To find the optimal SSIM threshold, we calculated the average grayscale value of each image and used it as an image brightness category. We used an interval of 25 for the SSIM threshold and recorded the TP, TN, FN, and FP counts for different thresholds under various brightness levels. This helped us calculate the precision and recall of SSIM at different thresholds (elaborated in Figures 9 and 10). Since we aimed to collect images that Yolo failed to recognize using SSIM for model training, we selected the threshold with the highest recall value as the optimal SSIM threshold for each brightness level in the experiment. If there were multiple highest recall values for a particular brightness level, we chose the threshold with the best precision among them. If duplicates remained, we selected the median as the optimal threshold.

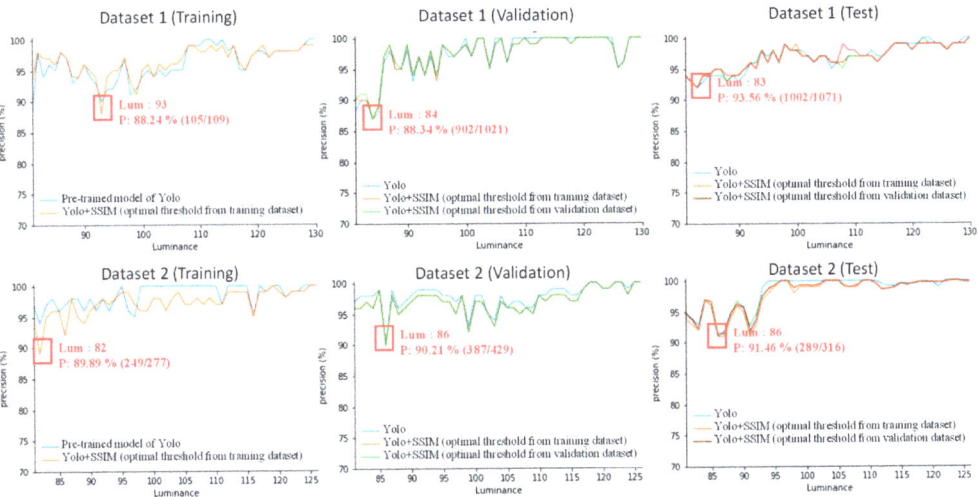

Figure 9. FenceTalk precision performance under various luminance (brightness) levels.

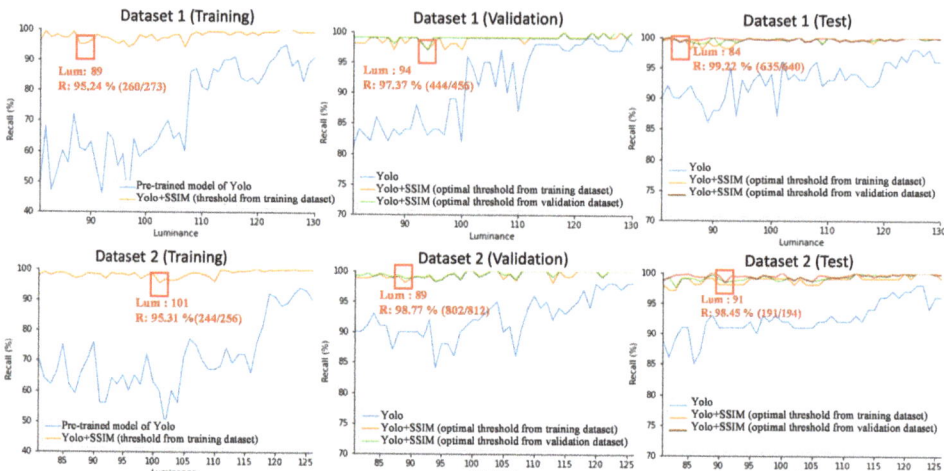

Figure 10. FenceTalk recall performance under various luminance (brightness) levels.

We employed the modified Yolo model trained on the training dataset to test the validation dataset and find the optimal SSIM threshold for different brightness levels in this dataset. Using the optimal SSIM threshold from the validation dataset, we identified images containing moving objects and images detected by the Yolo model. After manual labeling, these images were combined with all images containing humans from the training dataset, serving as training data for the Yolo model, which was then applied to the testing dataset for inferencing.

Table 2 presents the precision and recall of Yolo and SSIM using the optimal thresholds in Dataset 1. In the training phase, the recall for Yolo was relatively low at 77.16%. However, through the FenceTalk mechanism, the FenceTalk precision (Yolo + SSIM) was 97.71% and the recall (Yolo + SSIM) was 98.68%. The validation phase showed that FenceTalk's precision exceeded 97% and FenceTalk's recall (Yolo + SSIM) was above 99%. We re-trained the Yolo model after the validation phase. Therefore, the validation phase was a second training phase. Then, in the testing phase, FenceTalk's precision was 97.65% and its recall was 99.75%.

Table 2. The output measures for Dataset 1 (red font signifies the best performance).

Dataset 1	Yolo			SSIM			Yolo + SSIM		
	Precision	Recall	F1 Score	Precision	Recall	F1 Score	Precision	Recall	F1 Score
Training dataset (optimal SSIM thresholds from training dataset)	97.71%	77.16%	86.21%	97.99%	94.09%	96.00%	97.71%	98.68%	98.19%
Validation dataset (optimal SSIM thresholds from training dataset)	97.72%	92.42%	93.63%	96.74%	90.71%	93.63%	97.66%	99.30%	98.47%
Validation dataset (optimal SSIM thresholds from validation dataset)				97.59%	95.19%	96.37%	97.72%	99.64%	98.67%
Test dataset (optimal SSIM thresholds from training dataset)	97.67%	93.53%	95.59%	96.15%	92.93%	94.51%	97.64%	99.54%	98.58%
Test dataset (optimal SSIM thresholds from validation dataset)				96.29%	93.81%	95.03%	97.65%	99.75%	98.69%

Yolo's core technology enables efficient and accurate real-time object detection, making it a fundamental tool in various computer vision applications. Yolo revolutionized object detection by introducing the concept of single-shot detection, meaning it can detect and classify objects in an image in a single forward pass of a neural network. Yolo places a strong emphasis on maintaining good precision, which in turn results in a lower recall. Yolo provides a confidence threshold to adjust the level of recall. A lower confidence threshold can achieve higher recall but may lead to lower precision. In Tables 2 and 3, we show fine-tuning of the confidence threshold of Yolo to achieve a similar precision level as that of FenceTalk. Subsequently, we compared the differences in recall between Yolo and FenceTalk (Yolo + SSIM).

Table 3. The output measures for Dataset 2 (red font signifies the best performance).

Dataset 2	Yolo			SSIM			Yolo + SSIM		
	Precision	Recall	F1 Score	Precision	Recall	F1 Score	Precision	Recall	F1 Score
Training dataset (optimal SSIM thresholds from training dataset)	98.81%	72.63%	83.72%	95.87%	94.12%	94.99%	98.81%	98.39%	98.2%
Validation dataset (optimal SSIM thresholds from training dataset)	97.89%	92.83%	95.65%	87.97%	89.62%	88.79%	97.89%	99.26%	98.57%
Validation dataset (Optimal SSIM thresholds from validation dataset)				88.66%	92.00%	90.29%	97.89%	99.43%	98.67%
Test dataset (optimal SSIM thresholds from training dataset)	98.55%	94.41%	96.59%	92.54%	91.06%	91.80%	98.54%	99.50%	99.02%
Test dataset (optimal SSIM thresholds from validation dataset)				93.03%	90.40%	91.70%	98.57%	99.67%	99.11%

Table 3 presents the precision and recall of Yolo and SSIM using the optimal thresholds for Dataset 2. In the testing phase, FenceTalk's precision was 98.57% and its recall was 99.67%. Both cases (Tables 2 and 3) indicated that integrating SSIM into FenceTalk led to a further improvement in the overall F1 score compared to using only the Yolo model for recognition.

Figure 9 presents the precision accuracy of Yolo and SSIM under different sub-datasets and brightness levels. For Dataset 1, in the training phase, the lowest precision was 88.24% at brightness level 93. In the validation phase, the lowest precision was 87.34% at brightness level 84. In the testing phase, the lowest precision was 93.56% at brightness level 83. For Dataset 2, in the training phase, the lowest precision was 89.89% at brightness level 82. In the validation phase, the lowest precision was 90.21% at brightness level 86. In the testing phase, the lowest precision was 91.46% at brightness level 86.

Figure 10 depicts the recall accuracy of Yolo and SSIM under different sub-datasets and brightness levels. For Dataset 1, in the training dataset, the lowest recall was 95.24% at brightness level 89. In the validation phase, the lowest recall was 97.37% at brightness level 94. In the testing dataset, the lowest recall was 99.22% at brightness level 84. For Dataset 2, in the training phase, the lowest recall was 95.31% at brightness level 101. In the validation phase, the lowest recall was 98.77% at brightness level 89. In the testing phase, the lowest recall was 98.45% at brightness level 98. Figures 9 and 10 display the lowest precision and recall values across various sub-datasets and brightness levels. These lowest precision and recall metrics represent the baseline performance of FenceTalk. In the majority of cases, FenceTalk's performance exceeded these lower bound values.

We also conducted a comparison of how ViT and SSIM performed in detection of moving objects. We applied ViT and SSIM in greenhouse equipment operation status detection. For instance, when a user turned on the exhaust fan (Figure 11), we checked whether the fan started correctly to determine the equipment's normal operation.

(a) Fan is OFF.

(b) Fan is ON.

Figure 11. Greenhouse exhaust fan.

Table 4 displays the performance of ViT in recognizing equipment operation status. ViT achieved a recall of 1 and an F1 score of 0.999. Table 5 presents the results of SSIM, which were slightly lower than those of ViT. Specifically, SSIM achieved a recall of 0.98 and an F1 score of 0.989. These results indicate that SSIM can provide satisfactory performance when applied to moving object detection. However, when compared to ViT, SSIM requires significantly fewer computational resources.

Table 4. Performance of ViT.

ViT	Accuracy	Precision	Recall	F1-Score
Training (94,848 images)	0.999962	0.999860	1.0	0.999930
Validation (11,856 images)	0.999938	0.999716	1.0	0.999858
Testing (11,858 images)	0.999936	0.999642	1.0	0.999820

Table 5. Performance of SSIM.

SSIM	Accuracy	Precision	Recall	F1-Score
Testing (23,712 images)	0.993158	0.989927	0.989314	0.989620

Figure 12 illustrates the processing speed and GPU utilization of the embedded device Jetson Nano when executing FenceTalk. Each instance of FenceTalk was capable of performing image recognition for an RTSP streaming camera. Yolo (FPS) and Yolo (GPU) represent the execution speed and memory usage of Jetson Nano during object recognition. Yolo + r/w (FPS) and Yolo + r/w (GPU represent the execution speed and memory usage when Jetson Nano performs object recognition and reads/writes images. Yolo + r/w + SSIM (FPS) and Yolo + r/w + SSIM (GPU) represent the execution speed and memory usage when Jetson Nano performs object recognition, reads/writes images, and employs SSIM for motion detection.

Our study indicates that Jetson Nano can simultaneously run up to three instances of FenceTalk (i.e., the sources of video streaming came from three cameras). When the number of FenceTalk instances was 1, Jetson Nano achieved a processing speed of 14.5 FPS during object recognition, utilizing 0.79 GB of memory. However, with 3 FenceTalk instances, the processing speed dropped to 10.7 FPS during object recognition, and the memory usage increased to 2.56 GB. As the number of FenceTalk instances increased, Jetson Nano's processing speed decreased linearly rather than exponentially, while GPU utilization exhibited a linear increase.

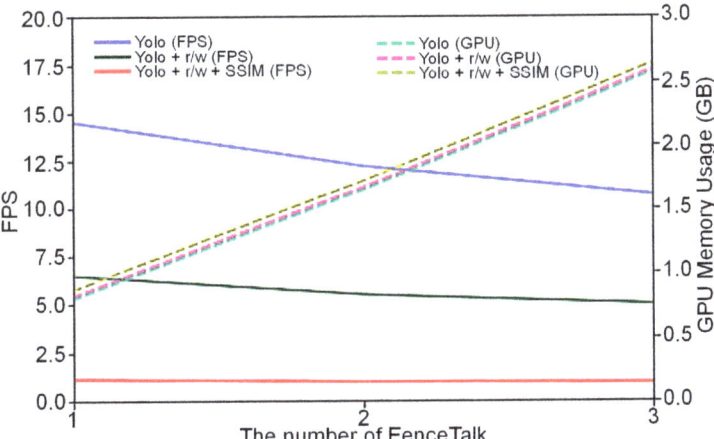

Figure 12. FenceTalk's processing speed and GPU usage on Jetson Nano.

6. Conclusions

A reliable image recognition model is crucial for security surveillance. Using image data containing moving objects from the specific area as training data can significantly enhance the model's accuracy in recognizing the area. To simplify the process of selecting target data from a large amount of image data, FenceTalk categorizes data based on brightness and utilizes SSIM and optimal thresholds to compare differences between current images and background images. It automatically selected suspicious images with moving objects that the Yolo model failed to recognize. This approach enables the model to learn the features of moving objects in the area using more training data. In the experimental results of FenceTalk, the recall values (Yolo + SSIM) surpassed 99%, demonstrating the universality of SSIM optimal thresholds. This also confirmed that FenceTalk effectively captures moving objects with motion characteristics.

FenceTalk can be deployed on the embedded device Jetson Nano, ensuring smooth system operation while reducing hardware costs. In the future, we will continue researching methods to minimize misjudgments of moving objects due to shadow and lighting changes. FenceTalk has a detection box to visualize recognized individuals; in the future, we will add the visualization feature (YOLOv7-gradCAM) [27] to FenceTalk. Additionally, GUI packages related to FenceTalk will be developed, making the system more user friendly and reducing deployment complexity.

Until August 2023, we have technically transferred FenceTalk to the Department of Education of Keelung City Government, Accton Technology Inc., Quanta Computer Inc., China Medical University, Asia University, and National Cheng Kung University.

Author Contributions: Conceptualization, Y.-W.L. and Y.-B.L.; methodology, Y.-W.L. and Y.-B.L.; software, J.-C.H.; validation, Y.-H.L.; data curation, J.-C.H.; writing—original draft preparation, Y.-B.L.; writing—review and editing, Y.-W.L.; supervision, Y.-W.L.; project administration, Y.-B.L.; funding acquisition, Y.-B.L. All authors have read and agreed to the published version of the manuscript.

Funding: This research was funded by National Science and Technology Council (NSTC) 112-2221-E-033-023, 110-2622-8-A49-022, NSTC112-2221-E-A49-049, NCKU Miin Wu School of Computing, Research Center for Information Technology Innovation.

Conflicts of Interest: The authors declare no conflict of interest.

References

1. Ahmad, M.B.; Abdullahi, A.; Muhammad, A.S.; Saleh, Y.B. The Various Types of sensors used in the Security Alarm system. *Int. J. New Comput. Archit. Their Appl.* **2020**, *10*, 50–59.
2. Chauhan, R.; Ghanshala, K.K.; Joshi, R.C. Convolutional Neural Network (CNN) for Image Detection and Recognition. In Proceedings of the 2018 First International Conference on Secure Cyber Computing and Communication (ICSCCC), Jalandhar, India, 15–17 December 2018; pp. 278–282.
3. Putra, M.H.; Yussof, Z.M.; Lim, K.C.; Salim, S.I. Convolutional neural network for person and car detection using yolo framework. *J. Telecommun. Electron. Comput. Eng. (JTEC)* **2018**, *10*, 67–71.
4. Bochkovskiy, A.; Wang, C.Y.; Liao, H.Y.M. Yolov4: Optimal speed and accuracy of object detection. *arXiv* **2020**, arXiv:2004.10934.
5. Raju, K.N.; Reddy, K.S.P. Comparative study of Structural Similarity Index (SSIM) by using different edge detection approaches on live video frames for different color models. In Proceedings of the 2017 International Conference on Intelligent Computing, Instrumentation and Control Technologies (ICICICT), Kannur, India, 6–7 July 2017; pp. 932–937.
6. Han, K.; Wang, Y.; Chen, H.; Chen, X.; Guo, J.; Liu, Z.; Tang, Y.; Xiao, A.; Xu, C.; Xu, Y.; et al. A Survey on Vision Transformer. *IEEE Trans. Pattern Anal. Mach. Intell.* **2023**, *45*, 87–110. [CrossRef] [PubMed]
7. Lin, Y.-B.; Chen, W.-E.; Chang, T.C.-Y. Moving from Cloud to Fog/Edge: The Smart Agriculture Experience. *IEEE Commun. Mag.* (Early Access) **2023**, 1–7. [CrossRef]
8. Priyadharshini, G.; Dolly, D.R.J. Comparative Investigations on Tomato Leaf Disease Detection and Classification Using CNN, R-CNN, Fast R-CNN and Faster R-CNN. In Proceedings of the International Conference on Advanced Computing and Communication Systems (ICACCS), Coimbatore, India, 17–18 March 2023; pp. 1540–1545.
9. Juyal, P.; Kundaliya, A. Multilabel Image Classification using the CNN and DC-CNN Model on Pascal VOC 2012 Dataset. In Proceedings of the International Conference on Sustainable Computing and Smart Systems (ICSCSS), Coimbatore, India, 14–16 June 2023; pp. 452–459.
10. Hmidani, O.; Alaoui, E.M.I. A Comprehensive Survey of the R-CNN Family for Object Detection. In Proceedings of the International Conference on Advanced Communication Technologies and Networking (CommNet), Marrakech, Morocco, 12–14 December 2022; pp. 1–6.
11. Lin, Y.-B.; Liu, C.-Y.; Chen, W.-L.; Chang, C.-H.; Ng, F.-L.; Yang, K.; Hsung, J. IoT-based Strawberry Disease Detection with Wall-mounted Monitoring Cameras. *IEEE Internet Things J.* **2023**, *1*. [CrossRef]
12. Wang, C.-Y.; Bochkovskiy, A.; Liao, H.Y.M. Scaled-yolov4: Scaling cross stage partial network. In Proceedings of the IEEE/CVF Conference on Computer Vision and Pattern Recognition, Nashville, TN, USA, 20–25 June 25 2021.
13. Lin, T.-Y.; Maire, M.; Belongie, S.; Hays, J.; Perona, P.; Ramanan, D.; Dollár, P.; Zitnick, C.L. Microsoft coco: Common objects in context. In Proceedings of the European Conference on Computer Vision, Zurich, Switzerland, 6–12 September 2014; Springer: Cham, Switzerland, 2014; pp. 740–755.
14. Li, C.; Li, L.; Jiang, H.; Weng, K.; Geng, Y.; Li, L.; Ke, Z.; Li, Q.; Cheng, M.; Nie, W.; et al. YOLOv6: A Single-Stage Object Detection Framework for Industrial Applications. *arXiv* **2022**, arXiv:2209.02976.
15. Zuo, X.; Li, J.; Huang, J.; Yang, F.; Qiu, T.; Jiang, Y. Pedestrian detection based on one-stage YOLO algorithm. *J. Phys. Conf. Ser.* **2021**, *1871*, 012131. [CrossRef]
16. Korhonen, J.; You, J. Peak signal-to-noise ratio revisited: Is simple beautiful? In Proceedings of the 2012 Fourth International Workshop on Quality of Multimedia Experience, Melbourne, VIC, Australia, 5–7 July 2012; pp. 37–38.
17. Wang, Z.; Bovik, A.C. Mean squared error: Love it or leave it? A new look at signal fidelity measures. *IEEE Signal Process. Mag.* **2009**, *26*, 98–117. [CrossRef]
18. Wibowo, H.T.; Wibowo, E.P.; Harahap, R.K. Implementation of Background Subtraction for Counting Vehicle Using Mixture of Gaussians with ROI Optimization. In Proceedings of the 2021 Sixth International Conference on Informatics and Computing (ICIC), Jakarta, Indonesia, 3–4 November 2021; pp. 1–6.
19. Rukundo, O.; Wu, K.; Cao, H. Image interpolation based on the pixel value corresponding to the smallest absolute difference. In Proceedings of the Fourth International Workshop on Advanced Computational Intelligence, Wuhan, China, 19–21 October 2011; pp. 432–435.
20. Hore, A.; Ziou, D. Image quality metrics: PSNR vs. SSIM. In Proceedings of the 2010 20th International Conference on Pattern Recognition, Istanbul, Turkey, 23–26 August 2010; pp. 2366–2369.
21. Wang, Z.; Bovik, A.C.; Sheikh, H.R.; Simoncelli, E.P. Image quality assessment: From error visibility to structural similarity. *IEEE Trans. Image Process.* **2004**, *13*, 600–612. [CrossRef] [PubMed]
22. Loza, A.; Wang, F.; Yang, J.; Mihaylova, L. Video object tracking with differential Structural SIMilarity index. In Proceedings of the 2011 IEEE International Conference on Acoustics, Speech and Signal Processing (ICASSP), Prague, Czech Republic, 22–27 May 2011; pp. 1405–1408.
23. Chen, G.; Shen, Y.; Yao, F.; Liu, P.; Liu, Y. Region-based moving object detection using SSIM. In Proceedings of the 2015 4th International Conference on Computer Science and Network Technology (ICCSNT), Harbin, China, 19–20 December 2015; Volume 1.
24. Zhou, M.; Wang, G.; Wang, J.; Hui, C.; Yang, W. Defect detection of printing images on cans based on SSIM and chromatism. In Proceedings of the 2017 3rd IEEE International Conference on Computer and Communications (ICCC), Chengdu, China, 13–16 December 2017.

25. Khalaf, H.A.; Tolba, A.; Rashid, M. Event triggered intelligent video recording system using MS-SSIM for smart home security. *AIN Shams Eng. J.* **2018**, *9*, 1527–1533. [CrossRef]
26. Xia, B.; Cao, J.; Wang, C. SSIM-NET: Real-time PCB defect detection based on SSIM and MobileNet-V3. In Proceedings of the 2019 2nd World Conference on Mechanical Engineering and Intelligent Manufacturing (WCMEIM), Shanghai, China, 22–24 November 2019.
27. Xiao-Dragon, yolov7-GradCAM. 2022. Available online: https://gitee.com/xiao-dragon/yolov7-GradCAM (accessed on 7 October 2023).

Disclaimer/Publisher's Note: The statements, opinions and data contained in all publications are solely those of the individual author(s) and contributor(s) and not of MDPI and/or the editor(s). MDPI and/or the editor(s) disclaim responsibility for any injury to people or property resulting from any ideas, methods, instructions or products referred to in the content.

Article

Parkinson's Disease Classification Framework Using Vocal Dynamics in Connected Speech

Sai Bharadwaj Appakaya *, Ruchira Pratihar * and Ravi Sankar

Department of Electrical Engineering, College of Engineering, University of South Florida, Tampa, FL 33620, USA; sankar@usf.edu
* Correspondence: saibharadwaj.appakaya@gmail.com (S.B.A.); ruchirapratihar@usf.edu (R.P.)

Abstract: Parkinson's disease (PD) classification through speech has been an advancing field of research because of its ease of acquisition and processing. The minimal infrastructure requirements of the system have also made it suitable for telemonitoring applications. Researchers have studied the effects of PD on speech from various perspectives using different speech tasks. Typical speech deficits due to PD include voice monotony (e.g., monopitch), breathy or rough quality, and articulatory errors. In connected speech, these symptoms are more emphatic, which is also the basis for speech assessment in popular rating scales used for PD, like the Unified Parkinson's Disease Rating Scale (UPDRS) and Hoehn and Yahr (HY). The current study introduces an innovative framework that integrates pitch-synchronous segmentation and an optimized set of features to investigate and analyze continuous speech from both PD patients and healthy controls (HC). Comparison of the proposed framework against existing methods has shown its superiority in classification performance and mitigation of overfitting in machine learning models. A set of optimal classifiers with unbiased decision-making was identified after comparing several machine learning models. The outcomes yielded by the classifiers demonstrate that the framework effectively learns the intrinsic characteristics of PD from connected speech, which can potentially offer valuable assistance in clinical diagnosis.

Keywords: Parkinson's disease; speech processing; pitch synchronous segmentation; MFCC

Citation: Appakaya, S.B.; Pratihar, R.; Sankar, R. Parkinson's Disease Classification Framework Using Vocal Dynamics in Connected Speech. *Algorithms* **2023**, *16*, 509. https://doi.org/10.3390/a16110509

Academic Editors: Chih-Lung Lin, Bor-Jiunn Hwang, Shaou-Gang Miaou, Yuan-Kai Wang and Frank Werner

Received: 15 September 2023
Revised: 24 October 2023
Accepted: 31 October 2023
Published: 4 November 2023

Copyright: © 2023 by the authors. Licensee MDPI, Basel, Switzerland. This article is an open access article distributed under the terms and conditions of the Creative Commons Attribution (CC BY) license (https://creativecommons.org/licenses/by/4.0/).

1. Introduction

Approximately 7.5 million people all over the world are diagnosed with Parkinson's disease (PD). Since its first description put forth by James Parkinson in 1817, PD has been a neurodegenerative disease that affects motor functioning [1]. With a prevalence of 572 per 100,000 individuals in North America, is the second most common neurodegenerative disorder caused by the degeneration and dysfunction of dopaminergic neurons in the substantia nigra [2]. A 2022 Parkinson's Foundation-backed study revealed that nearly 90,000 people are diagnosed with PD in the U.S. each year [3]. Projections show that the number of people with PD (45 years) will rise to approximately 1,238,000 by 2030 [4]. PD is primarily characterized by motor symptoms like muscle weakness, rigidity, tremor, bradykinesia (slow movement) that includes hypomimia, movement variability, and freezing of gait, and nonmotor symptoms like olfactory dysfunction, anxiety, depression, cognitive deficits, dementia, sleep disorders, and even melanoma [5]. The onset age ranges anywhere between 35 years and 60 years, and over 90% of people with PD are known to develop nonmotor manifestations [6]. However, the disease course and symptom manifestations are known to vary considerably from individual to individual. Such heterogeneity is also observed in response to levodopa medication. Many people with PD experience a drastic decrease in quality of life due to the substantial degeneration of dopaminergic neurons [7,8].

For over twenty-five years, clinical diagnosis of this highly heterogeneous disease has seen little improvement in terms of accuracy. Before the late 1980s, formal diagnostic criteria for this illness were nonexistent. Today, clinicians rely heavily on the Movement

Disorder Society-Sponsored Revision of the Unified Parkinson's Disease Rating Scale (MDS-UPDRS) and the Hoehn and Yahr (HY) rating scales to evaluate various aspects and primary motor symptoms in patients' daily lives. However, these subjective ratings can result in incorrect diagnoses due to atypical Parkinsonian conditions, essential tremors, and other dementias [9], which is why research in Parkinson's primarily focuses on disease detection, symptom research, and sustaining people's quality of life.

Reports indicate that the onset of prodromal Parkinson's disease is a slow and gradual process, lasting from three to fifteen years [10]. Consequently, the motor effects of the condition are often too subtle to be evident, making it difficult to diagnose. One of the most frequent and early manifestations of PD is speech impairment, which is experienced by approximately 90% of people living with the disorder [10–13]. Speech in people with PD is subject to several vocal impairments, such as tremors, lowered loudness, pitch alteration, hoarseness, imprecise articulation, delayed onset time, and decreased intelligibility [14–16]. These factors have provided researchers with an array of opportunities to conduct in-depth studies on the subject using signal processing and machine learning. As a method of day-to-day patient assessment, speech recording is an effective and non-invasive tool since it is fast, affordable, and generates objective data.

Over the past decade, speech research for PD has garnered increasing interest within the research community. Numerous studies have analyzed speech samples from individuals with PD and healthy controls (HC) to conduct classification and monitoring investigations. These studies can be broadly categorized into those based on ratings provided by trained listeners according to their perception [17–22], acoustic analysis [23–27], and/or computational methods involving mathematical modeling with signal processing and machine learning techniques [28–34]. Computational methods employ distinct sets of acoustic features extracted from various speech samples or tasks. The choice of speech stimuli and the specific features extracted from them depend on the research question and the type of speech impairment (phonatory or articulatory) targeted for classification.

Three types of speech tasks are commonly employed to elicit the effects of PD. These tasks include sustained vowel phonations, diadochokinetic (DDK; repetition of syllables), and connected speech (word/sentence/paragraph readings, monologues, etc.) tasks. Studies using sustained vowel phonations [32,35–37] focus on regularities or irregularities in phonation to train efficient classification models. Sustained phonations are relatively easy to analyze and do not suffer from language or accent barriers other than connected speech. However, researchers have reported the higher suitability of connected speech over sustained phonations to study pathological speech [38,39] due to their closeness to everyday speech. Although the DDK task is speech-like, it is less natural, and research consensus is mixed on the effectiveness of DDK in PD classification. Some studies have shown no significant differences between PD and HC groups [40], and other more recent studies have shown that people with PD have lower DDK rates than healthy speakers [41].

The individual diversities in cognitive processing behind speech production reflect misread texts, variations in pause durations, and other such manifestations. These manifestations make connected speech more complicated for analysis because it often requires manual corrections to maintain consistency in the data, which is essential for unbiased classification experiments. Hence, fewer research studies like [28] aim to automate PD classification/monitoring using connected speech. Rusz et al. showed significant progress in utilizing connected speech for automatic PD classification [42,43]. They used multiple types of speech tasks and observed that features extracted from monologue (connected speech) were sensitive enough and tasks like sustained phonations and DDK were not optimal to capture impairments due to prodromal PD. Studies like [44–46] used passage reading tasks to evaluate continuous speech and observe the variability between PD and HC using temporal and spectral features. Skodda et al. focused on specific vowel sounds and observed reductions in vowel space in PD speech [47]. In another study [48], various speech features, including NHR, fundamental frequency, relative shimmer, and jitters, were evaluated and compared. Lower values for these features are considered desirable for good speech signal quality. Additionally, the study explored the variants of vowel space area (triangular VSA—tVSA and quadrilateral VSA—qVSA), and FCR (Formant Centralization Ratio), which are indicators of potential dysarthria-related conditions that could lead to compression in VSA.

Furthermore, some recent studies [49] have discussed the application of different automatic speech recognition (ASR) services (such as Amazon, Google, and IBM) to differentiate between healthy controls and PD patients. Nonlinear mixed effects models (nLMEM) accounted for the unequal variance between healthy controls and individuals with speech disorders.

In recent years, there has been a notable shift towards utilizing deep learning models for speech processing in PD classification. Convolutional neural networks (CNNs) and recurrent neural networks (RNNs) [50–53] have gained prominence in detecting and classifying PD based on speech signals. These models have the capability to automatically extract features from speech signals [54,55] and learn to classify them based on underlying patterns within the data. However, it is essential to acknowledge that the use of deep learning techniques often has limitations that include data dependency, computationally intensive, lack of interpretability, and high training time. The positive aspect is that these opportunities provide us with a range of future prospects.

In our previous study [56], a novel classification protocol that deviates from the conventional process primarily in terms of signal segmentation was introduced. Syllabic-level feature variances were used instead of features themselves for training classifiers. The data used in that study consisted of three different word utterances repeated multiple times at various places in the paragraphs read by 40 different speakers.

Research on connected speech and sustained phonations has shown different strengths individually in capturing the effects of PD. In this study, a novel framework that draws on and combines these strengths of connected speech was developed and evaluated. The development and evaluation of this framework is a step toward establishing an efficient process that can be used to aid in diagnosis and telemonitoring applications. The proposed methodology closely follows the procedure adopted in our earlier study [56], with significant modifications. Here, different options for methodological steps representing the conventional and proposed frameworks were identified. The optimal choice for each step was identified through a comprehensive analysis. Features were extracted only from all the voiced segments in paragraph recordings collected from people with PD and HCs. A variety of classifiers were trained on the features themselves using hold-out validation. Through this approach, pitch-synchronous (PS) segmentation was proven to capture vocalic dynamics more efficiently, resulting in better and more reliable PD classification. To test the efficiency of the proposed framework, two separate datasets consisting of paragraph recordings in different languages were used, with each dataset being utilized in the training and testing phases individually. The features from our earlier study [56] were divided into two groups: mel-frequency cepstral coefficients (MFCCs) and pitch-synchronous features (PSFs), and they were compared. MFCCs have been widely used for speech recognition and speaker identification tasks. They have also been part of many studies aimed at PD classification and stood out as some of the best features yielding good performance [32,34,35,37,57]. The methodology adopted to evaluate the MFCCs for PD classification here has been designed to replicate these research works, where the results show a classification accuracy close to 90%. Most of these studies employ linear SVM kernels for classification, and we included the same in our classifier set. PSFs, designed to work with PS segmentation of voiced speech, had superior performance among the two feature groups.

In summary, the benefits of this study are as follows:

1. Efforts to establish an analytical framework that can be automated to classify between PD and HC using vocalic dynamics.
2. Providing evidence as to the robustness of the framework for the language being spoken.
3. Evaluation of PSFs for PD classification.

4. Providing evidence for the shortcomings of using MFCCs for PD classification due to their inherent nature of embedding patient identifiable information.

2. Dataset Description

This research utilized data from two different databases. The use of multiple databases helps verify the reproducibility of results and comprehend the effect of dataset size on classification performance. In this study, connected speech is evaluated using passages in Italian (Database 1) and English (Database 2) languages read by people with PD and healthy controls. The database descriptions are as follows:

2.1. Database 1

The first database was accessed from the IEEE DataPort [58], which is an Italian Parkinson's voice and speech collected by Dimauro and Girardi for the assessment of speech intelligibility in PD using a speech-to-text system. It contains two phonetically balanced Italian passages read by 50 subjects: 28 PD (19 male and 9 female) and 22 HC (10 male and 12 female). The recordings were created at a sampling rate of 44.1 kHz with 16 bits/sample. The duration of each passage recording varied between 1 and 4 min, with a mean of 1.3 min. The recordings were performed in an echo-free room, with the distance between the speaker and microphone varying between 15 and 25 cm. According to the authors of [59], none of the patients reported speech or language disorders unrelated to their PD symptoms prior to their study and were receiving antiparkinsonian treatment. The HY scale ratings were <4 for all the patients except for two patients with stage 4 and 1 patient with stage 5. Only the passage reading speech task from this database was used for this study.

2.2. Database 2

Database 2 was selected from a larger database collected at the Movement Disorders Center, University of Florida [60]. It comprises speech tasks such as passage reading ("The Rainbow Passage", Fairbanks, 1940) and sustained vowel phonations, though only the former was used in this study. This dataset is more balanced compared to Database 1 with 10 age and gender-matched data in both PD and HC groups. Recordings were taken with a Marantz portable recorder (Marantz America, LLC, Mahwah, NJ, USA) and stored digitally with a sampling rate of either 44.1 kHz or 22.05 kHz and 16 bits/sample. The duration of recordings from this database varied between 25 and 90 s, with a mean of 41 s. Despite the existence of two different sampling rates in Database 2, the procedure was not affected since the segmentation and feature extraction were designed to be robust and independent of the sampling frequency. Additionally, the sampling rates are not class-specific, i.e., both HC and PD classes consist of data that have both sampling frequencies. The institutional review board approved all test procedures at the University of Florida, and testing was completed following an informed consent process.

3. Materials and Methods

3.1. Methodology Block Description

The experimental methodology was conducted to assess the different alternatives identified for each block depicted in Figure 1. The final framework remains unchanged from the methodology block diagram, considering only the optimal choices identified from the results for each block.

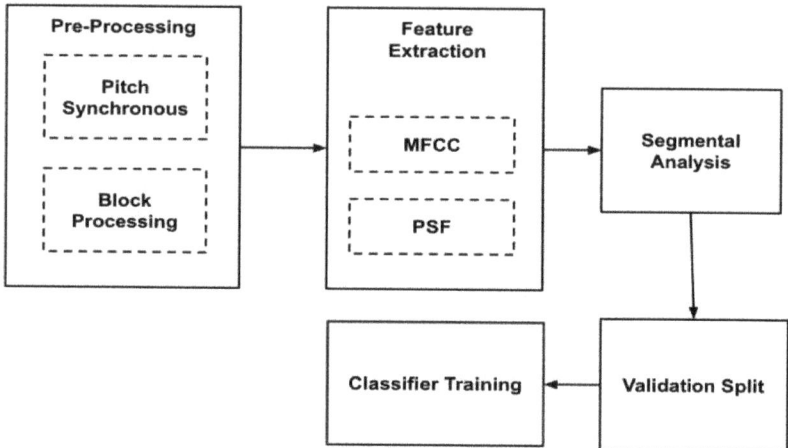

Figure 1. Methodology Block Diagram.

The methodology blocks, along with the different choices, are explained as follows:

3.1.1. Preprocessing: Block Processing and Pitch Synchronous Segmentation

After categorizing voiced and unvoiced portions from the speech signals, super-segments were recognized as the voiced components surrounded by unvoiced/silent portions on both sides. These super-segments are subsequently segmented into blocks or pitch-synchronous segments. Typical speech analysis follows a block processing approach where the block size can vary between 20 and 50 ms, depending on the application. These blocks are known to retain the necessary statistical stationarity in the data. In pitch-synchronous (PS) segmentation, the pitch cycles in voiced portions of speech are segmented out and processed.

Figure 2 shows a speech sample segmented using block processing (top) with a 25 ms window size and 50% overlap. The red vertical lines show the window edges for the first, third, and fifth windows, and the black lines show the same for the second, fourth, and sixth windows. The 50% overlap can be seen between red windows and black windows. In PS segmentation (bottom), the length of each segment is correlated with the fundamental frequency of the speaker and intonational variations and thus can vary from cycle to cycle. The PS segmentation was executed through an automated algorithm that detects voiced sections and subsequently segments them in accordance with the pitch-synchronous method outlined in [61].

In Database 1, a total of 378,000 pitch-synchronous cycles were identified, whereas in Database 2, approximately 87,000 pitch-synchronous segments were extracted. In Database 1, a total of 12,600 super segments were picked out. From each paragraph reading an average of 134 ± 24 super-segments have been identified. Each super segment contained an average of 30 cycles, with a standard deviation of 27 cycles. In Database 2, a total of 1357 super-segments were extracted, with an average of 68 ± 20 super segments per paragraph. Each super segment contained an average of 64 ± 58 cycles.

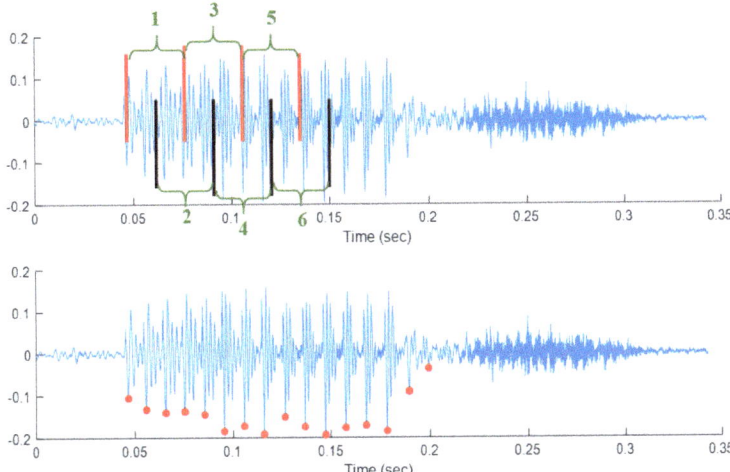

Figure 2. Segmentations: Block processing (top) with vertical (red/black) lines showing block limits with block indices (green) and pitch synchronous (bottom). Y–axis shows the normalized signal amplitude.

3.1.2. Feature Extraction: MFCCs and PSFs

MFCCs and custom-designed PSFs are used in this study. MFCCs are widely utilized features in speaker recognition applications due to their effectiveness. In this study, 15 MFCCs were extracted from each speech segment and used for classification. A set of 15 PSFs was extracted from each PS segment in voiced portions of the recordings. These features were explained in our previous work [56]. Table 1 provides the feature names along with the corresponding number of elements obtained from each feature. Since the speech samples underwent two distinct segmentation methods, MFCCs were separately extracted for each segmentation type. However, PSFs were exclusively extracted using the PS segmentation protocol. These features target cycle-to-cycle perturbations, which cannot be adequately captured by traditional measures like jitter and shimmer.

Table 1. Pitch Synchronous Features.

Feature Name	Size
Pitch Period	1
LOC—Length of Curve	1
Quarter Segment Energy	4
Total Energy	1
Correlation Canceller Efficiency	1
Correlation Canceller MSE	1
Peak Frequency (Hz)	1
Quarter Band Magnitude	4
Spectral Factor	1

Moreover, PSFs emphasize a wide range of attributes in both the time and frequency domains, enabling them to capture perturbations in both the vocal cords and the vocal tract. This comprehensive approach contributes to their effectiveness in the classification process.

3.1.3. Feature Preparation

In segmental analysis, the features extracted from each speech segment (block processing or PS) were treated as individual data points while training the classifiers. The available samples (feature vectors) were randomized and divided into training and testing sets. For PSFs, each feature vector except Quarter Segment Energy and Quarter Band Magnitude is transformed using min-max normalization at a super segment level to maintain consistency in their scales. As those two features contain four values, they are normalized as a group, i.e., All four vectors were grouped together and then the min and max values across the group of four vectors were used as the min and max for the group for normalization. This normalization allows the maintenance of variability within the features while matching the scales across the super segments for each feature.

3.1.4. Validation Split: Hold-Out

It is common practice to randomize data and extract 80% of it for training and test the model performance on the remaining 20% in machine learning problems with each data point treated as unique and independent.

In this study, due to data imbalance, after creating the training and testing datasets, they were further reduced following a specific protocol that mitigates the class imbalances while preserving generalization. After performing the 80–20% split, 'N' random samples from each class (PD/HC) in the original training set were pooled to have a total of 2N samples in the training set. N was chosen to be equivalent to 50% of the minority class size in the initial train set. This protocol ensures that the train and test sets have no shared samples and that the class distributions remain similar. The selection of the 50% factor also ensured sufficient training samples.

3.1.5. Classifier Training

A total of 17 classifiers available in MATLAB were employed for conducting classification experiments in this study. It included variants of Trees, Support Vector Machines (SVMs), k-nearest Neighbors (KNNs), Ensemble learners, and logistic regression. All the classifiers were trained with 5-fold cross-validation. Using a collection of classifiers, the suitability of different classifiers for this application was evaluated with unbiased decision-making. The arrangement of all the classifiers into their respective groups can be observed in Figure 3. Notably, all the classifiers were utilized with their default hyperparameter settings, as provided by MATLAB's Classification Learner Application.

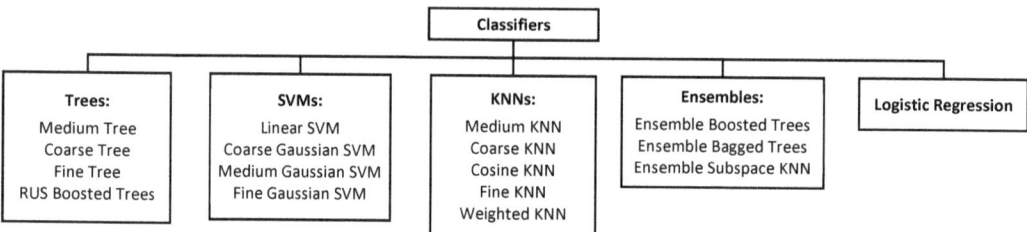

Figure 3. Organization of classifiers used in this study.

3.2. Experimental Design

Classification experiments were targeted to systematically identify the optimal choice for each methodology block in Figure 1. This experimental process included various steps, focused on the identification of optimal choices for segmentation methods, feature sets, and analysis techniques, subsequently, evaluating the appropriateness of each classifier for each application.

Finally, the performance was assessed when the optimal classifier set was trained and tested using data from different databases while employing the optimal segmentation method and feature set.

Individual step descriptions and their goals are described below in detail.

3.2.1. Importance of Block Size in Conventional Block Processing

Identification of the right block size becomes imperative when PS segmentation must be compared against block processing. In this step, performances for the block sizes ranging from 20 to 50 ms with 5 ms increments and 50% overlap were tested using MFCCs under the Hold-Out validation protocol for both genders individually. Results from this step were used to pick an optimal candidate for block size. In subsequent steps, whenever block processing was compared against PS segmentation, results from using the optimal block size choice identified in this step were used.

3.2.2. Identification of Optimal Choice for Segmentation Method

A novel comparison technique was adopted in this step to identify the optimal choice for segmentation and analysis methods. This technique was designed to measure the efficiency of classifiers in learning the effects of PD rather than their ability to remember the speakers in each class. The process is twofold: first, classification performance was determined under regular circumstances, and then speakers were randomly assigned PD/HC labels, and classification experiments were repeated without making any other changes to the protocol. As the speakers were randomly labeled, a drop in classification performance was anticipated across all cases. Since the segmentation and analysis technique combined with the ability to remember speakers in each category, comparable results were expected regardless of the original and random PD/HC labels. Therefore, the magnitude of the performance reduction was likely to be lower for such combinations.

$$\text{Relative Reduction} = \frac{OA - RA}{OA} \times 100 \qquad (1)$$

where OA & RA are accuracies with original random labels.

The relative reduction in accuracy defined by the drop in performance was calculated as per the equation. The experiments involved utilizing MFCCs from Database1 separately for each gender, employing all available classifiers. In the case of experimental parameters favoring speaker identification over recognition of PD effects, comparable results were expected regardless of the original and random PD/HC labels. Therefore, the magnitude of the performance reduction was likely to be lower for such combinations.

3.2.3. Identification of Optimal Choice for Feature Types

During this step, the selection between features MFCCs and PSFs, was made using the same protocol as in the previous step. The evaluation of feature combinations for each gender was carried out independently by assessing the relative reduction in classification accuracy resulting from random label assignment.

3.2.4. Evaluation of Classifiers

The 17 classifiers used for the experiments as listed in Figure 3, were evaluated for over-fitting by comparing the overfit factor and test accuracies. The overfit factor, calculated according to the given Equation (2), quantifies the relative difference between the training and testing accuracies. A higher overfit factor value indicates that the classifier performs well on the data seen during training but may be overfitting to that specific training data.

$$\text{Overfit factor} = \frac{\text{Train Acc} - \text{Test Acc}}{\text{Train Acc}} \qquad (2)$$

The test accuracy with random label assignment serves as an indicator of the classifier's tendency to memorize speakers more effectively than learning the patterns associated with PD. These two metrics, obtained from the results of experiments using the optimal choices determined in the previous steps, were utilized to identify, and eliminate classifiers that exhibited a significant inclination to overfitting.

3.2.5. Testing Using Different Databases

The optimal feature choice and a reduced list of classifiers were selected for the final framework development. The efficacy of this framework was tested by training the classifiers on a larger dataset and testing them on a different dataset. The latter dataset, except for the speech task, differed in terms of speakers, language, and acquisition environment, ensuring a more comprehensive assessment of the classifiers' performance and robustness. Furthermore, the influence of gender over the framework was examined by conducting replications of the experiment without any gender-based filters to the data. This allowed for a broader analysis of how the classifiers performed across both genders without any gender-specific constraints.

Owing to the variations in data acquisition conditions, such as differences in equipment and variations in speaker-to-microphone distances, the features extracted from different databases exhibited discrepancies in their numerical ranges. Normalization has been used to address these differences and maintain uniformity between the features from both databases before initiating the experimentation where training and testing were performed using different databases. z-scores with zero mean and unit standard deviation were extracted from feature data using Equation (3) and used for training and testing.

$$z_{ij} = \frac{x_{ij} - \mu_i}{\sigma_i} \quad (3)$$

where, x-Feature value, i-Feature index, j-segment index, μ-Mean, and σ-Standard deviation.

In addition to the normalization, a syllabic analysis approach is designed to emphasize more of the perturbations. The features extracted were transformed into covariances at a super segment level to do this. In the speech data, super segments were identified as the voiced components bordered by unvoiced/silent portions on both ends as explained earlier. Features extracted from all segments within a super segment were grouped and covariances of these feature groups were used for training classifiers. Figure 4 shows the method adopted to extract these covariances. C1 to Cn represent the feature covariances and subsequently used feature vectors from each super segment.

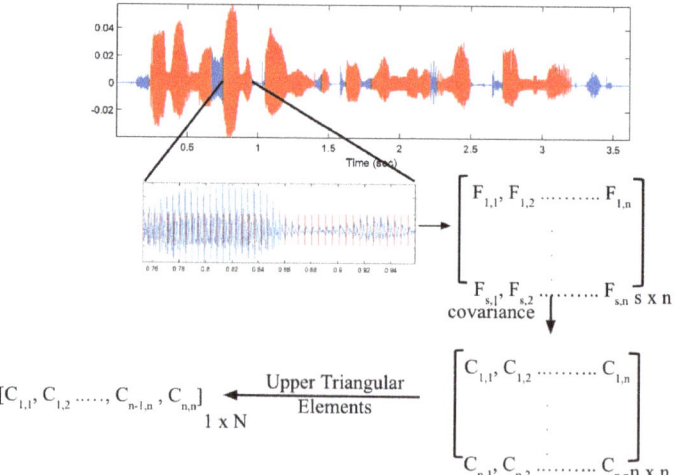

Figure 4. Syllabic learning approach showing the original recording (blue) overlaid with voiced sections (red) in the figure at the top. One of the voiced sections, segmented pitch synchronously, is zoomed into and shown at the center.

4. Results and Discussion

In this section, results for each one of the five steps mentioned in the previous section were presented. As results contain classification accuracies from multiple classifiers, box plots were used to show their distribution.

MFCCs with block segmentation with various block sizes ranging from 10 ms to 40 ms were extracted and utilized for PD/HC classification. The result did not demonstrate any significant impact due to block sizes and for brevity, only results from a block size of 25 ms were used whenever block processing results were discussed to be investigated from hereon.

The results from PS and block processing (25 ms block size, 50% overlap) with original labels and randomly assigned labels along with their relative differences were used. The outcomes from MFCCs using block and PS segmentations for both genders in Database 1 using holdout validation are provided in Figure 5. It was observed that both segmentation methods performed similarly across all genders.

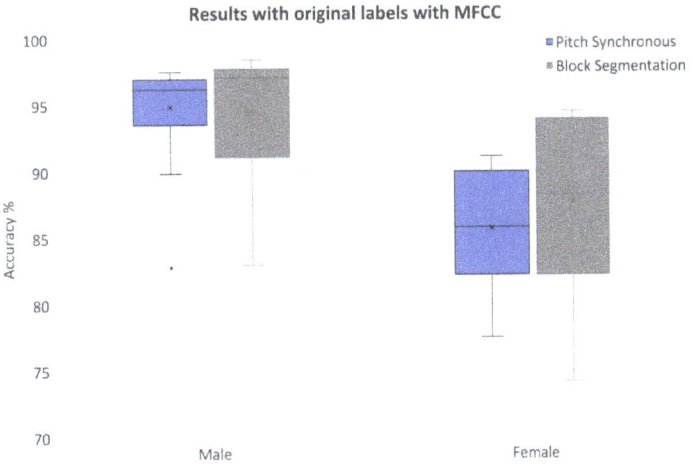

Figure 5. Classification results with different segmentations using original labels (MFCCs from Database 1).

These results indicated that except for females with block processing, the PS segmentation led to comparable or even greater degradation in all other cases. Notably, some of the classifiers exhibited negative values for percentage decrease, indicating better classifications with randomized labels. Upon closer examination, it was revealed that classifiers with fine kernels were primarily responsible for such effects. These classifiers were also found to be prone to overfitting issues, as discussed in detail later in this paper.

Overall, block segmentation yielded superior accuracies compared to PS segmentation when MFCCs are used.

When classifiers were trained using data with random labels, performance decreased in all cases as expected. The relative accuracy reduction due to random label assignment is shown in Figure 6.

The relative reduction in accuracy was also more pronounced in the case of block segmentation for females. In males, both segmentation methods were comparable when data was labeled randomly. In conclusion, it was observed that when trained with random labels, the degradation in accuracy was more pronounced, highlighting a lack of robustness in the classifiers while using block segmentation.

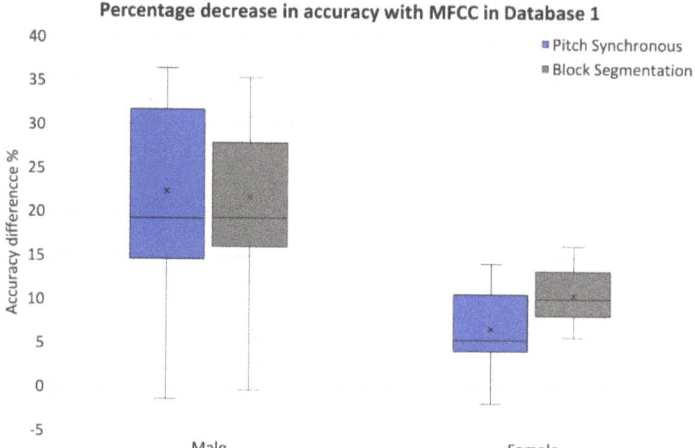

Figure 6. Percentage reduction in classification performance with original and random labels (MFCCs from Database 1).

The comparison between feature types was made using classification accuracies with original labels and the relative decline in performance due to random label assignment. The performance comparison between MFCCs from PS segmentation and PSFs using Database 1 for both genders is shown in Figure 7. We observe that in female, PSFs has better median accuracies than MFCCs. The significantly higher relative reduction in performance for PSFs along with higher classification accuracies for MFCC as depicted in Figure 6, indicates that MFCCs contain more speaker-identifiable information while PSFs exhibit a higher capability to capture the impact of PD on vocalic dynamics. Conversely, PSFs exhibited a higher capability to capture the impact of PD on vocalic dynamics compared to MFCCs.

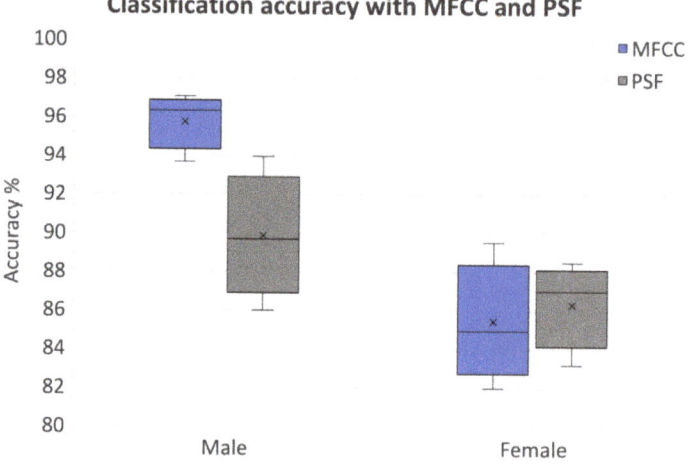

Figure 7. Classification performance comparison between MFCCs and PSFs (Database 1).

The percentage reduction in classification performance with original labels and randomly assigned labels was consistently higher for PSFs than MFCCs in all cases, as demonstrated in Figure 8. In conclusion, these findings suggest that PSFs are more effective than MFCCs in containing the effects of PD, while MFCCs are better suited for speaker identification and verification applications. As PSFs are extracted using PS segmentation, between block processing and PS segmentation, the latter becomes the optimal choice.

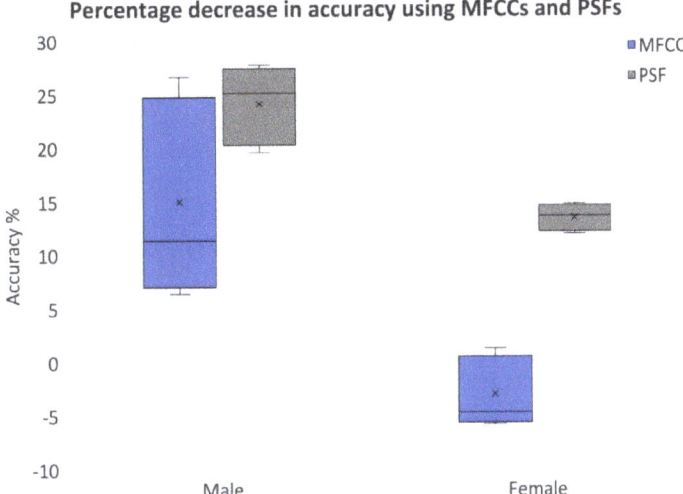

Figure 8. Percentage reduction in classification performances with original and random labels—classification between MFCCs and PSF (Database1).

Next, to recognize the optimal set of classifiers, each one of the 17 classifiers was evaluated using two different metrics:

1. Overfit factors are displayed in Table 2;
2. Test accuracy using random label assignment is shown in Table 3.

Table 2. Overfit Factor with Original Label.

Classifier No.	Classifier Name	Overfit Factor with Original Label			
		Male		Female	
		MFCC	PSF	MFCC	PSF
1	Medium KNN	−0.003	0.021	0.023	0.022
2	Coarse KNN	−0.012	0.017	0.014	0.003
3	Cosine KNN	0.005	0.015	0.016	0.021
4	Linear SVM	−0.001	−0.005	0.009	0
5	Coarse Tree	0.012	0.037	0.017	−0.001
6	Coarse Gaussian SVM	−0.002	0.006	0.013	0.001
7	Medium Tree	0.016	0.031	0.015	0.003
8	Ensemble Boosted Tree	0.018	0.026	0.025	0.004
9	RUS Boosted Tree	0.016	0.031	0,015	0.003
10	Logistic Regression	0	−0.012	0.007	−0.001
11	Fine Tree	0.039	0.03	0.061	0.008
12	Medium Gaussian SVM	0.003	0.006	0.02	0.001
13	Fine KNN	0.035	0.057	0.109	0.111
14	Weighted KNN	0.026	0.055	0.099	0.1
15	Ensemble Bagged Trees	0.046	0.051	0.1	0.051
16	Ensemble Subspace KNN	0.026	0.139	0.089	0.13
17	Fine Gaussian SVM	0.173	0.017	0.137	0.008

Table 3. Median Test Accuracies with Random Labels in Both Genders for Both Features.

Classifier No.	Classifier Name	Median Test Accuracy			
		Male		Female	
		MFCC	PSF	MFCC	PSF
1	Medium KNN	90.18	75.77	93.73	78.725
2	Coarse KNN	85.33	72.17	89.11	74.855
3	Cosine KNN	89.46	75.965	93.14	79.31
4	Linear SVM	67.69	58.8	80.44	53.65
5	Coarse Tree	66.52	59.52	77.01	69.605
6	Coarse Gaussian SVM	75.12	61.715	83.92	64.525
7	Medium Tree	68.83	63.09	76.81	72.405
8	Ensemble Boosted Tree	74.04	67.96	82.76	75.415
9	RUS Boosted Tree	68.83	63.09	76.81	72.35
10	Logistic Regression	67.62	59.335	79.92	59.38
11	Fine Tree	72.18	66.965	80.62	75.93
12	Medium Gaussian SVM	86.63	70.21	90.78	75.865
13	Fine KNN	91.27	75.32	93.91	76.43
14	Weighted KNN	91.97	77.4	94.29	78.66
15	Ensemble Bagged Trees	88.67	80.98	92.39	85.26
16	Ensemble Subspace KNN	90.62	66.125	94.08	73.915
17	Fine Gaussian SVM	92.14	78.415	95.86	80.62

In both Tables 2 and 3, the classifiers were arranged in ascending order of the overfit factor observed in males using MFCC for training. Index numbers were assigned to each classifier in the leftmost column of both tables for easy identification during discussions. To ensure robustness and bias-free analysis, all classifier experiments, including training, testing, and result acquisition, were repeated 10 times and the median values of overfit factors and test accuracies were then presented in Tables 2 and 3, respectively.

In Table 2, for both MFCCs and PSFs in both genders, classifiers 1 to 10 exhibited significantly lower overfit factors compared to the rest. Conversely, from classifier 11 to classifier 17, the overfit factors increased, indicating a greater tendency to overfit the training data. Table 3 displays the median test accuracies for each classifier under pitch-synchronous segmentation when labels were randomized. Among the classifiers with lower overfit values identified in Table 2 (classifiers 1 to 10), the first three classifiers exhibited significantly higher accuracy values even with random labels. This finding indicates that kNNs are more suitable for speaker verification applications than PD classification. They tend to rely on the proximity of samples in the feature space and learn minimal information related to the effects of PD. Comparably, the results in Table 3 revealed that MFCCs achieved much higher accuracy values than PSFs, suggesting that MFCCs are better suited for speaker identification tasks compared to PD classification. Based on these observations, only classifiers 4–10 were identified as optimal classifiers.

For the final step, training and testing were carried out using different databases. Database 1 and Database 2 consist of speakers with varying characteristics, including nationality, spoken language, paragraph length, and sampling frequency (16 kHz for Database 1 and 44.1 kHz for Database 2).

Figure 9 presents the results obtained when the optimal set of classifiers is trained and tested over different databases after applying normalization and using the z-scores before training the classifiers. The normalization process ensured that the classifiers were trained and tested on data with consistent scales, facilitating fair and reliable performance comparisons across the two databases. Figure 10 shows the results for the same grouping as Figure 9, where the syllabic level covariances are used for classification instead of the z-scores. Comparison between both figures shows the MFCCs cannot contain the effects of PD as well as PSFs. Covariances computed over PSFs at the super-segment level are more reliable than MFCCs. While it can clearly be improved further, these results provide evidence for the need to develop features that capture the vocalic dynamics more than the features that are reliable in identifying the speakers.

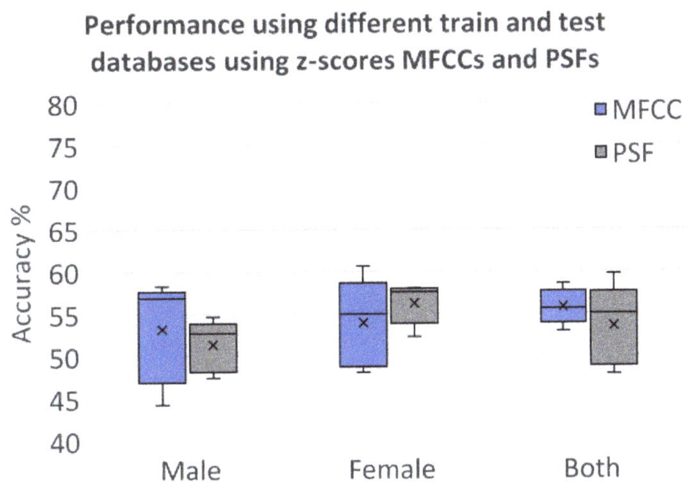

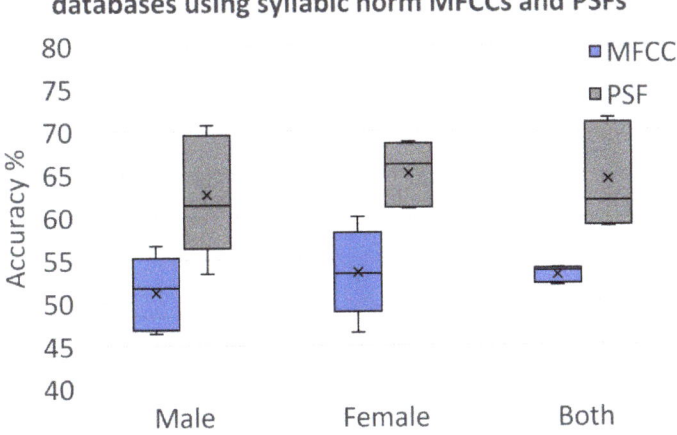

Figure 9. Classification accuracies with Database 1 used for training and Database 2 used for testing using z-scores for MFCCs and PSF.

Figure 10. Classification accuracies with Database 1 used for training and Database 2 used for testing using covariances from syllabic analysis z-scores for MFCCs and PSF.

Besides accuracy, additional metrics were used to evaluate the performance. The descriptions are as follows:

True Positives (*TP*): Number of PD samples predicted as PD.
True Negatives (*TN*): Number of HC samples predicted as HC.
False Positives (*FP*): Number of HC samples predicted as PD.

False Negatives (FN): Number of PD samples predicted as HC.
Accuracy: Proportion of test samples correctly predicted.

$$\text{Accuracy} = \frac{(TP+TN)}{(TP+TN+FP+FN)} \qquad (4)$$

Precision (P): Proportion of PD predictions that were correct.

$$\text{Precision} = \frac{TP}{(TP+FP)} \qquad (5)$$

Recall (R): Proportion of all PD samples correctly predicted.

$$\text{Recall} = \frac{TP}{(TP+FN)} \qquad (6)$$

F1-Score: Harmonic mean of precision and recall.

$$F1-\text{Score} = \frac{2 \times P \times R}{(P+R)} \qquad (7)$$

Matthews Correlation Coefficient (MCC): An improvement over F1-Score as it includes the TN in its computation.

$$\text{MCC} = \frac{TP \times TN - FP \times FN}{\sqrt{(TP+FP)(TP+FN)(TN+FP)(TN+FN)}} \qquad (8)$$

ROC-AUC: Area under Receiver Operating Characteristic (ROC) curve.

All the metrics except MCC have values between 0 and 1, with 1 being the best possible value. MCC can have values between −1 and 1, with 1 being the best possible value.

Under gender-independent grouping (black) for MFCCs, Coarse Gaussian SVM and Logistic Regression demonstrated relatively higher performance across all metrics, as illustrated in Figure 11a For PSFs, Medium Trees also exhibited comparable performance to Coarse Gaussian SVM and Logistic Regression, as depicted in Figure 11b Therefore, Coarse Gaussian SVM and Logistic Regression emerge as superior classifiers for PD classification, whether using MFCCs or PSFs. with PSFs, Coarse Gaussian SVM, and Ensemble Boosted Trees achieved classification accuracies close to 75% without any indication of bias in F1-scores and MCC. Upon a more detailed analysis of the results in Figure 5, it was observed that under the same conditions, these two classifiers achieved an average accuracy of 85% when exclusively using Database 1 for training and testing.

In this paper, all the results presented so far were obtained by treating each segment as an individual sample. Using logistic regression with PSFs, the percentage of correctly predicted segments from each participant was determined. The results from coarse Gaussian SVM closely followed the outcomes from logistic regression.

Remarkably, the proposed framework demonstrated a high recall and good precision for detecting PD; this was supported by the result that more than 80% of PD speaker segments and over 50% of HC speaker segments were correctly predicted with their corresponding labels. The proposed methodology is being further developed to improve the precision by decreasing false positives. It involves capturing the cycle-to-cycle perturbations much better and evaluating each syllable holistically. In our latest studies, a syllabic analysis protocol is being developed and preliminary results show signs of PSFs consistently outperforming MFCCs. These insights will be incorporated into our upcoming report.

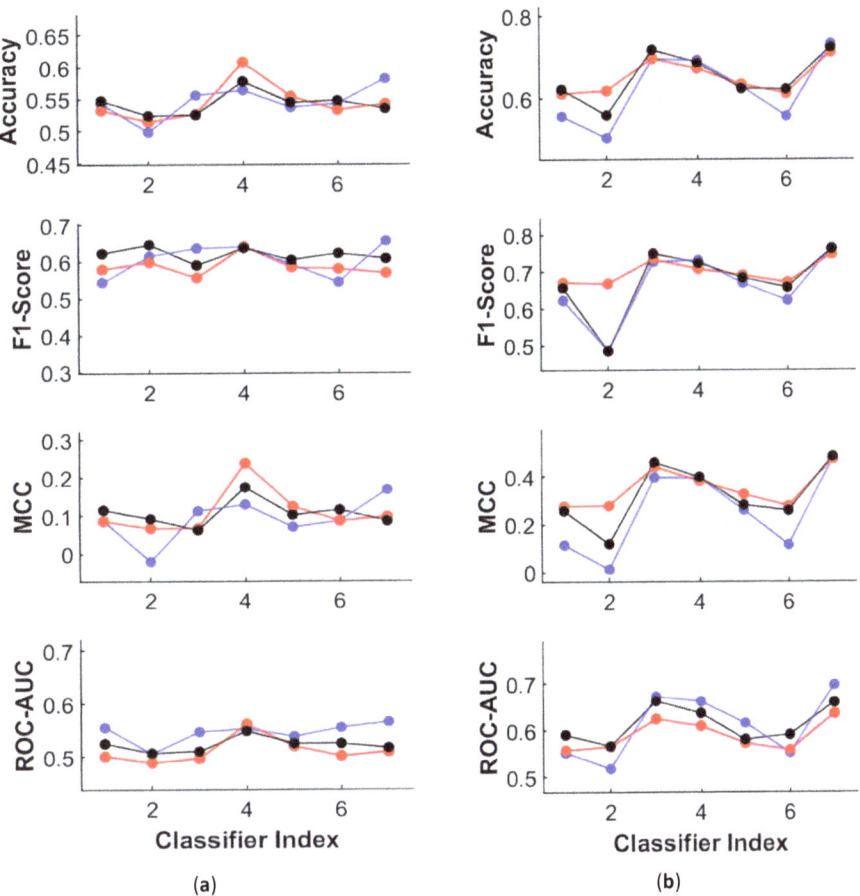

Figure 11. Performance Metrics with Database 1 for training and Database 2 for testing with (**a**) MFCCs in three groups; and (**b**) PSFs in three groups. The three groups are namely group 1 (blue line), group 2 (red line), and group 3 (black line).

5. Conclusions

In this study, a novel protocol for PD classification using connected speech has been proposed and thoroughly tested under various experimental conditions. Different choices representing proposed and existing frameworks for some of the methodological blocks were identified and systematically evaluated. In the analysis of block processing with varying block sizes using MFCCs, it was revealed that the block size had no significant impact on the classification performance. On comparing the two segmentation protocols, PS segmentation exhibited superior performance, which could be attributed to its ability to maintain consistent resolution across speakers with different and varying fundamental frequencies. Furthermore, when considering results with randomized labels, MFCCs demonstrated their strength in providing speaker-identifiable information to classifiers, rendering them more suitable for speaker-identification applications.

Between MFCCs and PSFs for PS segmental analysis, PSF provides reliability due to its ability to capture the patterns in speech affected by Parkinson's disease. The study demonstrated the positive impact of reducing identifying information on classifier reliability. A total of 17 classifiers were utilized for testing, with each one being individually assessed for reliability and overfitting capability. Among the 17 classifiers, 10 displayed tendencies towards overfitting. The remaining seven classifiers were employed to test the proposed gender framework independently using different databases. The performance results showed that coarse Gaussian SVMs, ensemble boosted trees, and logistic regression are all well-suited for this application. These results are promising, considering the magnitude of the differences in both data sets.

Future work is aimed at identifying the effects of various factors, like variants of features used, and improving the analysis methods presented here by employing advanced techniques like autoencoders to further improve the ability to capture the effects of PD on speech production. We plan to compare performances utilizing sustained phonations with varying vowels and diverse linear and nonlinear features. Our objective is to create a comprehensive automated classification procedure capable of overcoming challenges such as data availability, ease of implementation, and enhanced generalizability. Such a protocol can very well be used beyond PD for other ailments that affect speech production.

Author Contributions: Conceptualization, S.B.A. and R.S.; methodology, S.B.A. and R.S.; software, S.B.A.; validation, S.B.A. and R.S.; formal analysis, S.B.A.; investigation, S.B.A. and R.S.; resources, S.B.A.; data curation, R.S. and S.B.A.; writing—original draft preparation, S.B.A., R.P. and R.S.; writing—review and editing, S.B.A., R.P. and R.S.; visualization, S.B.A. and R.P.; supervision, R.S.; project administration, R.S. All authors have read and agreed to the published version of the manuscript.

Funding: This research received no external funding.

Institutional Review Board Statement: Not applicable for Database 1. The institutional review board approved all test procedures at the University of Florida for Database 2.

Informed Consent Statement: Informed consent was obtained from all test subjects involved in the study.

Data Availability Statement: Database 1 presented in the study is openly available in "IEEE DataPort" at http://dx.doi.org/10.21227/aw6b-tg17, accessed on 3 April 2020.

Acknowledgments: The authors would like to thank Dimauro and Girardi for sharing their Italian dataset via IEEE DataPort and the researchers at the Movement Disorder Center at the University of Florida for sharing the English dataset. We would like to thank Supraja Anand at the University of South Florida College of Behavioral and Community Sciences for providing help in curating and analyzing the dataset.

Conflicts of Interest: The authors declare no conflict of interest.

References

1. Parkinson, J. An essay on the shaking palsy. *J. Neuropsychiatry Clin. Neurosci.* **2002**, *14*, 223–236. [CrossRef]
2. Tohgi, H.; Abe, T.; Takahashi, S. Parkinson's disease: Diagnosis, treatment and prognosis. *Nihon Ronen Igakkai Zasshi. Jpn. J. Geriatr.* **1996**, *33*, 911–915. [CrossRef] [PubMed]
3. New Study Shows the Incidence of Parkinson's Disease in the U.S. Is 50% Higher than Previous Estimates. Available online: https://www.parkinson.org/about-us/news/incidence-2022 (accessed on 3 April 2020).
4. Marras, C.; Beck, J.C.; Bower, J.H.; Roberts, E.; Ritz, B.; Ross, G.W.; Abbott, R.D.; Savica, R.; Van Den Eeden, S.K.; Willis, A.W.; et al. Prevalence of Parkinson's disease across North America. *NPJ Park. Dis.* **2018**, *4*, 21. [CrossRef]
5. Lima, M.S.M.; Martins, E.F.; Marcia Delattre, A.; Proenca, M.B.; Mori, M.A.; Carabelli, B.; Ferraz, A.C. Motor and non-motor features of Parkinson's disease–A review of clinical and experimental studies. *CNS Neurol. Disord. Drug Targets Former. Curr. Drug Targets CNS Neurol. Disord.* **2012**, *11*, 439–449.
6. Lang, A.E. A critical appraisal of the premotor symptoms of Parkinson's disease: Potential usefulness in early diagnosis and design of neuroprotective trials. *Mov. Disord.* **2011**, *26*, 775–783. [CrossRef] [PubMed]
7. Schrag, A.; Jahanshahi, M.; Quinn, N. How does Parkinson's disease affect quality of life? A comparison with quality of life in the general population. *Mov. Disord. Off. J. Mov. Disord. Soc.* **2000**, *15*, 1112–1118. [CrossRef]
8. Lang, A.E.; Obeso, J.A. Time to move beyond nigrostriatal dopamine deficiency in Parkinson's disease. *Ann. Neurol. Off. J. Am. Neurol. Assoc. Child Neurol. Soc.* **2004**, *55*, 761–765. [CrossRef]
9. Rizzo, G.; Copetti, M.; Arcuti, S.; Martino, D.; Fontana, A.; Logroscino, G. Accuracy of clinical diagnosis of Parkinson disease: A systematic review and meta-analysis. *Neurology* **2016**, *86*, 566–576. [CrossRef] [PubMed]
10. Postuma, R.; Lang, A.; Gagnon, J.; Pelletier, A.; Montplaisir, J. How does parkinsonism start? Prodromal parkinsonism motor changes in idiopathic REM sleep behaviour disorder. *Brain* **2012**, *135*, 1860–1870. [CrossRef] [PubMed]

11. Harel, B.T.; Cannizzaro, M.S.; Cohen, H.; Reilly, N.; Snyder, P.J. Acoustic characteristics of Parkinsonian speech: A potential biomarker of early disease progression and treatment. *J. Neurolinguist.* **2004**, *17*, 439–453. [CrossRef]
12. Ho, A.K.; Iansek, R.; Marigliani, C.; Bradshaw, J.L.; Gates, S. Speech impairment in a large sample of patients with Parkinson's disease. *Behav. Neurol.* **1998**, *11*, 131–137. [CrossRef] [PubMed]
13. Kent, R.D.; Kent, J.F.; Weismer, G.; Duffy, J.R. What dysarthrias can tell us about the neural control of speech. *J. Phon.* **2000**, *28*, 273–302. [CrossRef]
14. Rudzicz, F. Articulatory knowledge in the recognition of dysarthric speech. *IEEE Trans. Audio Speech Lang. Process.* **2010**, *19*, 947–960. [CrossRef]
15. Canter, G.J. Speech characteristics of patients with Parkinson's disease: III. Articulation, diadochokinesis, and over-all speech adequacy. *J. Speech Hear. Disord.* **1965**, *30*, 217–224. [CrossRef]
16. Darley, F.L.; Aronson, A.E.; Brown, J.R. *Motor Speech Disorders*; Saunders: Philadelphia, PA, USA, 1975.
17. Tjaden, K.; Kain, A.; Lam, J. Hybridizing conversational and clear speech to investigate the source of increased intelligibility in speakers with Parkinson's disease. *J. Speech Lang. Hear. Res.* **2014**, *57*, 1191–1205. [CrossRef]
18. Anand, S.; Stepp, C.E. Listener perception of monopitch, naturalness, and intelligibility for speakers with Parkinson's disease. *J. Speech Lang. Hear. Res.* **2015**, *58*, 1134–1144. [CrossRef]
19. Chiu, Y.-F.; Neel, A. Predicting Intelligibility Deficits in Parkinson's Disease With Perceptual Speech Ratings. *J. Speech Lang. Hear. Res.* **2020**, *63*, 433–443. [CrossRef]
20. Miller, N.; Allcock, L.; Jones, D.; Noble, E.; Hildreth, A.J.; Burn, D.J. Prevalence and pattern of perceived intelligibility changes in Parkinson's disease. *J. Neurol.* **2007**, *78*, 1188–1190. [CrossRef]
21. Cannito, M.P.; Suiter, D.M.; Beverly, D.; Chorna, L.; Wolf, T.; Pfeiffer, R.M. Sentence intelligibility before and after voice treatment in speakers with idiopathic Parkinson's disease. *J. Voice* **2012**, *26*, 214–219. [CrossRef]
22. Cannito, M.P.; Suiter, D.M.; Chorna, L.; Beverly, D.; Wolf, T.; Watkins, J. Speech intelligibility in a speaker with idiopathic Parkinson's disease before and after treatment. *J. Med. Speech Lang. Pathol.* **2008**, *16*, 207–213.
23. Goberman, A.M.; Elmer, L.W. Acoustic analysis of clear versus conversational speech in individuals with Parkinson disease. *J. Commun. Disord.* **2005**, *38*, 215–230. [CrossRef] [PubMed]
24. Chenausky, K.; MacAuslan, J.; Goldhor, R. Acoustic analysis of PD speech. *Park. Dis.* **2011**, *2011*, 435232. [CrossRef] [PubMed]
25. Kuo, C.; Tjaden, K. Acoustic variation during passage reading for speakers with dysarthria and healthy controls. *J. Commun. Disord.* **2016**, *62*, 30–44. [CrossRef] [PubMed]
26. Fletcher, A.R.; McAuliffe, M.J.; Lansford, K.L.; Liss, J.M. Assessing vowel centralization in dysarthria: A comparison of methods. *J. Speech Lang. Hear. Res.* **2017**, *60*, 341–354. [CrossRef]
27. Burk, B.R.; Watts, C.R. The effect of Parkinson disease tremor phenotype on cepstral peak prominence and transglottal airflow in vowels and speech. *J. Voice* **2019**, *33*, 580.e11–580.e19. [CrossRef]
28. Orozco-Arroyave, J.R.; Hönig, F.; Arias-Londoño, J.D.; Vargas-Bonilla, J.F.; Skodda, S.; Rusz, J.; Nöth, E. Automatic detection of Parkinson's disease from words uttered in three different languages. In Proceedings of the Fifteenth Annual Conference of the International Speech Communication Association, Singapore, 14–18 September 2014.
29. Orozco-Arroyave, J.; Hönig, F.; Arias-Londoño, J.; Vargas-Bonilla, J.; Daqrouq, K.; Skodda, S.; Rusz, J.; Nöth, E. Automatic detection of Parkinson's disease in running speech spoken in three different languages. *J. Acoust. Soc. Am.* **2016**, *139*, 481–500. [CrossRef]
30. Rusz, J.; Cmejla, R.; Tykalova, T.; Ruzickova, H.; Klempir, J.; Majerova, V.; Picmausova, J.; Roth, J.; Ruzicka, E. Imprecise vowel articulation as a potential early marker of Parkinson's disease: Effect of speaking task. *J. Acoust. Soc. Am.* **2013**, *134*, 2171–2181. [CrossRef]
31. Skodda, S.; Grönheit, W.; Schlegel, U. Intonation and speech rate in Parkinson's disease: General and dynamic aspects and responsiveness to levodopa admission. *J. Voice* **2011**, *25*, e199–e205. [CrossRef]
32. Tsanas, A.; Little, M.A.; McSharry, P.E.; Spielman, J.; Ramig, L.O. Novel speech signal processing algorithms for high-accuracy classification of Parkinson's disease. *IEEE Trans. Biomed. Eng.* **2012**, *59*, 1264–1271. [CrossRef]
33. Rusz, J.; Cmejla, R.; Ruzickova, H.; Ruzicka, E. Quantitative acoustic measurements for characterization of speech and voice disorders in early untreated Parkinson's disease. *J. Acoust. Soc. Am.* **2011**, *129*, 350–367. [CrossRef]
34. Little, M.; McSharry, P.; Hunter, E.; Spielman, J.; Ramig, L. Suitability of dysphonia measurements for telemonitoring of Parkinson's disease. *Nat. Preced.* **2008**, 1. [CrossRef]
35. Almeida, J.S.; Rebouças Filho, P.P.; Carneiro, T.; Wei, W.; Damaševičius, R.; Maskeliūnas, R.; de Albuquerque, V.H.C. Detecting Parkinson's disease with sustained phonation and speech signals using machine learning techniques. *Pattern Recognit. Lett.* **2019**, *125*, 55–62. [CrossRef]
36. Benba, A.; Jilbab, A.; Hammouch, A. Detecting patients with Parkinson's disease using Mel frequency cepstral coefficients and support vector machines. *Int. J. Electr. Eng. Inform.* **2015**, *7*, 297.
37. Rahn, D.A., III; Chou, M.; Jiang, J.J.; Zhang, Y. Phonatory impairment in Parkinson's disease: Evidence from nonlinear dynamic analysis and perturbation analysis. *J. Voice* **2007**, *21*, 64–71. [CrossRef] [PubMed]
38. Klingholtz, F. Acoustic recognition of voice disorders: A comparative study of running speech versus sustained vowels. *J. Acoust. Soc. Am.* **1990**, *87*, 2218–2224. [CrossRef]

39. Zraick, R.I.; Dennie, T.M.; Tabbal, S.D.; Hutton, T.J.; Hicks, G.M.; O'Sullivan, P.S. Reliability of speech intelligibility ratings using the Unified Parkinson Disease Rating Scale. *J. Med. Speech Lang. Pathol.* **2003**, *11*, 227–241.
40. Ackermann, H.; Konczak, J.; Hertrich, I. The temporal control of repetitive articulatory movements in Parkinson's disease. *Brain Lang.* **1997**, *56*, 312–319. [CrossRef]
41. Kumar, S.; Kar, P.; Singh, D.; Sharma, M. Analysis of diadochokinesis in persons with Parkinson's disease. *J. Datta Meghe Inst. Med. Sci. Univ.* **2018**, *13*, 140. [CrossRef]
42. Hlavnička, J.; Čmejla, R.; Tykalová, T.; Šonka, K.; Růžička, E.; Rusz, J. Automated analysis of connected speech reveals early biomarkers of Parkinson's disease in patients with rapid eye movement sleep behaviour disorder. *Sci. Rep.* **2017**, *7*, 12. [CrossRef]
43. Rusz, J.; Hlavnička, J.; Tykalová, T.; Novotný, M.; Dušek, P.; Šonka, K.; Růžička, E. Smartphone allows capture of speech abnormalities associated with high risk of developing Parkinson's disease. *IEEE Trans. Neural Syst. Rehabil. Eng.* **2018**, *26*, 1495–1507. [CrossRef]
44. Whitfield, J.A.; Goberman, A.M. Articulatory–acoustic vowel space: Application to clear speech in individuals with Parkinson's disease. *J. Commun. Disord.* **2014**, *51*, 19–28. [CrossRef] [PubMed]
45. MacPherson, M.K.; Huber, J.E.; Snow, D.P. The intonation–syntax interface in the speech of individuals with Parkinson's disease. *J. Speech Lang. Hear. Res.* **2011**, *54*, 19–32. [CrossRef] [PubMed]
46. Forrest, K.; Weismer, G.; Turner, G.S. Kinematic, acoustic, and perceptual analyses of connected speech produced by Parkinsonian and normal geriatric adults. *J. Acoust. Soc. Am.* **1989**, *85*, 2608–2622. [CrossRef]
47. Skodda, S.; Visser, W.; Schlegel, U. Vowel articulation in Parkinson's disease. *J. Voice* **2011**, *25*, 467–472. [CrossRef] [PubMed]
48. Vizza, P.; Tradigo, G.; Mirarchi, D.; Bossio, R.B.; Lombardo, N.; Arabia, G.; Quattrone, A.; Veltri, P. Methodologies of speech analysis for neurodegenerative diseases evaluation. *Int. J. Med. Inform.* **2019**, *122*, 45–54. [CrossRef]
49. Schultz, B.G.; Tarigoppula, V.S.A.; Noffs, G.; Rojas, S.; van der Walt, A.; Grayden, D.B.; Vogel, A.P. Automatic speech recognition in neurodegenerative disease. *Int. J. Speech Technol.* **2021**, *24*, 771–779. [CrossRef]
50. Shi, X.; Wang, T.; Wang, L.; Liu, H.; Yan, N. Hybrid Convolutional Recurrent Neural Networks Outperform CNN and RNN in Task-state EEG Detection for Parkinson's Disease. In Proceedings of the 2019 Asia-Pacific Signal and Information Processing Association Annual Summit and Conference (APSIPA ASC), Lanzhou, China, 18–21 November 2019; pp. 939–944.
51. Quan, C.; Ren, K.; Luo, Z. A Deep Learning Based Method for Parkinson's Disease Detection Using Dynamic Features of Speech. *IEEE Access* **2021**, *9*, 10239–10252. [CrossRef]
52. Gunduz, H. Deep Learning-Based Parkinson's Disease Classification Using Vocal Feature Sets. *IEEE Access* **2019**, *7*, 115540–115551. [CrossRef]
53. Nagasubramanian, G.; Sankayya, M. Multi-Variate vocal data analysis for Detection of Parkinson disease using Deep Learning. *Neural Comput. Appl.* **2020**, *33*, 4849–4864. [CrossRef]
54. Karaman, O.; Çakın, H.; Alhudhaif, A.; Polat, K. Robust automated Parkinson disease detection based on voice signals with transfer learning. *Expert Syst. Appl.* **2021**, *178*, 115013. [CrossRef]
55. Ali, L.; Zhu, C.; Zhang, Z.; Liu, Y. Automated Detection of Parkinson's Disease Based on Multiple Types of Sustained Phonations Using Linear Discriminant Analysis and Genetically Optimized Neural Network. *IEEE J. Transl. Eng. Health Med.* **2019**, *7*, 2000410. [CrossRef] [PubMed]
56. Appakaya, S.B.; Sankar, R. Classification of Parkinson's disease Using Pitch Synchronous Speech Analysis. In Proceedings of the 2018 40th Annual International Conference of the IEEE Engineering in Medicine and Biology Society (EMBC), Honolulu, HI, USA, 17–21 July 2018; pp. 1420–1423.
57. Benba, A.; Jilbab, A.; Hammouch, A. Discriminating between patients with Parkinson's and neurological diseases using cepstral analysis. *IEEE Trans. Neural Syst. Rehabil. Eng.* **2016**, *24*, 1100–1108. [CrossRef] [PubMed]
58. Giovanni, D.; Francesco, G. Italian Parkinson's Voice and Speech. *IEEE Dataport* **2019**. [CrossRef]
59. Dimauro, G.; Di Nicola, V.; Bevilacqua, V.; Caivano, D.; Girardi, F. Assessment of speech intelligibility in Parkinson's disease using a speech-to-text system. *IEEE Access* **2017**, *5*, 22199–22208. [CrossRef]
60. Skowronski, M.D.; Shrivastav, R.; Harnsberger, J.; Anand, S.; Rosenbek, J. Acoustic discrimination of Parkinsonian speech using cepstral measures of articulation. *J. Acoust. Soc. Am.* **2012**, *132*, 2089. [CrossRef]
61. Appakaya, S.B.; Sankar, R. Parkinson's Disease Classification using Pitch Synchronous Speech Segments and Fine Gaussian Kernels based SVM. In Proceedings of the 2020 42nd Annual International Conference of the IEEE Engineering in Medicine & Biology Society (EMBC), Montreal, QC, Canada, 20–24 July 2020; pp. 236–239.

Disclaimer/Publisher's Note: The statements, opinions and data contained in all publications are solely those of the individual author(s) and contributor(s) and not of MDPI and/or the editor(s). MDPI and/or the editor(s) disclaim responsibility for any injury to people or property resulting from any ideas, methods, instructions or products referred to in the content.

Article

Relational Fisher Analysis: Dimensionality Reduction in Relational Data with Global Convergence [†]

Li-Na Wang [1], Guoqiang Zhong [2,*], Yaxin Shi [2] and Mohamed Cheriet [3]

[1] Qingdao Vocational and Technical College of Hotel Management, Qingdao 266100, China; alinagq@163.com
[2] College of Computer Science and Technology, Ocean University of China, Qingdao 266100, China; syxzhuanshu@163.com
[3] Synchromedia Laboratory for Multimedia Communication in Telepresence, École de Technologie Supérieure, Montréal, QC H3C 1K3, Canada; mohamed.cheriet@etsmtl.ca
* Correspondence: gqzhong@ouc.edu.cn
[†] This paper is an extended version of our paper published in Zhong, G.; Shi, Y.; Cheriet, M. Relational Fisher analysis: A general framework for dimensionality reduction. In Proceedings of the 2016 International Joint Conference on Neural Networks (IJCNN), Vancouver, BC, Canada, 24–29 July 2016.

Abstract: Most of the dimensionality reduction algorithms assume that data are independent and identically distributed (i.i.d.). In real-world applications, however, sometimes there exist relationships between data. Some relational learning methods have been proposed, but those with discriminative relationship analysis are lacking yet, as important supervisory information is usually ignored. In this paper, we propose a novel and general framework, called relational Fisher analysis (RFA), which successfully integrates relational information into the dimensionality reduction model. For nonlinear data representation learning, we adopt the kernel trick to RFA and propose the kernelized RFA (KRFA). In addition, the convergence of the RFA optimization algorithm is proved theoretically. By leveraging suitable strategies to construct the relational matrix, we conduct extensive experiments to demonstrate the superiority of our RFA and KRFA methods over related approaches.

Keywords: relational learning; dimensionality reduction; graph embedding; trace ratio; document understanding and recognition; face recognition

1. Introduction

In some applications, such as pattern recognition and data mining, dimensionality reduction methods are often used since they can reduce space-time complexity, denoise, and make the model more robust. Principal component analysis (PCA) [1–3] and linear discriminant analysis (LDA) [2,4–7] are two typical linear algorithms. Following them, researchers have proposed many variants, such as kernel PCA [8], generalized discriminant analysis (GDA) [9], and linear discriminant analysis for robust dimensionality reduction (RLDA) [10].

For nonlinear dimensionality reduction problems, manifold learning provides an effective solution. By supposing that data are located on a low-dimensional manifold, data samples observed in high-dimensional space can be represented in a low-dimensional space. Some representative manifold learning algorithms are ISOMAP [11], locally linear embedding (LLE) [12] and Laplacian Eigenmaps (LE) [13].

From the algorithmic perspective, algorithms mentioned above can be categorized as global methods or local methods. Global methods learn the low-dimensional representations by using global information of data. PCA and LDA are all global methods. The global methods are often effective and efficient, such that they are widely used in many real world applications. However, when dealing with non-linear data, using the global method cannot capture the genuine distribution of data very well. Local methods using the manifold learning idea, such as LLE and LE, pay special attention to the intrinsic structure of data. Nevertheless, most of these manifold learning methods disregard label information when recovering the low-dimensional manifold structure, in that they are inherently unsupervised.

Although the above-mentioned methods are defined from different perspectives of dimensionality reduction, graph embedding provide a unified framework for understanding and comparing them [14]. Furthermore, by integrating label information in the computation of intrinsic and penalty graphs within the graph embedding framework, a supervised dimensionality reduction method called marginal Fisher analysis (MFA) is proposed [14].

In traditional dimensionality reduction algorithms as described above, a data distribution assumption is generally applied that data are independent and identically distributed (i.i.d.). However, in real-world applications, there are often certain relativity or links between certain data, for instance, geometrical or semantic similarity, links among web pages, citation relations between scientific papers. Relationships usually indicate that these related samples are likely to have similarities or belonging to the same class. Nevertheless, although some dimensionality reduction methods consider to preserve the locality of data [15–18], the useful relationships are often simply ignored during the learning process of most existing dimensionality reduction methods.

Recently, relational learning has often been used in practical applications, for instance, web mining [19] and social network analysis [20]. In addition, relational information is also considered in social network discovery, document classification, sequential data analysis and semi-supervised graph embedding [21–23].

In the domain of dimensionality reduction, some algorithms have already been proposed in which the relational information is integrated into the representation learning process. In [24], Duin et al. propose the relational discriminant analysis (RDA). In RDA, relationships among data are measured by the Euclidean distance between objects and prototypes or support objects of each class. However, RDA uses mean squared error as the objective function, and cannot perform well on multi-class learning problems. In [25], Li et al. propose the probabilistic relational PCA (PRPCA) to build a probabilistic model associated with PCA and relational learning. RDA and PRPCA effectively integrate relational learning into the dimensionality reduction algorithms. Nevertheless, how to better inject relationships into traditional dimensionality reduction models is still worth exploring.

Recently, a few deep relational learning algorithms have been proposed. Specifically, Gao et al. design a deep learning model based on relational network for hyperspectral image few-shot classification [26], Chen et al. apply local relation learning for face forgery detection [27], and Cho et al. develop a weakly supervised anomaly detection method via context-motion relational learning [28]. In addition, some relational learning methods are used in the one-shot [29] or zero-shot [30] learning scenarios. However, many real-world applications have only very few data. Hence, shallow relational learning algorithms are still needed to be proposed and utilized.

In this paper, we propose a novel and general framework for dimensionality reduction, called relational Fisher analysis (RFA) [31]. Besides the intrinsic and penalty graph in the graph embedding framework, we further construct a relational graph which captures the relational information encoded within data. Through this graph, the proposed RFA takes into account the impact of the relational information in the presentation learning process. An effective iterative trace ratio algorithm is proposed to optimize RFA. Futhermore, we use the kernel trick to extend RFA to its kernelized version—KRFA. Additiionally, we theoretically prove that the optimization algorithm of RFA converges. To evaluate the effectiveness of RFA, we conduct extensive experiments in many real-world applications. The results demonstrate that the proposed RFA outperforms most of the classic dimensionality reduction algorithms on the datasets we use. The effectiveness of KRFA is also tested.

This paper is based on one of our previous conference papers [31], with significant improvements. For concreteness, we propse the KRFA algorithm and add more exclusive experiments with comparison to the related approaches. The rest of this paper is organized as follows: In Section 2, we introduce several related ideas, including graph embedding, trace ratio problem and relational learning, which are highly relevant to our work. In Section 3, we focus on our proposed method RFA, including the notation, the formulation and the iterative optimization method of RFA. Section 4 includes the proof of the convergence of RFA and we present how to extend RFA to a kernel version—KRFA. In Section 5, we compare our methods RFA and KRFA to other commonly used dimensionality reduction methods with extensive experiments, which demonstrate the effectiveness of our proposed methods. Finally, we summarize this paper in Section 6.

2. Related Work

In this section, we first briefly introduce some traditional dimensionality reduction methods, and then offer a detailed description about relevant ideas including graph embedding, trace ratio problem and relational learning, respectively. Finally, we specify how those ideas are used in this work.

2.1. Traditional Dimensionality Reduction Methods

Some basic ideas of traditional dimensionality reduction methods are presented in this subsection, such as PCA, LDA as well as several locality based manifold learning methods including LLE and LE. Advantages and drawbacks of these methods are also presented in this part.

2.1.1. PCA

The main idea of PCA [1] is to seek projection directions with maximal variances of the low-dimensional embeddings. It effectively extract and retain the principle components of the original data. However, as PCA is an unsupervised dimensionality reduction method, low-dimensional embeddings obtained from this method cannot perfectly maintain the discrimination between data of different classes.

2.1.2. LDA

LDA [7], well known as a supervised dimensionality reduction method, aims to seek projection directions to minimize the intraclass scattering and maximize the interclass scattering for the low-dimensional embeddings. However, for LDA, if the dimensionality of data is far greater than the data size, the intraclass scattering matrix may suffer from the singularity problem and thus it influences the solution of this dimensionality reduction algorithm. Furthermore, since the rank of the interclass scattering matrix is at most $C-1$, the number of available projection directions of LDA is at most $C-1$, where C is the number of classes.

2.1.3. Manifold Learning Methods

PCA and LDA are all global methods which use global information to project the original data into a subspace and obtain the low-dimensional data representations. However, for highly nonlinear data structure, these linear methods cannot learn the nonlinear relationships between data and thus the results are not ideal. By assuming that the high-dimensional data have a low-dimensional manifold structure, manifold learning algorithms can nonlinearly map the high-dimensional data onto their low-dimensional manifold. Among manifold learning methods, local geometric information-based methods, such as LLE, LE and local preserving projection (LPP) [32] are widely used. The ideas behind them are as follows.

LLE [12] preserves the linear reconstruction characteristics in a local neighborhood of each datum. Hence, the low-dimensional embeddings obtained by LLE presents the local geometrical structure of the data manifold. LE [13] preserves the similarities of the neighboring data points based on an adjacency matrix and a graph Laplacian matrix. However, for LLE and LE, as the nonlinear mapping function between the high-dimensional and low-dimensional spaces is not learned, we cannot easily obtain the low-dimensional representations of new data. To the end, LPP [32] performs a linear approximation of LE, and successfully overcomes its drawback as mentioned above.

2.2. Graph Embedding

In [14], Yan et al. show that some commonly used dimensionality reduction algorithms could be transformed into a unified framework despite their different motivations, and the unified framework is called graph embedding. This framework derives a low-dimensional feature space, which preserves the adjacency relationship between sample pairs. The general objective function of this framework is presented in Equation (1), where $\mathbf{W}$ denotes the similarity matrix of the undirected weighted graph $\mathbf{G} = \{\mathbf{X}, \mathbf{W}\}$ and $\mathbf{B}$ is the constraint matrix defined to avoid a trivial solution of the objective function,

$$\mathbf{V}^* = \operatorname*{argmin}_{\mathbf{V}^T \mathbf{S}_B \mathbf{V} = \mathbf{I}} \mathbf{V}^T \mathbf{S}_W \mathbf{V} = \operatorname*{argmin}_{\mathbf{V}} \frac{\mathbf{V}^T \mathbf{S}_W \mathbf{V}}{\mathbf{V}^T \mathbf{S}_B \mathbf{V}}, \tag{1}$$

where $\mathbf{S}_W$ and $\mathbf{S}_B$ are matrices constructed with respect to $\mathbf{X}$, $\mathbf{W}$ and $\mathbf{B}$, respectively.

We note that this unified framework graph embedding also provides a new idea for researchers to propose new dimensionality reduction algorithms. In particular, Yan et al. propose a novel dimensionality reduction method by defining an intrinsic graph which characterizes the intraclass compactness and a penalty graph which characterizes the interclass separability in the graph embedding framework, and call it marginal Fisher analysis (MFA).

2.3. Trace Ratio Problems

As presented in the above subsection, within the context of graph embedding, the dimensionality reduction methods can be viewed as trying to obtain the transformation matrix $\mathbf{W}$ that makes $Tr(\mathbf{W}^T \mathbf{S}_p \mathbf{W})$ maximum and $Tr(\mathbf{W}^T \mathbf{S}_l \mathbf{W})$ minimum. This is often formulated as a trace ratio optimization problem, that is $\max_\mathbf{W} Tr(\mathbf{W}^T \mathbf{S}_p \mathbf{W})/Tr(\mathbf{W}^T \mathbf{S}_l \mathbf{W})$ [33]. Generally, there are two kinds of solutions for this problem: (1) Simplifying the problem into a ratio trace problem: $\max_\mathbf{W} Tr[(\mathbf{W}^T \mathbf{S}_l \mathbf{W})^{-1}(\mathbf{W}^T \mathbf{S}_p \mathbf{W})]$, then using generalized eigenvalue decomposition (GED) to obtain the transformation matrix $\mathbf{W}$; (2) Directly optimizing the objective function through an iterative procedure, with each step presented as a trace difference problem: $Tr[(\mathbf{W}^T (\mathbf{S}_p - \lambda^n \mathbf{S}_l) \mathbf{W})]$. However, for the first solution, the optimization of ratio trace formulation may deviate from the original objective, which results in a closed-form but inexact solution and may subsequently lead to uncertainty in subsequent classification or clustering problems. For the second solution, Wang et al. [33] propose an efficient iterative procedure by solving the trace difference problem in each iterative step, named iterative procedure (ITR). It is proven that ITR could converge to the optimal solution and solve the trace ratio problem. With the orthogonal assumption on the projection matrix, objective function of the ITR optimization can be defined as

$$\mathbf{W} = \operatorname*{argmax}_{\mathbf{W}^T \mathbf{W} = \mathbf{I}} \frac{tr(\mathbf{W}^T \mathbf{S}_p \mathbf{W})}{tr(\mathbf{W}^T \mathbf{S}_l \mathbf{W})}. \tag{2}$$

In [34], Nie et al. address the graph-based feature selection framework using the iterative process of the trace ratio problem. In [35], Zhong et al. analyze the iterative procedures for the trace ratio problem and prove necessary and sufficient conditions of the existence of the optimal solution of trace ratio problems, which are that there is a sequence $\{\lambda_1^*, \lambda_2^*, \ldots, \lambda_n^*\}$ that converges to λ^* as $n \to +\infty$, where λ^* is the optimal value of Equation (2). Based on these previous works, we also formulate RFA as a trace ratio problem and theoretically prove the convergence of its optimization algorithm.

2.4. Relational Learning

In many real-world applications, data generally share some kinds of relations, such as geometrical or semantic similarity, links or citations. This relation information encoded inside data provides valuable evidence for some issues, such as classification and retrieval. To the end, relational learning is generally integrated into the representation learning models.

In [36], Duin et al. prove that it is possible to use only proximity measure (distances or similarities) to represent the samples rather than mapping the feature vectors to the low-dimensional space. In addition, they propose a proximity description-based dimensionality reduction method called relational discriminant analysis (RDA) in [24]. Instead of data, RDA uses similarities to a subset of objects in the training data as features. In this case, dimensionality reduction can be conducted either by selection methods (such as random selection [37], systematic selection [38]), or by feature extraction methods (such as multi-dimensional scaling [39], Sammon mapping [40] and Niemann mapping [41]).

In [25], Li et al. model the covariance of data with the relationships between instances and propose a Gaussian latent variable model which successfully integrates relational information into the dimensionality reduction process, called probabilistic relational PCA (PRPCA). In PRPCA, relational information is defined by the relevance between data samples. We take the scientific paper citation as an example. If there is a quoting between the papers, it means that these papers most likely have similar topics. To take the inter-influence between cited papers into account, Li et al. further construct a matrix $\mathbf{\Phi} = \mathbf{\Delta}^{-1}$, which satisfy the condition that similar instances often have a lower probability density at the latent space. To the end, PRPCA, based on the relational covariance $\mathbf{\Phi}$, successfully applies the relational information to the dimensionality reduction algorithms.

Relational learning is also commonly used for data mining, information retrieval and other machine learning-related applications. Paccanaro et al. [42] propose a method, called linear relational embedding, for the distributed representations of data, where data consist of the relationship of concepts. Wang et al. [43] utilize the characteristic that existing relations between items are often useful in recommendation systems and propose a model called relational collaborative topic regression (RCTR), which expand the traditional CTR model by integrating feedback information, item content information and relational information. Xuan et al. [44] propose a nonparametric relational topic model using stochastic processes instead of fixed-dimensional probability distributions.

Based on the classical works mentioned above, we propose a general and effective dimensionality reduction framework named relational Fisher analysis (RFA). This framework uses graph embedding [14] as theoretical foundation and integrates the relational information [24,25] encoded inside data into the dimensionality reduction process. Besides the intrinsic graph and the penalty graph as defined using graph embedding, we further construct a relational graph based on the existing relationships between data, which enables the desired low-dimensional space to preserve the intrinsic information, reduce the penalty information and further learn and preserve the relational information among the data samples. In addition, through the derivation and equivalent transformation operations, the objective function of our proposed method can be transformed into the trace ratio form for optimization. Based on a systematic analysis of two optimization method for trace ratio problems [33], we propose a novel iterative algorithm which uses the value of the trace ratio as criterion for the algorithmic convergence. In addition, by further introducing the ITR-Score defined in [34] into the iterative process, optimal projection directions are learned, which improves the effectiveness of the proposed RFA model.

3. Methodology

In this section, we first present some notations used in our work. The iterative steps, the optimization method and the proof of global convergence of RFA are then introduced in detail.

3.1. Notation

Matrices are represented in uppercase bold letters, for instance, $\mathbf{A}$, while vectors are represented in boldface lowercase letters, for instance, $\mathbf{a}$, and $\mathbf{a}_i$ is the ith element of $\mathbf{a}$. $\mathbf{A}_{i*}$ and $\mathbf{A}_{*j}$ denote the ith row and jth column of a matrix $\mathbf{A}$; therefore, the element of the

ith row and jth column of the matrix is represented by $\mathbf{A}_{ij}$. The trace of $\mathbf{A}$ is defined by $\mathrm{tr}(\mathbf{A})$ and the transpose of $\mathbf{A}$ is defined by $\mathbf{A}^T$. In addition, $|\mathbf{A}_{ij}|$ is the absolute value of $\mathbf{A}_{ij}$, $\|\mathbf{A}\|_F$ is the Frobenius norm of $\mathbf{A}$. If $\mathbf{A}$ is positive definite, we have $\mathbf{A} \succ 0$, while it is positive semi-definite (psd), we have $\mathbf{A} \succeq 0$.

In a learning task that contains multiple classes of data, we usually have dataset $\{\{\mathbf{X}_{*i}, \mathbf{y}_i\} \in \Re^D \times \Re^1, i = 1, 2, \ldots, N\}$, where each $\mathbf{X}_{*i}$ represents a sample and $\mathbf{y}_i \in \{1, 2, \ldots, C\}$ is the class of that sample, $C \geqslant 2$ is the total number of classes and N is the total number of samples. For a linear dimensionality reduction task, we hope to find a projection matrix $\mathbf{W}$ and obtain the d-dimensional representation $\mathbf{Z}_{*i}$ of $\mathbf{X}_{*i}$ by $\mathbf{Z}_{*i} = \mathbf{W} * \mathbf{X}_{*i}$, where $\mathbf{Z}_{*i} \in \Re^d, i = 1, 2, \ldots, N, d < D$ is the dimensionality of the output.

3.2. Formulation of RFA

As discussed in Section 2, graph embedding has already been proven to be a general framework for dimensionality reduction. However, there are two shortages of graph embedding. First, graph embedding does not obtain and preserve the relational information between data. Second, the graph embedding framework needs to be solved by generalized eigenvalue decomposition, which is only an approximate approach. Inspired by graph embedding, we propose a new dimensionality reduction framework called RFA, which integrates relationships among data into the dimensionality reduction model and can alleviate the two problems mentioned above. Formulation of the proposed RFA is described as follows.

We use $\mathbf{R} \in \Re^{N \times N}$ to denote the relational matrix. The dimensionality reduction framework RFA is modeled as

$$\mathcal{L} = \min_{\mathbf{W}} \frac{\mathrm{tr}(\mathbf{W}^T \mathbf{X} \mathbf{L}_I \mathbf{X}^T \mathbf{W})}{\mathrm{tr}(\mathbf{W}^T \mathbf{X} \mathbf{L}_P \mathbf{X}^T \mathbf{W})} + \lambda \mathrm{tr}(\mathbf{W}^T \mathbf{X} \mathbf{R} \mathbf{X}^T \mathbf{W}), \tag{3}$$

where $\mathbf{L}_I$ and $\mathbf{L}_P$ define the intrinsic and penalty graphs, respectively, and $\lambda \geqslant 0$ is a hyperparameter. Specifically, we only consider undirected graph and assume that $\mathbf{L}_I$, $\mathbf{L}_P$ and $\mathbf{R}$ are symmetric and psd. Based on this formulation and these assumptions, the generality of RFA can be explained from the following two points:

(1) If $\lambda = 0$, our algorithm can be simplified to a basic graph embedding model, so that some commonly used dimensionality reduction algorithms can be regarded as special cases of RFA;

(2) Otherwise, if $\mathcal{L}$ only contains relational information, RFA can be considered to use relational learning to reduce the dimensionality of data. For instance, the MDS algorithm is a special RFA algorithm under this condition.

3.3. Optimization of RFA

We reformulate Problem (3) as

$$\mathcal{L} = \min_{\mathbf{W}} \frac{\mathrm{tr}(\mathbf{W}^T \mathbf{S}_I \mathbf{W})}{\mathrm{tr}(\mathbf{W}^T \mathbf{S}_P \mathbf{W})} + \lambda \mathrm{tr}(\mathbf{W}^T \mathbf{S}_R \mathbf{W}), \tag{4}$$

where $\mathbf{S}_I = \mathbf{X} \mathbf{L}_I \mathbf{X}^T$, $\mathbf{S}_P = \mathbf{X} \mathbf{L}_P \mathbf{X}^T$ and $\mathbf{S}_R = \mathbf{X} \mathbf{R} \mathbf{X}^T$.

As $\mathbf{R}$ is psd, $\mathbf{S}_R$ is as well. We suppose $\mathbf{S}_R = \mathbf{U} \mathbf{\Lambda} \mathbf{U}^T$. We have

$$\mathcal{L} = \min_{\mathbf{V}} \frac{\mathrm{tr}(\mathbf{V}^T \widetilde{\mathbf{S}}_I \mathbf{V})}{\mathrm{tr}(\mathbf{V}^T \widetilde{\mathbf{S}}_P \mathbf{V})} + \lambda \mathrm{tr}(\mathbf{V}^T \mathbf{V}), \tag{5}$$

where $\mathbf{W} = \mathbf{U} \mathbf{\Lambda}^{-1/2} \mathbf{V}$, $\widetilde{\mathbf{S}}_I = \mathbf{\Lambda}^{-1/2} \mathbf{U}^T \mathbf{S}_I \mathbf{U} \mathbf{\Lambda}^{-1/2}$ and $\widetilde{\mathbf{S}}_P = \mathbf{\Lambda}^{-1/2} \mathbf{U}^T \mathbf{S}_P \mathbf{U} \mathbf{\Lambda}^{-1/2}$.
Furthermore,

$$\frac{\partial \mathcal{L}}{\partial \mathbf{V}} = \frac{2 \mathrm{tr}(\mathbf{V}^T \widetilde{\mathbf{S}}_P \mathbf{V}) \widetilde{\mathbf{S}}_I \mathbf{V} - 2 \mathrm{tr}(\mathbf{V}^T \widetilde{\mathbf{S}}_I \mathbf{V}) \widetilde{\mathbf{S}}_P \mathbf{V}}{\mathrm{tr}(\mathbf{V}^T \widetilde{\mathbf{S}}_P \mathbf{V})^2} + 2 \lambda \mathbf{V}. \tag{6}$$

We let $\frac{\partial \mathcal{L}}{\partial \mathbf{V}} = \mathbf{0}$. We have

$$\frac{\text{tr}(\mathbf{V}^T\widetilde{\mathbf{S}}_P\mathbf{V})\widetilde{\mathbf{S}}_I\mathbf{V} - \text{tr}(\mathbf{V}^T\widetilde{\mathbf{S}}_I\mathbf{V})\widetilde{\mathbf{S}}_P\mathbf{V}}{\text{tr}(\mathbf{V}^T\widetilde{\mathbf{S}}_P\mathbf{V})^2} = -\lambda\mathbf{V}. \tag{7}$$

Equation (7) can be rewritten as

$$(\widetilde{\mathbf{S}}_I - \frac{\text{tr}(\mathbf{V}^T\widetilde{\mathbf{S}}_I\mathbf{V})}{\text{tr}(\mathbf{V}^T\widetilde{\mathbf{S}}_P\mathbf{V})}\widetilde{\mathbf{S}}_P)\mathbf{V} = -\lambda\text{tr}(\mathbf{V}^T\widetilde{\mathbf{S}}_P\mathbf{V})\mathbf{V}. \tag{8}$$

We let $\eta = \frac{\text{tr}(\mathbf{V}^T\widetilde{\mathbf{S}}_I\mathbf{V})}{\text{tr}(\mathbf{V}^T\widetilde{\mathbf{S}}_P\mathbf{V})}$ and $\widetilde{\lambda} = -\lambda\text{tr}(\mathbf{V}^T\widetilde{\mathbf{S}}_P\mathbf{V})$. We obtain

$$(\widetilde{\mathbf{S}}_I - \eta\widetilde{\mathbf{S}}_P)\mathbf{V} = \widetilde{\lambda}\mathbf{V}. \tag{9}$$

It can be seen from Equation (9) that the columns of matrix $\mathbf{V}$ are the eigenvectors of $\widetilde{\mathbf{S}}_I - \eta\widetilde{\mathbf{S}}_P$, where η is a parameter related to $\mathbf{V}$.

Without loss of generality, we assume $\mathbf{V}^T\mathbf{V} = \mathbf{I}_d$, where $\mathbf{I}_d$ is an identity matrix. Hence, we have the following constrained trace ratio problem [33,45]:

$$\widetilde{\mathcal{L}} = \min_{\mathbf{V}^T\mathbf{V}=\mathbf{I}_d} \frac{\text{tr}(\mathbf{V}^T\widetilde{\mathbf{S}}_I\mathbf{V})}{\text{tr}(\mathbf{V}^T\widetilde{\mathbf{S}}_P\mathbf{V})}. \tag{10}$$

Problem (10) can be solved with the iterative method similar to that in [33,45]. The specific steps are as follows:

(1) **Removing the null space of $\mathbf{S}_t = \widetilde{\mathbf{S}}_I + \widetilde{\mathbf{S}}_P$ [46].** We assume that $\mathbf{S}_t = \widetilde{\mathbf{U}}\widetilde{\mathbf{\Lambda}}\widetilde{\mathbf{U}}^T$, where $\widetilde{\mathbf{\Lambda}}$ is a diagonal matrix and $\widetilde{\mathbf{U}}$ contains the eigenvectors of $\mathbf{S}_t$ corresponding to nonzero eigenvalues. Therefore, Formula (10) can be transformed to

$$\widetilde{\mathcal{L}} = \min_{\widetilde{\mathbf{W}}^T\widetilde{\mathbf{W}}=\mathbf{I}_d} \frac{\text{tr}(\widetilde{\mathbf{W}}^T\widehat{\mathbf{S}}_I\widetilde{\mathbf{W}})}{\text{tr}(\widetilde{\mathbf{W}}^T\widehat{\mathbf{S}}_P\widetilde{\mathbf{W}})}, \tag{11}$$

where $\mathbf{V} = \widetilde{\mathbf{U}}\widetilde{\mathbf{W}}, \widetilde{\mathbf{W}} \in \Re^{r\times d}$, $\widehat{\mathbf{S}}_I = \widetilde{\mathbf{U}}^T\widetilde{\mathbf{S}}_I\widetilde{\mathbf{U}}$ and $\widehat{\mathbf{S}}_P = \widetilde{\mathbf{U}}^T\widetilde{\mathbf{S}}_P\widetilde{\mathbf{U}}$. We can further rewrite the problem (11) as

$$\widetilde{\mathcal{L}} = \min_{\widetilde{\mathbf{W}}^T\widetilde{\mathbf{W}}=\mathbf{I}_d} \frac{\text{tr}(\widetilde{\mathbf{W}}^T\widehat{\mathbf{S}}_I\widetilde{\mathbf{W}})}{\text{tr}(\widetilde{\mathbf{W}}^T\widehat{\mathbf{S}}_T\widetilde{\mathbf{W}})}, \tag{12}$$

where $\widehat{\mathbf{S}}_T = \widetilde{\mathbf{U}}^T(\widetilde{\mathbf{S}}_I + \widetilde{\mathbf{S}}_P)\widetilde{\mathbf{U}} = \widehat{\mathbf{S}}_I + \widehat{\mathbf{S}}_P$. Since $\widehat{\mathbf{S}}_T$ is positive definite, for any orthonormal matrix $\widetilde{\mathbf{W}}$, Problem (12) satisfies that the denominator is positive.

(2) **Efficient iterative optimization.** The original trace ratio problem (12) can be rewritten as a trace difference problem:

$$\widetilde{\mathbf{W}}^* = \operatorname*{argmin}_{\widetilde{\mathbf{W}}^T\widetilde{\mathbf{W}}=\mathbf{I}_d} \text{tr}(\widetilde{\mathbf{W}}^T(\widehat{\mathbf{S}}_I - \widetilde{\eta}\widehat{\mathbf{S}}_T)\widetilde{\mathbf{W}}), \tag{13}$$

where $\widetilde{\eta}$ is a parameter which can be calculated in the iterative process. In the iterative process, we first randomly initialize the target matrix $\widetilde{\mathbf{W}}$ to be an arbitrary orthogonal matrix as $\widetilde{\mathbf{W}}_0 \in \Re^{r\times d}$, and then calculate $\widetilde{\eta}_0 = \frac{\text{tr}(\widetilde{\mathbf{W}}_0^T\widehat{\mathbf{S}}_I\widetilde{\mathbf{W}}_0)}{\text{tr}(\widetilde{\mathbf{W}}_0^T\widehat{\mathbf{S}}_T\widetilde{\mathbf{W}}_0)}$. By using the calculated $\widetilde{\eta}_0$, we can obtain $\widetilde{\mathbf{W}}_1$ by solving Problem (13). In the end, through several iterations, we obtain $\widetilde{\mathbf{W}}_T$, where T is the number of iterations and it's satisfied $|\widetilde{\eta}_T - \widetilde{\eta}_{T-1}| < \epsilon$ ($\epsilon = 10^{-5}$ is used in our experiments). Then, $\widetilde{\mathbf{W}}_T$ is the optimal solution of Problem (12). In next section, we prove that RFA owns the global convergence. In order to improve the effectiveness of our method, we select some superior projection directions for each $\widetilde{\mathbf{W}}_t$, as performed in [47]. Our selection criterion is

$$\widetilde{\mathbf{W}}_t = \underset{\widetilde{\mathbf{W}} \in \Phi}{\mathrm{argmin}} \, \frac{\mathrm{tr}(\widetilde{\mathbf{W}}^T \widehat{\mathbf{S}}_I \widetilde{\mathbf{W}})}{\mathrm{tr}(\widetilde{\mathbf{W}}^T \widehat{\mathbf{S}}_T \widetilde{\mathbf{W}})}, \tag{14}$$

where Φ is a set of $r \times d$ matrices with columns formed by eigenvectors of $\widehat{\mathbf{S}}_I - \widetilde{\eta}_{t-1} \widehat{\mathbf{S}}_T$. We use the eigenvectors corresponding to d smallest ITR-score [48] to initialize the selection.

Algorithm 1 specifically describes the iterative procedure of Problem (12).

Algorithm 1 Optimization of Problem (12)

1: **Initialization:** Initialize $\widetilde{\mathbf{W}}$ as an orthonormal matrix $\widetilde{\mathbf{W}}_0 \in \mathfrak{R}^{r \times d}$ and Let $\widetilde{\eta}_0 = 0$.
2: **Iterations:**
3: **for** $t = 0$ **to** *MaxIt* **do**
4:     (1) Compute $\widetilde{\eta}_t$ as

$$\widetilde{\eta}_t = \frac{\mathrm{tr}(\widetilde{\mathbf{W}}_{t-1}^T \widehat{\mathbf{S}}_I \widetilde{\mathbf{W}}_{t-1})}{\mathrm{tr}(\widetilde{\mathbf{W}}_{t-1}^T \widehat{\mathbf{S}}_T \widetilde{\mathbf{W}}_{t-1})}.$$

5:     (2) Solve the eigenvalue decomposition problem:

$$(\widehat{\mathbf{S}}_I - \widetilde{\eta} \widehat{\mathbf{S}}_T) \widehat{\mathbf{W}}_{*i} = \widetilde{\lambda}_i \widehat{\mathbf{W}}_{*i}.$$

6:     (3) Compute

$$s_i = \frac{\mathrm{tr}(\widehat{\mathbf{W}}_{*i}^T \widehat{\mathbf{S}}_I \widehat{\mathbf{W}}_{*i})}{\mathrm{tr}(\widehat{\mathbf{W}}_{*i}^T \widehat{\mathbf{S}}_T \widehat{\mathbf{W}}_{*i})}, \, i = 1, \ldots, r.$$

7:     (4) Use $\{\widehat{\mathbf{W}}_{*i}\}_{i=1}^r$ to initialize $\widetilde{\mathbf{W}}_t$ and solve the following problem:

$$\widetilde{\mathbf{W}}_t^* = \underset{\widetilde{\mathbf{W}} \in \Phi}{\mathrm{argmin}} \, \frac{\mathrm{tr}(\widetilde{\mathbf{W}}^T \widehat{\mathbf{S}}_I \widetilde{\mathbf{W}})}{\mathrm{tr}(\widetilde{\mathbf{W}}^T \widehat{\mathbf{S}}_T \widetilde{\mathbf{W}})},$$

where Φ is a set of matrices with columns formed by $\{\widehat{\mathbf{W}}_{*i}\}_{i=1}^r$.
8:     **if** $|\widetilde{\eta}_t - \widetilde{\eta}_{t-1}| < \epsilon$ ($\epsilon = 10^{-5}$ is used in our experiments) **then**
9:         Break.
10:     **end if**
11: **end for**
12: **Output:** $\widetilde{\mathbf{W}}_t$.

---

4. Global Convergence of RFA and Extensions

In this section, we first prove that RFA can converge to the global optimal solution, and then, we apply the kernel trick to RFA for nonlinear relational dimensionality reduction.

4.1. Global Convergence of RFA

Theorem 1 states the convergence of RFA.

Theorem 1. *The RFA algorithm converges to a global optimal solution of Problem (3).*

Proof. Considering that all the formulas, from Problem (3) to Problem (12) in the previous section, are all transformed equivalently, we only prove the convergence of Problem (12) here. Specifically, we show that $\widetilde{\mathcal{L}}$ of Problem (12) has a lower bound, which gradually decreases with the iterative process.

We can easily see that for any $\widetilde{\mathbf{W}}$, it satisfies that $0 \leq \frac{\mathrm{tr}(\widetilde{\mathbf{W}}^T \widehat{\mathbf{S}}_I \widetilde{\mathbf{W}})}{\mathrm{tr}(\widetilde{\mathbf{W}}^T \widehat{\mathbf{S}}_T \widetilde{\mathbf{W}})} \leq 1$, so that the lower bound of $\widetilde{\mathcal{L}}$ is 0.

Next, we prove that the objective value of Problem (12) gradually decreases with the iterative process of RFA. Defining

$$\widetilde{\eta}_t = \frac{\text{tr}(\widetilde{\mathbf{W}}_{t-1}^T \widehat{\mathbf{S}}_I \widetilde{\mathbf{W}}_{t-1})}{\text{tr}(\widetilde{\mathbf{W}}_{t-1}^T \widehat{\mathbf{S}}_T \widetilde{\mathbf{W}}_{t-1})}, \tag{15}$$

we then have

$$\text{tr}(\widetilde{\mathbf{W}}_{t-1}^T (\widehat{\mathbf{S}}_I - \widetilde{\eta}_t \widehat{\mathbf{S}}_T) \widetilde{\mathbf{W}}_{t-1}) = 0. \tag{16}$$

However,

$$\widetilde{\mathbf{W}}_t = \underset{\widetilde{\mathbf{W}}^T \widetilde{\mathbf{W}} = \mathbf{I}_d}{\arg\min}\, \text{tr}(\widetilde{\mathbf{W}}^T (\widehat{\mathbf{S}}_I - \widetilde{\eta}_t \widehat{\mathbf{S}}_T) \widetilde{\mathbf{W}}). \tag{17}$$

Therefore,

$$\text{tr}(\widetilde{\mathbf{W}}_t^T (\widehat{\mathbf{S}}_I - \widetilde{\eta}_t \widehat{\mathbf{S}}_T) \widetilde{\mathbf{W}}_t) \leq 0, \tag{18}$$

and

$$\widetilde{\eta}_{t+1} = \frac{\text{tr}(\widetilde{\mathbf{W}}_t^T \widehat{\mathbf{S}}_I \widetilde{\mathbf{W}}_t)}{\text{tr}(\widetilde{\mathbf{W}}_t^T \widehat{\mathbf{S}}_T \widetilde{\mathbf{W}}_t)} \leq \widetilde{\eta}_t. \tag{19}$$

Thus, we prove that $\widetilde{\eta}_t$ gradually decreases with the iterative process, and Theorem 1 holds. □

According to Theorem 1, we can see that RFA can converge to the global optimal solution. In addition, for given intrinsic and penalty graphs and the relational matrix, the computational complexity of the RFA algorithm is only $\Theta(N)$, where N is the total number of data. That also illustrates the efficiency of our algorithm.

4.2. Kernel Extension

In this subsection, we apply kernel trick to our RFA method and present the kernel RFA (KRFA) method, which can be used to nonlinear dimensionality reduction problems.

The KRFA optimization problem is basically the same as Problem (3), except that the data point $\mathbf{X}_{*i}$ needs to be mapped to a reproducing kernel Hilbert space and obtain $\phi(\mathbf{X}_{*i})$, where $\phi(\cdot)$ denotes the mapping function. In addition, the corresponding intrinsic and penalty graphs and the relational matrix should also be mapped to the reproducing kernel Hilbert space. The kernel function $\mathbf{K}(\mathbf{X}_{*i}, \mathbf{X}_{*j}) = \phi(\mathbf{X}_{*i})^T \phi(\mathbf{X}_{*j})$.

We suppose the projection matrix $\mathbf{W} = \mathbf{\Phi}\mathbf{\Gamma}$. We have $\mathbf{W}^T \mathbf{\Phi} = \mathbf{\Gamma}^T \mathbf{K}$.

Normalizing and centering the data in the high-dimensional feature space, we substitute $\mathbf{K}$ with

$$\hat{\mathbf{K}} = \mathbf{K} - \mathbf{1}_N \mathbf{K} - \mathbf{K} \mathbf{1}_N + \mathbf{1}_N \mathbf{K} \mathbf{1}_N. \tag{20}$$

In this way, the learning task of RFA is able to be described as

$$\mathcal{L}_{\hat{\mathbf{K}}} = \min_{\mathbf{\Gamma}} \frac{\text{tr}(\mathbf{\Gamma}^T \hat{\mathbf{K}} \widetilde{\mathbf{L}}_I \hat{\mathbf{K}}^T \mathbf{\Gamma})}{\text{tr}(\mathbf{\Gamma}^T \hat{\mathbf{K}} \widetilde{\mathbf{L}}_P \hat{\mathbf{K}}^T \mathbf{\Gamma})} + \lambda \text{tr}(\mathbf{\Gamma}^T \hat{\mathbf{K}} \widetilde{\mathbf{R}} \hat{\mathbf{K}}^T \mathbf{\Gamma}), \tag{21}$$

where $\widetilde{\mathbf{L}}_I$, $\widetilde{\mathbf{L}}_P$ and $\widetilde{\mathbf{R}}$ are the intrinsic graph, the penalty graph and the relational matrix, respectively. The distance of the samples in the reproducing kernel Hilbert space can be calculated by

$$D(x_i, x_j) = \sqrt{\mathbf{K}(x_i, x_i) + \mathbf{K}(x_i, x_j) - 2\mathbf{K}(x_i, x_j)}. \tag{22}$$

With reference to the derivation and transformation procedure of RFA, KRFA can be eventually transformed into

$$\widetilde{\mathcal{L}}_{\mathbf{K}} = \min_{\mathbf{V}_{\mathbf{K}}^T \mathbf{V}_{\mathbf{K}} = \mathbf{I}_d} \frac{\text{tr}(\mathbf{V}_{\mathbf{K}}^T \widetilde{\mathbf{S}}_I^{\mathbf{K}} \mathbf{V}_{\mathbf{K}})}{\text{tr}(\mathbf{V}_{\mathbf{K}}^T \widetilde{\mathbf{S}}_P^{\mathbf{K}} \mathbf{V}_{\mathbf{K}})}. \tag{23}$$

As KRFA's optimization process is similar to RFA, we still use iterative methods to solve it.

5. Experiments

In this section, extensive experiments are conducted to validate the effectiveness of our RFA and KRFA methods. For the linear case, we conduct experiments on document analysis, handwritten digits recognition, face recognition and webpage classification problems. For KRFA, we test its performance on several benchmark datasets. The results of comparative experiments are presented below.

5.1. Performance of RFA

To evaluate the performance of RFA, we selected several related dimensionality reduction methods for comparison with RFA. These methods were LDA, MFA, RDA, and PRPCA, respectively. Among them, LDA, MFA and RDA are supervised methods, while PRPCA is unsupervised. We note that, due to the flexibility in the design of the relational matrix, RFA could be either global or local. We describe this detail in the following part.

We followed MFA to construct the intrinsic and penalty graphs [14]. The numbers of nearest neighbors for constructing the intrinsic graph (k_i) and the penalty graph (k_p) were set to 5 and 20, respectively, for all the datasets.

We applied RFA to document understanding, face recognition and several other recognition tasks. In order to test the performance of RFA on document recognition tasks, we used the Ibn Sina ancient Arabic document dataset [49], USPS handwritten digits dataset (http://www.cs.nyu.edu/~roweis/data.html, accessed on 19 October 2023), two handwritten digits datasets (Optdigits and Pendigits) and one English letter dataset (Letter) from the UCI machine learning repository [50]. For face recognition tasks, we used two face datasets [51–54], the CMU PIE (http://www.face-rec.org/databases/, accessed on 19 October 2023) and the YaleB (http://www.cad.zju.edu.cn/home/dengcai/Data/FaceData.html, accessed on 19 October 2023) datasets. At the same time, we used some UCI datasets, including Shuttle, Thyroid, Vowel and Waveform21, to evaluate RFA.

In our experiments, we used the classical graph Laplacian matrix, $\mathbf{L} = \mathbf{D} - \mathbf{M}$, to define the relational matrix, whose weight matrix $\mathbf{M}$ is shown as below:

$$\mathbf{M}_{ij} = \begin{cases} 1, & \text{if } \mathbf{x}_i \in \mathcal{N}_k(\mathbf{x}_j) \text{ or } \mathbf{x}_j \in \mathcal{N}_k(\mathbf{x}_i), \\ 0, & \text{otherwise}, \end{cases} \tag{24}$$

where $\mathcal{N}_k(\mathbf{x}_i)$ is the set consisting of the k nearest neighborForof $\mathbf{x}_i$. For each dataset, the value of k was selected based on 5-fold cross-validation. Moreover, $\mathbf{D}$ is the diagonal degree matrix with $\mathbf{D}_{ii} = \sum_j \mathbf{M}_{ij}$.

We note that the Laplacian weight matrix formed by the above formula contains the relationship information between the sample and a certain number of its neighbors, which allows the iintegration of the local relationships between data into the supervised representation learning algorithm. At the same time, as a general model, the relationship matrix $\mathbf{R}$ in RFA can be of various forms. For example, $\mathbf{R}$ can be the centralization matrix $\mathbf{H} = \mathbf{I}_N - \frac{1}{N}\mathbf{1}\mathbf{1}^T$, where N denotes the data size and $\mathbf{1}$ is a column vector of length N with all ones.

For the PRPCA algorithm, we performed this experiment using the codes provided by the authors. For the RDA algorithm, we randomly selected the prototypes [24]. For algorithms other than RDA, we used the 1-nearest neighbor classifier to evaluate the classification performance of them.

5.1.1. Comparison on Hand-Writing Datasets

We first tested the performance of RFA on the hand-writing datasets. For clarity, the details of the used datasets are shown in Table 1.

Table 1. Statistics of the used datasets.

Dataset	Classes	Samples	Dimensions
Ibn Sina	174	20,688	200
USPS	10	9298	256
Pendigits	10	5620	16
Optdigits	10	10,992	64
Letter	26	20,000	16

Ibn Sina dataset: This dataset [55] is an ancient manuscript dataset, one image of which is shown in Figure 1, and we tried to identify the Arabic subwords on this dataset. In the experiment, we used a 50-page manuscript as the training set and 10 pages as the test set. We extracted the square-root velocity (SRV) representation [56] of the Arabic subwords. Then, we removed the outlier classes, including the classes that had less than 10 samples. Finally, we obtained a 174-class Arabic subword dataset with 17,543 samples for training and 3125 samples for test.

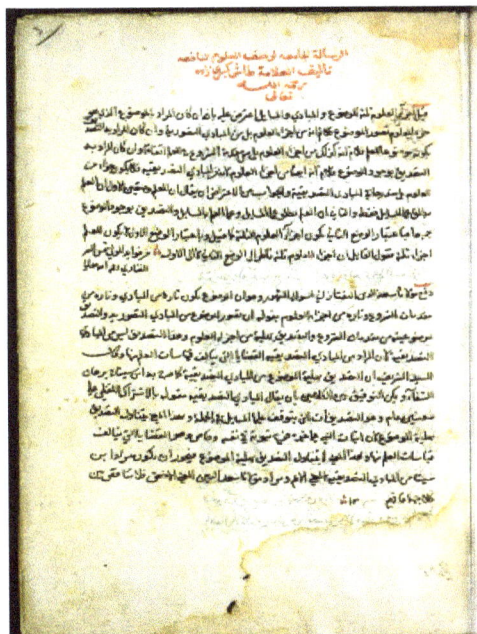

Figure 1. One image of the Ibn Sina dataset.

In this experiment, we set the number of the nearest neighbors – k of the RFA to 8 and compared RFA with the LDA and MFA dimensionality reduction algorithms. The classification accuracies after using these three dimensionality reduction methods to map the data to different dimensionalities are shown in Figure 2. We can see that RFA is far better than the LDA algorithm and slightly better than the MFA algorithm. At the same time, when the dimensionality is from 50 to $C - 1$ (C is the number of classes), the correct rate of the MFA algorithm has a certain fluctuation and tends to decline, while the classification performance of our RFA algorithm is relatively stable, which means that our algorithm is more robust than MFA.

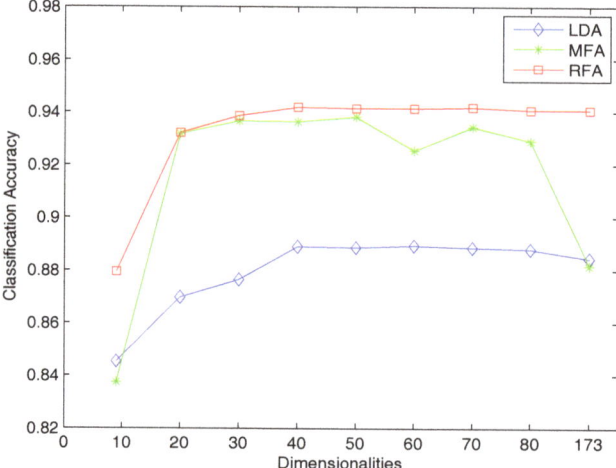

Figure 2. Classification accuracy obtained by LDA, MFA and RFA on the Ibn Sina dataset.

USPS dataset: The USPS is an U.S. post handwritten digits dataset that contains 7291 training data and 2007 test data from 10 classes and the dimensionality of the data features is 256. Some handwritten digits in the USPS dataset are shown in Figure 3.

In this experiment, we set the number of the nearest neighbors – k of RFA to 24. The classification accuracies obtained by RFA and the compared algorithms are shown in Figure 4. Because the LDA subspace has a maximum of $C - 1$ dimension (C is the number of classes), LDA is only presented with a black star in the figure. We can see that when the dimension is 9, RFA obtains a comparative results with LDA and MFA. When the dimension increases, the results of RFA are always optimal. At the same time, the classification accuracy obtained by RDA is low.

Figure 5 presents 3D visualization of the learned data representations by RFA, which shows the effect of RFA to obtain better classification boundaries between the classes. In Figure 6, we show the 2D projections of data learned by both RFA and MFA, to further show the effectiveness of RFA. It is easy to see that the samples processed by RFA are less likely to overlap at the boundary, indicating that compared to MFA, RFA preserves more properties that help distinguish the samples.

In addition, the robustness of RFA is tested with respect to k_i and k_p (used to construct the intrinsic and penalty graphs). From Figure 4, it can be seen that RFA obtained the best result when the subspace dimension was 35, with parameter settings $k_i = 5$, $k_p = 20$ and $k = 24$. We fixed k and one of k_i and k_p to obtain the results when another parameter took different values. Figures 7 and 8 show that RFA is very robust.

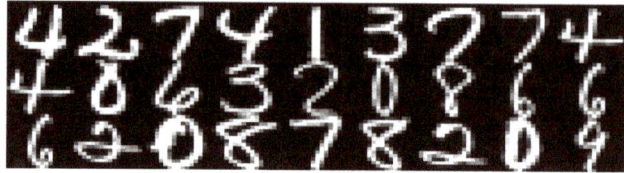

Figure 3. Sample images from the USPS dataset.

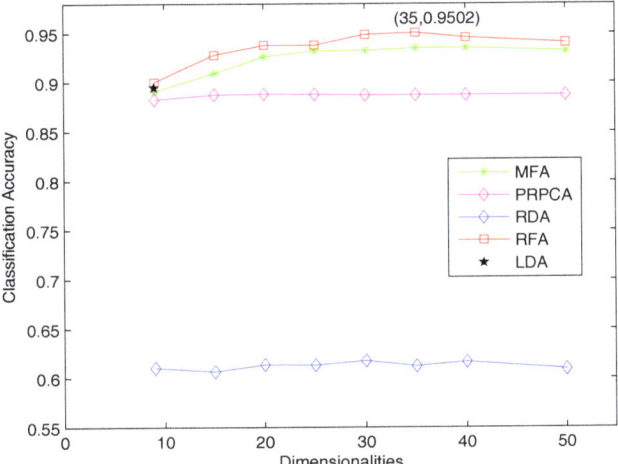

Figure 4. Classification results obtained by RFA and the compared methods on the USPS dataset.

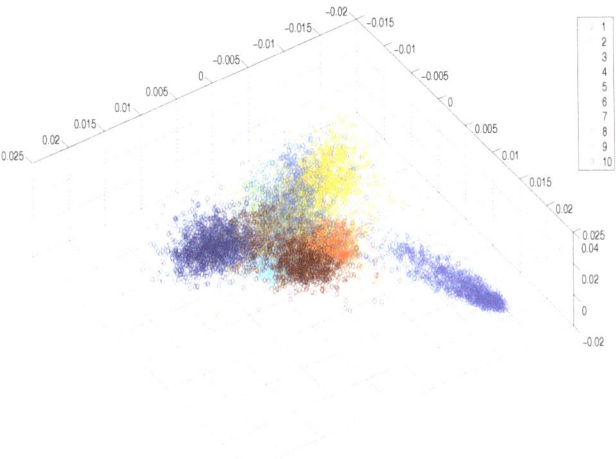

Figure 5. 3D visualization of the mapped data obtained by RFA. Samples of different classes are marked with different colors.

We also selected three document recognition-related datasets from the UCI machine learning repository to further test the effectiveness of RFA. They are Optdigits, Pendigits and Letter. Optdigits was preprocessed by NIST [57] programs to obtain 5620 instances in 8×8 dimensions. The Pendigits dataset contains a large number of preprocessed 16-dimensional samples written by 44 different authors, including 7494 training examples and 3498 test samples. Letter consists of 20,000 handwritten characters written by 20 fonts from 26 capital letters in the English alphabet. We used the 5-fold cross-validation for these experiments. Tables 2–4 show the classification accuracy and standard deviation obtained on these three datasets and the boldface results are the best ones. We can see that RFA performs consistently better than other compared methods.

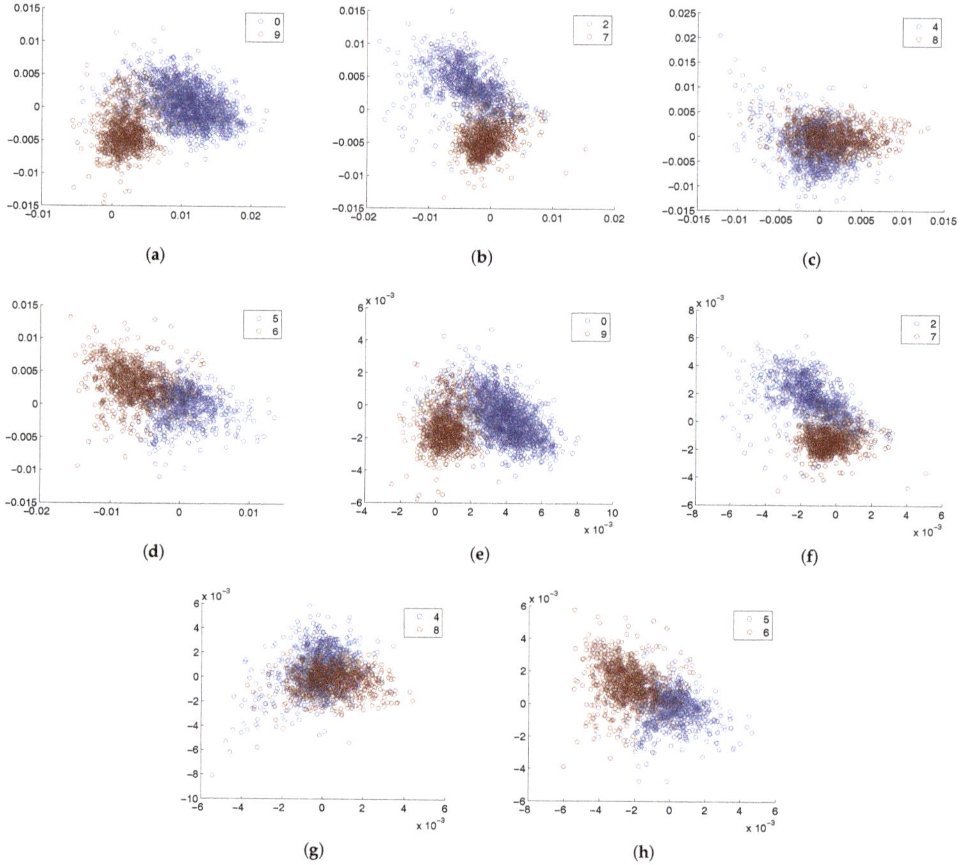

Figure 6. 2D visualization for data pairs. (**a**–**d**) are the results obtained by RFA, and (**e**–**h**) are the results obtained by MFA.

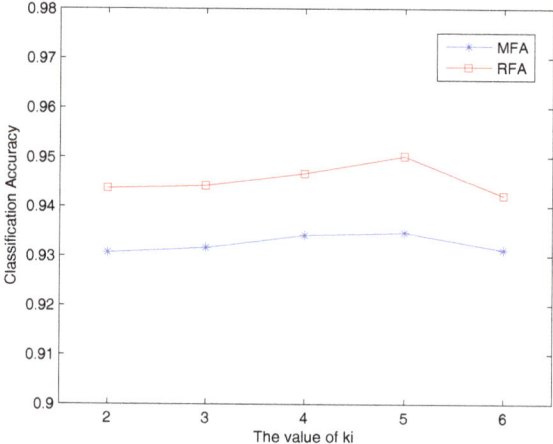

Figure 7. Classification results obtained by RFA and MFA with different values of k_i on the USPS dataset.

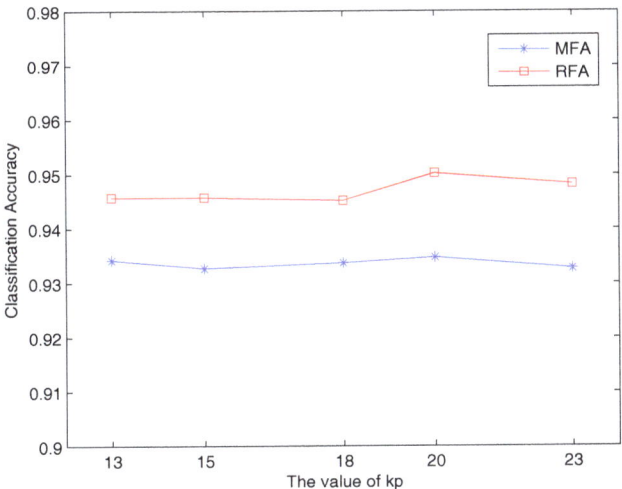

Figure 8. Classification results obtained by RFA and MFA with different values of k_p on the USPS dataset.

Table 2. Classification results obtained on the Optdigits dataset. The best results are highlighted in boldface.

Dimension	LDA	MFA	PRPCA	RDA	RFA
9	0.9219 ± 0.0117	0.9657 ± 0.0020	0.9687 ± 0.0053	0.7407 ± 0.0048	**0.9662 ± 0.0017**
15	-	0.9845 ± 0.0039	0.9690 ± 0.0087	0.7407 ± 0.0017	**0.9875 ± 0.0031**
20	-	0.9836 ± 0.0041	0.9696 ± 0.0075	0.7399 ± 0.0054	**0.9891 ± 0.0025**
25	-	0.9849 ± 0.0030	0.9696 ± 0.0071	0.7425 ± 0.0042	**0.9891 ± 0.0035**
30	-	0.9819 ± 0.0062	0.9698 ± 0.0071	0.7415 ± 0.0053	**0.9875 ± 0.0048**
35	-	0.9795 ± 0.0051	0.9703 ± 0.0067	0.7404 ± 0.0066	**0.9879 ± 0.0036**
40	-	0.9783 ± 0.0058	0.9710 ± 0.0079	0.7402 ± 0.0056	**0.9875 ± 0.0032**
50	-	0.9781 ± 0.0026	0.9701 ± 0.0071	0.7407 ± 0.0043	**0.9877 ± 0.0023**

Table 3. Classification results obtained on the Pendigits dataset.

Dimension	LDA	MFA	PRPCA	RDA	RFA
9	0.9843 ± 0.0020	0.9929 ± 0.0013	0.9797 ± 0.0027	0.9203 ± 0.0046	**0.9939 ± 0.0006**
13	-	0.9940 ± 0.0026	0.9604 ± 0.0028	0.8856 ± 0.0052	**0.9950 ± 0.0018**
15	-	0.9945 ± 0.0018	0.9887 ± 0.0008	0.9212 ± 0.0050	**0.9962 ± 0.0013**

Table 4. Classification results obtained on the Letter dataset.

Dimension	LDA	MFA	PRPCA	RDA	RFA
9	0.9108 ± 0.0070	0.9570 ± 0.0014	0.9230 ± 0.0018	0.3699 ± 0.0064	**0.9580 ± 0.0018**
13	0.9108 ± 0.0070	0.9725 ± 0.0022	0.9604 ± 0.0028	0.3727 ± 0.0065	**0.9753 ± 0.0022**
15	0.9575 ± 0.0023	0.9596 ± 0.0018	0.9567 ± 0.0022	0.3715 ± 0.0087	**0.9695 ± 0.0011**

5.1.2. Comparison on Face Datasets

Here, we tested RFA on the face recognition problems. The PIE and YaleB datasets were used. The details of these two datasets are shown in Table 5. For the corresponding experimental settings, we set the number of the nearest neighbors – k on the PIE dataset to 8 and k on the YaleB dataset to 18.

Considering that RDA cannot perform well on multi-class classification problems, we used LPP instead as a compared method in these experiments. We used the 5-fold cross-validation to decide the value of parameter k for graph construction in LPP. We performed the experiments in different low-dimensional spaces on the PIE and YaleB datasets, and the experimental results are shown in Tables 6 and 7.

Table 5. Statistics of the face recognition datasets.

Dataset	Classes	Samples	Feature Dimension
PIE	68	11,554	1024
YaleB	38	2414	1024

Table 6. Classification results obtained on the PIE dataset.

Dimension	LDA	MFA	PRPCA	LPP	RFA
9	0.9021 ± 0.0070	0.9173 ± 0.0045	0.7736 ± 0.0113	0.8156 ± 0.0089	**0.9198 ± 0.0026**
15	0.9430 ± 0.0038	0.9526 ± 0.0040	0.8810 ± 0.0063	0.8930 ± 0.0054	**0.9540 ± 0.0021**
20	0.9528 ± 0.0031	0.9592 ± 0.0027	0.9089 ± 0.0062	0.9122 ± 0.0044	**0.9603 ± 0.0030**
25	0.9597 ± 0.0040	0.9617 ± 0.0028	0.9224 ± 0.0056	0.9262 ± 0.0042	**0.9621 ± 0.0040**
30	0.9613 ± 0.0037	0.9639 ± 0.0042	0.9299 ± 0.0041	0.9320 ± 0.0031	**0.9644 ± 0.0027**
35	0.9623 ± 0.0025	0.9643 ± 0.0024	0.9342 ± 0.0037	0.9352 ± 0.0040	**0.9660 ± 0.0024**
40	0.9633 ± 0.0019	0.9638 ± 0.0021	0.9370 ± 0.0049	0.9383 ± 0.0036	**0.9655 ± 0.0021**
50	0.9639 ± 0.0012	0.9643 ± 0.0029	0.9394 ± 0.0052	0.9411 ± 0.0034	**0.9656 ± 0.0024**
C-1(67)	0.9641 ± 0.0048	0.9640 ± 0.0034	0.9356 ± 0.0045	0.9437 ± 0.0035	**0.9654 ± 0.0037**

Table 7. Classification results obtained on the YaleB dataset.

Dimension	LDA	MFA	PRPCA	LPP	RFA
9	0.7788 ± 0.0199	**0.8355 ± 0.0190**	0.3675 ± 0.0378	0.4859 ± 0.0437	0.8347 ± 0.0212
15	0.8774 ± 0.0092	0.9031 ± 0.0071	0.4793 ± 0.0572	0.5911 ± 0.0367	**0.9039 ± 0.0096**
20	0.9209 ± 0.0136	0.9273 ± 0.0104	0.5460 ± 0.0426	0.6276 ± 0.0171	**0.9279 ± 0.0067**
25	**0.9395 ± 0.0189**	0.9370 ± 0.0126	0.5427 ± 0.0377	0.6392 ± 0.0345	0.9387 ± 0.0109
30	**0.9507 ± 0.0157**	0.9490 ± 0.0081	0.5518 ± 0.0399	0.6736 ± 0.0219	0.9495 ± 0.0079
35	**0.9582 ± 0.0132**	0.9503 ± 0.0093	0.5481 ± 0.0385	0.6740 ± 0.0109	0.9503 ± 0.0116
40	-	0.9511 ± 0.0102	0.2084 ± 0.3698	0.6997 ± 0.0059	**0.9569 ± 0.0146**
50	-	0.9482 ± 0.0101	0.3698 ± 0.1481	0.7142 ± 0.0123	**0.9532 ± 0.0089**
C-1(37)	**0.9594 ± 0.0094**	0.9511 ± 0.0122	0.5306 ± 0.2147	0.6864 ± 0.0102	0.9548 ± 0.0107

We can see from the experimental results that RFA performs very well. Although LPP is an effective dimensionality reduction method for face recognition, RFA is significantly better than LPP. Moreover, RFA obtains comparable results with LDA and MFA. These results demonstrate the effectiveness of RFA in the face recognition applications.

Additionally, convergence of RFA is verified on these two datasets. As illustrated in Figures 9 and 10, the value of η (trace ratio) decreases through the iterative procedures until it reaches the global optimal value η^* on both of the two datasets, which clearly shows the convergence of RFA.

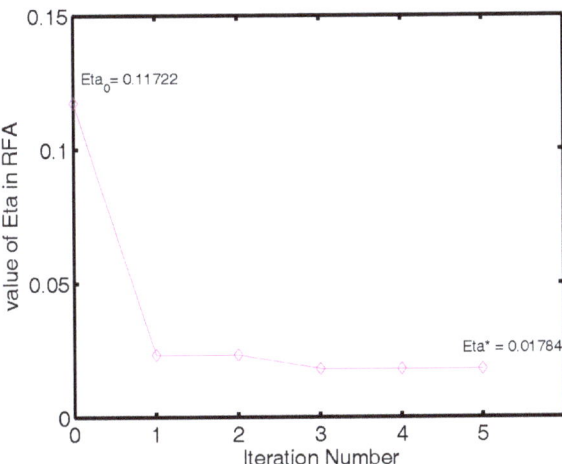

Figure 9. Changing cave of η over the iteration number on the PIE dataset.

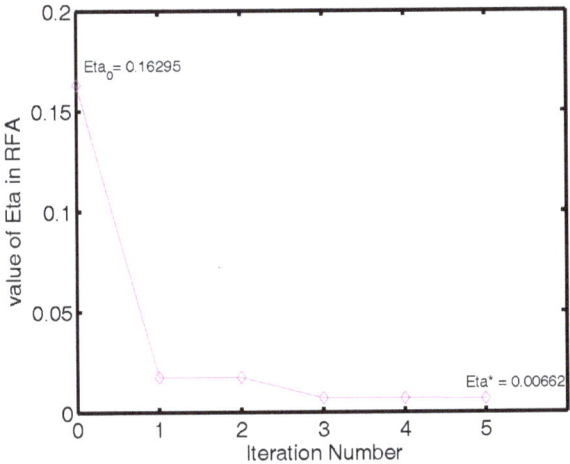

Figure 10. Changing cave of η over the iteration number on the YaleB dataset.

5.1.3. Comparison on Other UCI Datasets

To evaluate the generalization ability of RFA, we conducted experiments on UCI datasets of other fields. The details of the used datasets are shown in Table 8. For the corresponding experimental settings of these four dataset, we set the number of the nearest neighbors – k to 15. For the fairness of comparison, the subspace dimension of each method was set to $C - 1$.

Table 8. Statistics of the UCI datasets.

Dataset	Classes	Samples	Dimensions
Shuttle	7	14,516	9
Thyroid	3	215	5
Vowel	11	990	10
Waveform21	3	5000	21

The results shown in Table 9 demonstrate the advantage of RFA over the related approaches. It is very effective in a wide range of applications.

Table 9. Classification results obtained on several UCI datasets.

Dimension	LDA	MFA	PRPCA	RDA	RFA
Shuttle	0.9975 ± 0.0009	0.9979 ± 0.0012	0.9959 ± 0.0037	0.9162 ± 0.0132	**0.9981 ± 0.0003**
Thyroid	0.9395 ± 0.0265	0.9488 ± 0.0303	0.9488 ± 0.0195	0.9209 ± 0.0265	**0.9581 ± 0.0255**
Vowel	0.9859 ± 0.0109	0.9869 ± 0.0085	0.9859 ± 0.0090	0.3505 ± 0.0349	**0.9879 ± 0.0077**
Waveform21	0.8184 ± 0.0119	0.8258 ± 0.0197	0.8150 ± 0.0118	0.5972 ± 0.0279	**0.8278 ± 0.0175**

5.1.4. Comparison on Document Classification and Webpage Classification Problems

As a general dimensionality reduction framework, the relational matrix can be constructed with different strategies. In the previous sections, we considered the relationship between samples based on their class labels or similarity. However, in some complicated problems, relationships may presented in other forms. For example, as indicated in [25], if there is a reference relationship between two papers, they are likely to have the same topic. However, due to the sparse nature of the bag-of-words representation, the similarity between these two papers may be very low. Thus, to further testify RFA, we designed a relational matrix based on the citation relevance between data samples to model RFA, and tested its effectiveness on document classification and webpage classification problems.

For this experiment, we used two datasets, citeseer and WebKB (https://linqs-data.soe.ucsc.edu/public/lbc/, accessed on 19 October 2023). We note that the WebKB dataset contains four subsets: Cornell, Texas, Washington and Wisconsin, and we show the experimental results of these four subsets separately. Each dataset contains bag-of-words representation of documents or webpages and citation links between the instances. Citeseer contains 3312 scientific documents from 6 different classes, and there are 4732 citation relation between the documents. WebKB consists of 877 webpages from 5 different classes, and there are 1608 page links within this dataset. We adopted the same strategy as in PRPCA to construct the relational matrix:

(1) Constructing the adjacent graph $\mathbf{A}$ according to the relevance between data samples. If there was a citation or link between sample i and j, then $\mathbf{A}_{ij} = 1$; else, $\mathbf{A}_{ij} = 0$.

(2) Letting $\tilde{\mathbf{D}}_{ii} = \sum_j \mathbf{A}_{ij} = (\mathbf{AA})_{ij}$, then $\mathbf{B} = \mathbf{AA} - \tilde{\mathbf{D}}$,

$$\mathbf{B}_{ij} = \begin{cases} (\mathbf{AA})_{ij} = \sum_{k=1}^{N} \mathbf{A}_{ij}\mathbf{A}_{kj}, & \text{if } i \neq j, \\ 0, & \text{otherwise.} \end{cases} \quad (25)$$

(3) Defining $\mathbf{G} = 2\mathbf{A} + \mathbf{B}$ as the relational matrix in RFA.

We took PRPCA as the baseline method in this part. Experimental results are illustrated in Figure 11; we can see that RFA achieves comparable results with PRPCA in all these five datasets and is even better than PRPCA on some of the datasets.

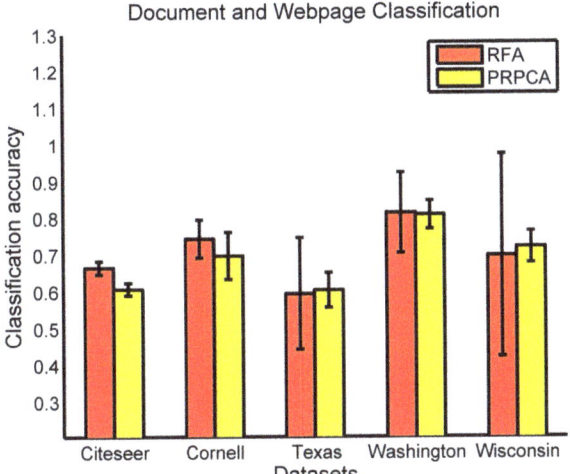

Figure 11. Classification accuracy and standard deviation obtained by RFA and PRPCA on the document and webpage classification problems.

5.2. Performance of KRFA

To evaluate the efficiency of KRFA, we tested its performance on several benchmark datasets from the UCI machine learning repository. The details of these datasets are shown in Table 10. For the corresponding experimental settings, we set the number of the nearest neighbors – k of these five dataset to 15. To avoid the singular value issue, we adopted KPCA to retain 98% of the variance before formally performing KMFA and KRFA. We used Gaussian kernel in the experiment and for the fairness of the comparison, the subspace dimension of each method was set to $C - 1$. Table 11 shows the comparison results obtained by KRFA, KMFA and RFA.

Table 10. Statistics of the UCI datasets.

Dataset	Classes	Samples	Dimensions
Ecoli	8	336	8
Satimage	7	6435	36
Vehicle	4	846	18
Waveform40	3	5000	40
Wine	3	178	13

Table 11. Classification accuracy and standard deviation obtained on several UCI datasets.

Dimension	KMFA	RFA	KRFA
Ecoli	0.4200 ± 0.0431	0.4230 ± 0.0423	**0.4511 ± 0.0648**
Satimage	0.7749 ± 0.0345	0.3915 ± 0.0834	**0.8182 ± 0.0343**
Vehicle	0.7694 ± 0.0152	0.6808 ± 0.0489	**0.7824 ± 0.0187**
Waveform40	0.7996 ± 0.0136	0.8216 ± 0.0097	**0.8256 ± 0.0062**
Wine	0.9716 ± 0.0350	0.9716 ± 0.0286	**0.9773 ± 0.0239**

As shown in Table 11, the proposed KRFA obtained a comparable and even better result than KMFA. Furthermore, experimental results of KRFA were all better than RFA on the used datasets. That superiority can be especially reflected on the Satimage dataset. The performance of RFA on Satimage was unsatisfactory. However, KRFA conducted effective nonlinear dimensionality reduction and thus obtained good result on the following classification problem. These two points clearly demonstrate the nonlinear dimensionality reduction ability of KRFA.

6. Conclusions

In this paper, we propose a novel and general framework named relational Fisher analysis (RFA) which integrates relational information into the dimensionality reduction models. RFA can be effectively optimized with an iterative method based on trace ratio. For nonlinear dimensionality reduction, we adopt kernel trick to RFA and design its kernel version named KRFA. Extensive experiments demonstrate that RFA and KRFA outperform other related dimensionality reduction algorithms in most cases. In future work, we plan to extend this research in the following aspects: (1) Exploiting efficient relationship metric for different relational data to further test the effectiveness of the proposed RFA model; (2) Further extending the formulation of RFA for semi-supervised learning; and (3) Extending RFA for tensor representation learning and applying it to tensor analysis problems.

Author Contributions: Conceptualization, L.-N.W. and G.Z.; methodology, L.-N.W. and M.C.; software, Y.S.; validation, L.-N.W. and M.C.; formal analysis, G.Z.; investigation, L.-N.W.; resources, G.Z.; data curation, Y.S.; writing—original draft preparation, L.-N.W., Y.S. and G.Z.; writing—review and editing, M.C.; visualization, Y.S.; supervision, G.Z.; project administration, M.C.; funding acquisition, G.Z. All authors have read and agreed to the published version of the manuscript.

Funding: This work was partially supported by the National Key Research and Development Program of China under Grant No. 2018AAA0100400, HY Project under Grant No. LZY2022033004, the Natural Science Foundation of Shandong Province under Grants No. ZR2020MF131 and No. ZR2021ZD19, the Science and Technology Program of Qingdao under Grant No. 21-1-4-ny-19-nsh, and Project of Associative Training of Ocean University of China under Grant No. 202265007.

Institutional Review Board Statement: Not applicable.

Informed Consent Statement: Not applicable.

Data Availability Statement: The data used in this work were public on the Internet and no new data were created.

Acknowledgments: We thank "Qingdao AI Computing Center" and "Eco-Innovation Center" for providing inclusive computing power and technical support of MindSpore during the completion of this paper.

Conflicts of Interest: The authors declare no conflict of interest.

References

1. Jolliffe, I. *Principal Component Analysis*; Wiley Online Library: Hoboken, NJ, USA, 2002.
2. Martínez, A.M.; Kak, A.C. Pca versus lda. *IEEE Trans. Pattern Anal. Mach. Intell.* **2001**, *23*, 228–233. [CrossRef]
3. Turk, M.A.; Pentland, A.P. Face recognition using eigenfaces. In Proceedings of the 1991 IEEE Computer Society Conference on Computer Vision and Pattern Recognition, Maui, HI, USA, 3–6 June 1991; pp. 586–591.
4. Fukunaga, K. *Introduction to Statistical Pattern Recognition*; Academic Press: Cambridge, MA, USA, 2013.
5. Ye, J.; Janardan, R.; Park, C.H.; Park, H. An optimization criterion for generalized discriminant analysis on undersampled problems. *IEEE Trans. Pattern Anal. Mach. Intell.* **2004**, *26*, 982–994. [PubMed]
6. Lu, J.; Plataniotis, K.N.; Venetsanopoulos, A.N. Face recognition using LDA-based algorithms. *IEEE Trans. Neural Netw.* **2003**, *14*, 195–200. [PubMed]
7. Fisher, R.A. The use of multiple measurements in taxonomic problems. *Ann. Eugen.* **1936**, *7*, 179–188. [CrossRef]
8. Schölkopf, B.; Smola, A.J.; Müller, K. Nonlinear Component Analysis as a Kernel Eigenvalue Problem. *Neural Comput.* **1998**, *10*, 1299–1319. [CrossRef]
9. Baudat, G.; Anouar, F. Generalized Discriminant Analysis Using a Kernel Approach. *Neural Comput.* **2000**, *12*, 2385–2404. [CrossRef]
10. Zhao, H.; Wang, Z.; Nie, F. A New Formulation of Linear Discriminant Analysis for Robust Dimensionality Reduction. *IEEE Trans. Knowl. Data Eng.* **2018**, *31*, 629–640. [CrossRef]
11. Tenenbaum, J.B.; de Silva, V.; Langford, J.C. A Global Geometric Framework for Nonlinear Dimensionality Reduction. *Science* **2000**, *290*, 2319–2323. [CrossRef]

12. Roweis, S.T.; Saul, L.K. Nonlinear Dimensionality Reduction by Locally Linear Embedding. *Science* **2000**, *290*, 2323–2326. [CrossRef]
13. Belkin, M.; Niyogi, P. Laplacian Eigenmaps for Dimensionality Reduction and Data Representation. *Neural Comput.* **2003**, *15*, 1373–1396. [CrossRef]
14. Yan, S.; Xu, D.; Zhang, B.; Zhang, H.J.; Yang, Q.; Lin, S. Graph Embedding and Extensions: A General Framework for Dimensionality Reduction. *IEEE Trans. Pattern Anal. Mach. Intell.* **2007**, *29*, 40–51. [CrossRef] [PubMed]
15. He, X.; Niyogi, P. Locality Preserving Projections. In Proceedings of the 16th International Conference on Neural Information Processing Systems, Whistler, BC, Canada, 9–11 December 2003; pp. 153–160.
16. Wong, W.K.; Zhao, H. Supervised optimal locality preserving projection. *Pattern Recognit.* **2012**, *45*, 186–197. [CrossRef]
17. Chen, J.; Liu, Y. Locally linear embedding: A survey. *Artif. Intell. Rev.* **2011**, *36*, 29–48. [CrossRef]
18. Wang, Q.; Chen, K. Zero-Shot Visual Recognition via Bidirectional Latent Embedding. *Int. J. Comput. Vis.* **2017**, *124*, 356–383. [CrossRef]
19. Chang, J.; Blei, D.M. Relational Topic Models for Document Networks. *J. Mach. Learn. Res. Proc. Track* **2009**, *5*, 81–88.
20. Chang, J.; Boyd-Graber, J.L.; Blei, D.M. Connections between the Lines: Augmenting Social Networks with Text. In Proceedings of the 15th ACM SIGKDD International Conference on Knowledge Discovery and Data Mining, Paris, France, 28 June–1 July 2009; pp. 169–178.
21. Peel, L. Graph-based semi-supervised learning for relational networks. *arXiv* **2016**, arXiv:1612.05001.
22. Yang, Z.; Cohen, W.; Salakhutdinov, R. Revisiting semi-supervised learning with graph embeddings. *arXiv* **2016**, arXiv:1603.08861.
23. Weston, J.; Ratle, F.; Mobahi, H.; Collobert, R. Deep learning via semi-supervised embedding. In *Neural Networks: Tricks of the Trade*; Springer: Berlin/Heidelberg, Germany, 2012; pp. 639–655.
24. Duin, R.P.W.; Pekalska, E.; de Ridder, D. Relational Discriminant Analysis. *Pattern Recognit. Lett.* **1999**, *20*, 1175–1181. [CrossRef]
25. Li, W.J.; Yeung, D.Y.; Zhang, Z. Probabilistic Relational PCA. In Proceedings of the 22nd International Conference on Neural Information Processing Systems, Vancouver, BC, Canada, 7–10 December 2009; pp. 1123–1131.
26. Gao, K.; Liu, B.; Yu, X.; Qin, J.; Zhang, P.; Tan, X. Deep Relation Network for Hyperspectral Image Few-Shot Classification. *Remote Sens.* **2020**, *12*, 923. [CrossRef]
27. Chen, S.; Yao, T.; Chen, Y.; Ding, S.; Li, J.; Ji, R. Local Relation Learning for Face Forgery Detection. In Proceedings of the Thirty-Fifth AAAI Conference on Artificial Intelligence, Virtually, 2–9 February 2021; pp. 1081–1088.
28. Cho, M.; Kim, M.; Hwang, S.; Park, C.; Lee, K.; Lee, S. Look Around for Anomalies: Weakly-Supervised Anomaly Detection via Context-Motion Relational Learning. In Proceedings of the 2023 IEEE/CVF Conference on Computer Vision and Pattern Recognition (CVPR), Vancouver, BC, Canada, 17–24 June 2023; pp. 12137–12146.
29. Ma, R.; Mei, B.; Ma, Y.; Zhang, H.; Liu, M.; Zhao, L. One-shot relational learning for extrapolation reasoning on temporal knowledge graphs. *Data Min. Knowl. Discov.* **2023**, *37*, 1591–1608. [CrossRef]
30. Li, X.; Ma, J.; Yu, J.; Zhao, M.; Yu, M.; Liu, H.; Ding, W.; Yu, R. A structure-enhanced generative adversarial network for knowledge graph zero-shot relational learning. *Inf. Sci.* **2023**, *629*, 169–183. [CrossRef]
31. Zhong, G.; Shi, Y.; Cheriet, M. Relational Fisher analysis: A general framework for dimensionality reduction. In Proceedings of the 2016 International Joint Conference on Neural Networks (IJCNN), Vancouver, BC, Canada, 24–29 July 2016; pp. 2244–2251.
32. He, X.; Yan, S.; Hu, Y.; Zhang, H.J. Learning a Locality Preserving Subspace for Visual Recognition. In Proceedings of the IEEE International Conference on Computer Vision, Nice, France, 13–16 October 2003; Volume 1, pp. 385–392.
33. Wang, H.; Yan, S.; Xu, D.; Tang, X.; Huang, T.S. Trace Ratio vs. Ratio Trace for Dimensionality Reduction. In Proceedings of the 2007 IEEE Computer Society Conference on Computer Vision and Pattern Recognition (CVPR 2007), Minneapolis, MN, USA, 18–23 June 2007.
34. Nie, F.; Xiang, S.; Jia, Y.; Zhang, C.; Yan, S. Trace Ratio Criterion for Feature Selection. In Proceedings of the Twenty-Third AAAI Conference on Artificial Intelligence, Chicago, IL, USA, 13–17 July 2008; Volume 2, pp. 671–676.
35. Zhong, G.; Ling, X. The necessary and sufficient conditions for the existence of the optimal solution of trace ratio problems. In Proceedings of the Chinese Conference on Pattern Recognition, Chengdu, China, 5–7 November 2016; Springer: Berlin/Heidelberg, Germany, 2016; pp. 742–751.
36. Duin, R.P.; de Ridder, D.; Tax, D.M. Experiments with a featureless approach to pattern recognition. *Pattern Recognit. Lett.* **1997**, *18*, 1159–1166. [CrossRef]
37. Duin, R.P. Relational discriminant analysis and its large sample size problem. In Proceedings of the 14th International Conference on Pattern Recognition, Washington, DC, USA, 16–20 August 1998; Volume 1, pp. 445–449.
38. Ypma, A.; Duin, R.P. Support objects for domain approximation. In Proceedings of the 8th International Conference on Artificial Neural Networks, Skövde, Sweden, 2–4 September 1998; Springer: Berlin/Heidelberg, Germany, 1998; pp. 719–724.
39. Borg, I.; Groenen, P.J. *Modern Multidimensional Scaling: Theory and Applications*; Springer Science & Business Media: Berlin/Heidelberg, Germany, 2005.
40. Sammon, J., Jr. A nonlinear structure analysis mapping for data. *IEEE Trans. Comp.* **1969**, *C-18*, 401–409.
41. Niemann, H. Linear and nonlinear mapping of patterns. *Pattern Recognit.* **1980**, *12*, 83–87. [CrossRef]
42. Paccanaro, A.; Hinton, G.E. Learning Distributed Representations of Concepts Using Linear Relational Embedding. *IEEE Trans. Knowl. Data Eng.* **2001**, *13*, 232–244. [CrossRef]

43. Wang, H.; Li, W. Relational Collaborative Topic Regression for Recommender Systems. *IEEE Trans. Knowl. Data Eng.* **2015**, *27*, 1343–1355. [CrossRef]
44. Xuan, J.; Lu, J.; Zhang, G.; Xu, R.Y.D.; Luo, X. Bayesian Nonparametric Relational Topic Model through Dependent Gamma Processes. *IEEE Trans. Knowl. Data Eng.* **2017**, *29*, 1357–1369. [CrossRef]
45. Jia, Y.; Nie, F.; Zhang, C. Trace Ratio Problem Revisited. *IEEE Trans. Neural Networks* **2009**, *20*, 729–735.
46. Xiang, S.; Nie, F.; Zhang, C. Learning A Mahalanobis Distance Metric for Data Clustering and Classification. *Pattern Recognit.* **2008**, *41*, 3600–3612. [CrossRef]
47. Nie, F.; Xiang, S.; Jia, Y.; Zhang, C. Semi-supervised Orthogonal Discriminant Analysis via Label Propagation. *Pattern Recognit.* **2009**, *42*, 2615–2627. [CrossRef]
48. Zhao, M.B.; Zhang, Z.; Chow, T.W.S. Trace Ratio Criterion Based Generalized Discriminative Learning for Semi-supervised Dimensionality Reduction. *Pattern Recognit.* **2012**, *45*, 1482–1499. [CrossRef]
49. Moghaddam, R.F.; Cheriet, M.; Milo, T.; Wisnovsky, R. A Prototype System for Handwritten Sub-word Recognition: Toward Arabic-manuscript Transliteration. In Proceedings of the 2012 11th International Conference on Information Science, Signal Processing and Their Applications (ISSPA), Montreal, QC, Canada, 2–5 July 2012; pp. 1198–1204.
50. Lichman, M. UCI Machine Learning Repository, 2013. Available online: https://archive.ics.uci.edu/ (accessed on 19 October 2023).
51. Gross, R. Face Databases. In *Handbook of Face Recognition*; Li, S.Z., Jain, A.K., Eds.; Springer: Berlin/Heidelberg, Germany, 2005.
52. Cai, D.; He, X.; Hu, Y.; Han, J.; Huang, T. Learning a Spatially Smooth Subspace for Face Recognition. In Proceedings of the 2007 IEEE Computer Society Conference on Computer Vision and Pattern Recognition (CVPR 2007), Minneapolis, MN, USA, 18–23 June 2007.
53. Cai, D.; He, X.; Han, J. Spectral Regression for Efficient Regularized Subspace Learning. In Proceedings of the 11th International Conference on Computer Vision, ICCV 2007, Rio de Janeiro, Brazil, 14–20 October 2007.
54. Cai, D.; He, X.; Han, J.; Zhang, H.J. Orthogonal Laplacianfaces for face recognition. *IEEE Trans. Image Process.* **2006**, *15*, 3608–3614. [CrossRef]
55. Zheng, Y.; Cai, Y.; Zhong, G.; Chherawala, Y.; Shi, Y.; Dong, J. Stretching Deep Architectures for Text Recognition. In Proceedings of the 2015 13th International Conference on Document Analysis and Recognition (ICDAR), Tunis, Tunisia, 23–26 August 2015; pp. 236–240.
56. Srivastava, A.; Klassen, E.; Joshi, S.H.; Jermyn, I.H. Shape Analysis of Elastic Curves in Euclidean Spaces. *IEEE Trans. Pattern Anal. Mach. Intell.* **2011**, *33*, 1415–1428. [CrossRef]
57. Garris, M.D.; Blue, J.L.; Candela, G.T.; Dimmick, D.L.; Geist, J.; Grother, P.J.; Janet, S.A.; Wilson, C.L. *NIST Form-Based Handprint Recognition System*; NIST Interagency/Internal Report (NISTIR); National Institute of Standards and Technology: Gaithersburg, MD, USA, 1994.

Disclaimer/Publisher's Note: The statements, opinions and data contained in all publications are solely those of the individual author(s) and contributor(s) and not of MDPI and/or the editor(s). MDPI and/or the editor(s) disclaim responsibility for any injury to people or property resulting from any ideas, methods, instructions or products referred to in the content.

Article

A New Algorithm for Detecting GPN Protein Expression and Overexpression of IDC and ILC Her2+ Subtypes on Polyacrylamide Gels Associated with Breast Cancer

Jorge Juarez-Lucero [1], Maria Guevara-Villa [2], Anabel Sanchez-Sanchez [1,*], Raquel Diaz-Hernandez [1] and Leopoldo Altamirano-Robles [1]

1. Instituto Nacional de Astrofísica Optica y Electronica, Luis Enrique Erro # 1, Tonantzintla, Puebla 72840, Mexico; jjlucero@inaoep.mx (J.J.-L.); raqueld@inaoep.mx (R.D.-H.); robles@inaoep.mx (L.A.-R.)
2. Faculty of Architecture, Meritorious Autonomous University of Puebla, 4 Sur 104 Centro Histórico, Puebla 72000, Mexico; maria.guevaravilla@correo.buap.mx
* Correspondence: anabel@inaoep.mx; Tel.: +52-222-266-3100 (ext. 2113)

Citation: Juarez-Lucero, J.; Guevara-Villa, M.; Sanchez-Sanchez, A.; Diaz-Hernandez, R.; Altamirano-Robles, L. A New Algorithm for Detecting GPN Protein Expression and Overexpression of IDC and ILC Her2+ Subtypes on Polyacrylamide Gels Associated with Breast Cancer. *Algorithms* **2024**, *17*, 149. https://doi.org/10.3390/a17040149

Academic Editors: Chih-Lung Lin, Bor-Jiunn Hwang, Shaou-Gang Miaou and Yuan-Kai Wang

Received: 24 January 2024
Revised: 22 March 2024
Accepted: 26 March 2024
Published: 2 April 2024

Copyright: © 2024 by the authors. Licensee MDPI, Basel, Switzerland. This article is an open access article distributed under the terms and conditions of the Creative Commons Attribution (CC BY) license (https://creativecommons.org/licenses/by/4.0/).

Abstract: Sodium dodecyl sulfate–polyacrylamide gel electrophoresis (SDS-PAGE) is used to identify protein presence, absence, or overexpression and usually, their interpretation is visual. Some published methods can localize the position of proteins using image analysis on images of SDS-PAGE gels. However, they cannot automatically determine a particular protein band's concentration or molecular weight. In this article, a new methodology to identify the number of samples present in an SDS-PAGE gel and the molecular weight of the recombinant protein is developed. SDS-PAGE images of different concentrations of pure GPN protein were created to produce homogeneous gels. Then, these images were analyzed using the developed methodology called Image Profile Based on Binarized Image Segmentation (IPBBIS). It is based on detecting the maximum intensity values of the analyzed bands and produces the segmentation of images filtered by a binary mask. The IPBBIS was developed to identify the number of samples in an SDS-PAGE gel and the molecular weight of the recombinant protein of interest, with a margin of error of 3.35%. An accuracy of 0.9850521 was achieved for homogeneous gels and 0.91736 for heterogeneous gels of low quality.

Keywords: sodium dodecyl sulfate–polyacrylamide gel electrophoresis; image analysis; protein band; molecular weight; image segmentation; binary mask

1. Introduction

Among the techniques used to identify the presence, absence, or overexpression of proteins of biological interest, sodium dodecyl sulfate–polyacrylamide gel electrophoresis (SDS-PAGE) is present. This technique separates proteins by applying an electric field to the gel. The proteins move through the gel and are retained in different positions according to molecular weight. The gel is stained using Coomassie blue. Each of the samples analyzed is represented in columns (lanes), and the horizontal bands correspond to the proteins detected in each piece. Thicker bands indicate higher protein concentration [1] (see Figure 1).

The gels obtained are used to visually find proteins and gene fragments in DNA gels [2–4] or as a method for disease diagnosis [3–6]. Unlike DNA gels, protein gels can contain many bands per sample, making them difficult for the human eye to interpret (see lane 4 in Figure 1). As a result, misinterpretations of protein gels can occur due to errors generated by optical illusions, visual sensitivity, or fatigue [3,5–7].

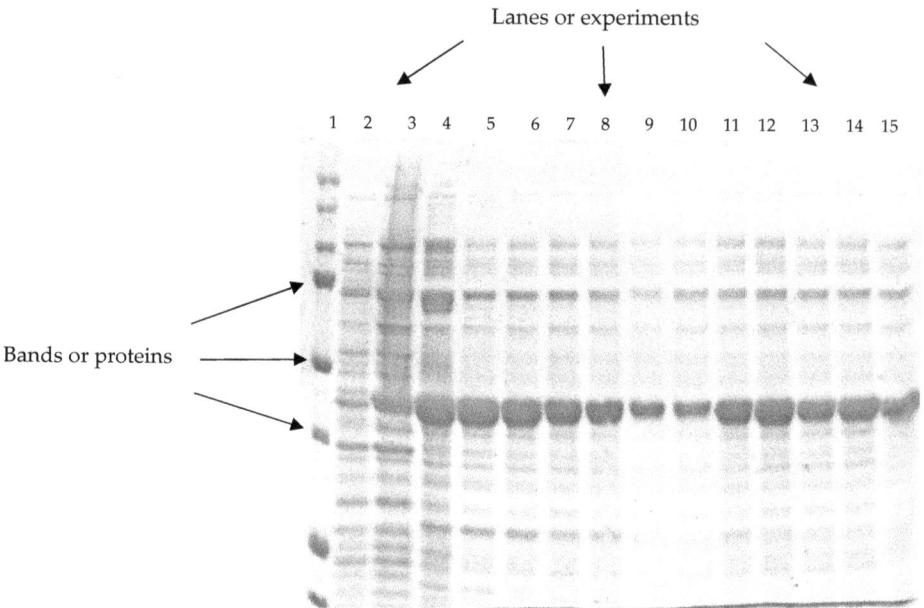

Figure 1. Image of a polyacrylamide protein gel. The vertical columns or lanes represent different experiments placed within the gel (numbered 1 to 15). The horizontal lines or bands represent the proteins identified per column.

Image analysis has been used to interpret the DNA and protein bands in the images of SDS-PAGE gels. These analyses include background noise removal, lane detection, and segmentation of the lanes to analyze the protein or gene sought [8,9].

Many techniques have been used to remove background noise, such as contourlet and wavelet transforms, top-hat transforms, Gaussian low-pass filters, normalization, intensity shifts, median filters, adaptive thresholding, non-linear Gaussians, Fourier analysis, fuzzy-c-means, and some convolution matrix and image slices to search for regions of interest. These methods have been used individually or in combination to obtain improved and noise-free image profiles to extract gel features [1,3–5,7,8,10–20].

Several tools have been employed to detect lanes and segmentation of the bands [1,3–5,7,8,10–16,18–22]. These techniques involve various methods to improve segmentation by selecting necessary pixels related to background contrast. They include edge detection using Bayesian approximations, thresholding, or Otsu segmentation. Users can choose a region of interest to reduce size, minimize noise, calculate standard deviation, and manually set a threshold to generate the gel profile. Division of the area between user-specified lanes can be achieved using Gaussian functions or templates. Sobel filters can be applied to identify lanes and bands in the gel.

Additionally, Gaussian processes or templates can split the region between user-specified lanes. Sobel filters can also detect peaks and troughs in the intensity profiles, aiding in identifying the gaps between lanes. Brightness changes can be measured, and the number of pixels in the lanes can be counted to determine their distances using variations in grey levels. Spectral density can be calculated to average the width of lanes, allowing for the selection of specific regions within the image. An analysis is also performed on the profile of the peaks through their areas to group them using techniques such as K-means, and the bands are delimited with ellipses to avoid their intersection and to determine the separation of the lanes. Among all the proposed methods, the ones that have shown the best results include the user's choice of regions of interest to reduce noise and generate more accurate profiles, which facilitates the identification of the minima related to the separation of lanes in the gel.

Currently, programs such as Scanalytics, GelcomparII, GelJ, Gel-Pro Analyzer, ImageJ, PyElph, TotalLab, PDQuest, Proteomweaver, Dcyder 2D, imageMaster, Melanie, BioNumerics, Redfin, Gel IQ, Z3, and Delta2D Flicker are used to analyze images of DNA or protein gels [1,10,23–29]. These programs employ semi-automatic filters to remove noise, meaning the operator must manually select the column or band of interest. The user also adjusts the intensity changes and decides the threshold that reduces the background noise. However, due to the dependence on analysts with little knowledge of intensities and thresholding, this often results in poor gel analysis.

In this article, the methodology described in reference [30] is used to obtain different concentrations of pure GPN protein added to a bacterial cell extract to create SDS-PAGE gels. The incorrect or excessive expression of some proteins may be associated with an imbalance in health. So, this research produces gels representing patient samples with different amounts of protein expression, including absence and different levels of overexpression, to emulate various stages of Invasive Ductal Carcinoma (IDC) and Invasive Lobular Carcinoma (ILC) Her2+ breast cancer. Therefore, identifying specific proteins can be helpful in disease diagnosis. Examples of such proteins are GTPases, such as Rho, Rab27, and GPN, which have been linked to the development of breast cancer [31]. Reference [30], an article by the same authors as the present article, describes a new methodology for obtaining pure GPN protein in high levels in homogeneous form. In the current research, an algorithm was proposed for a new method to identify the lanes and bands of samples present in an SDS-PAGE gel and the molecular weight, to identify the overexpressed protein related to breast cancer. For that, the obtained images were analyzed using the newly developed image analysis methodology to identify different levels of expression of the GPN protein expressed per sample. This methodology is based on detecting the pixel values of the white color of the histogram segmentation of images filtered by a binary mask.

1.1. Novelty

Current investigations to search proteins in SDS-PAGE gels require manually selecting the area corresponding to the protein of interest to locate the separation between proteins by processing. The methodology proposed in this article automatically identifies the protein of interest and automatically determines its overexpression levels, and has not been presented in other works.

1.2. Limitations and Challenges

Although the algorithm performs best in finding the protein of interest automatically, the gel image must be of a minimum good quality, otherwise, the algorithm will fail. A database of polyacrylamide gel does not exist to apply and train a neural network to find the overexpressed protein. It is difficult to obtain samples of various stages of IDC and ILC Her2+ breast cancer.

This article is segmented as follows: Section 1 shows the advances in SDS-PAGE gel image analysis to identify proteins and their overexpression. Section 2 describes the procedure IPBBIS for obtaining overexpression gels. The IPBBIS method includes preprocessing techniques, diagrams, and pseudocodes that determine the number of samples present in a gel, the protein bands, and their overexpression. Section 3 details the results obtained and their discussion. At the end for Section 4 of the article, conclusions and future work are described.

2. Materials and Methods

2.1. Creation of Samples with Different GPN Concentrations

SDS-PAGE images corresponding to samples of different concentrations were obtained to replicate or emulate the GPN protein overexpression in vivo during the involvement of IDC and ILC Her2+ breast cancer [31]. For this activity, the purification methodology shown in [30] was followed. Images of the gels are shown in Figure 2A,B.

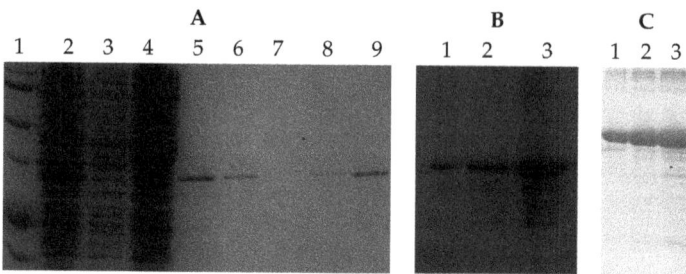

Figure 2. (**A**) Purification of recombinant GPN protein from *Escherichia coli* bacteria. Lane 1: Molecular weight control. Lane 2: Positive control of GPN protein expression. Lane 3: Negative expression control. Lane 4: Total protein extract. Lanes 5–9: purified GPN protein. (**B**) Different concentrations of purified GPN protein. (**C**) Different concentrations of BSA protein.

In addition, to obtain SDS-PAGE images from controlled concentrations, a dilution of bovine serum albumin (BSA) protein obtained from the Bio-Rad Protein Assay kit was prepared to get three samples with the following concentrations: 2 mg/mL, 1 mg/mL, and 0.5 mg/mL (lanes 1, 2, and 3, respectively, in Figure 2C). This experiment demonstrated that the methodology proposed here also achieves the set goal for other protein classes.

Samples numbering 2230 were used, including endogenous *Escherichia coli* proteins with random addition of GPN at different concentrations. Two minibatches were used from this dataset. The first one, 1561 samples, was called heterogeneous because the samples presented different degrees of staining, smiley face effects. or curved lines, and even SDS-PAGE breaks. The second minibatch comprised 669 homogeneous samples because the gel characteristics did not vary. For each minibatch, 70% was used for training and the rest for testing.

2.2. Image Acquisition

SDS-PAE images were obtained with a Gel Doc XR+ photo documenter system based on CCD high resolution, using image Lab Software to capture pictures following the specifications of the supplier Bio-Rad. The resolution of the images was 4 megapixels, and the pixel density was 4096 ppi [32].

2.3. Preprocessing and Feature Extraction

Preprocessing and feature extraction were carried out through the identification of the number of white pixels by an image segmentation histogram using a binary mask (profile-based image segmentation, algorithm showed in Figure 3) as follows:

Size adjustment: The SDS-PAGE images obtained from the Gel Doc XR+ photo documenter system, were resized to images of 600 by 400 pixels size. Image equalization was performed if the samples in the gels had a high protein concentration. Subsequently, all images were binarized and dilated, as shown in Figure 3. The images were inverted after an erosion operation was applied.

Analysis of lanes and bands: A binary mask with a dimension of 1 pixel wide (MAXWIDE variable) and 400 pixels high (MAXHIGH variable) was used for the lanes. For the study of the bands present per sample, the value of 1 pixel wide for the MAXWIDE variable was used, assigning 50 pixels for the MAXHIGH variable of the binary mask. The mask was displaced pixel by pixel across the entire image width, that is, 600 data (one for each pixel for lanes) or 400 data (one for each pixel for bands).

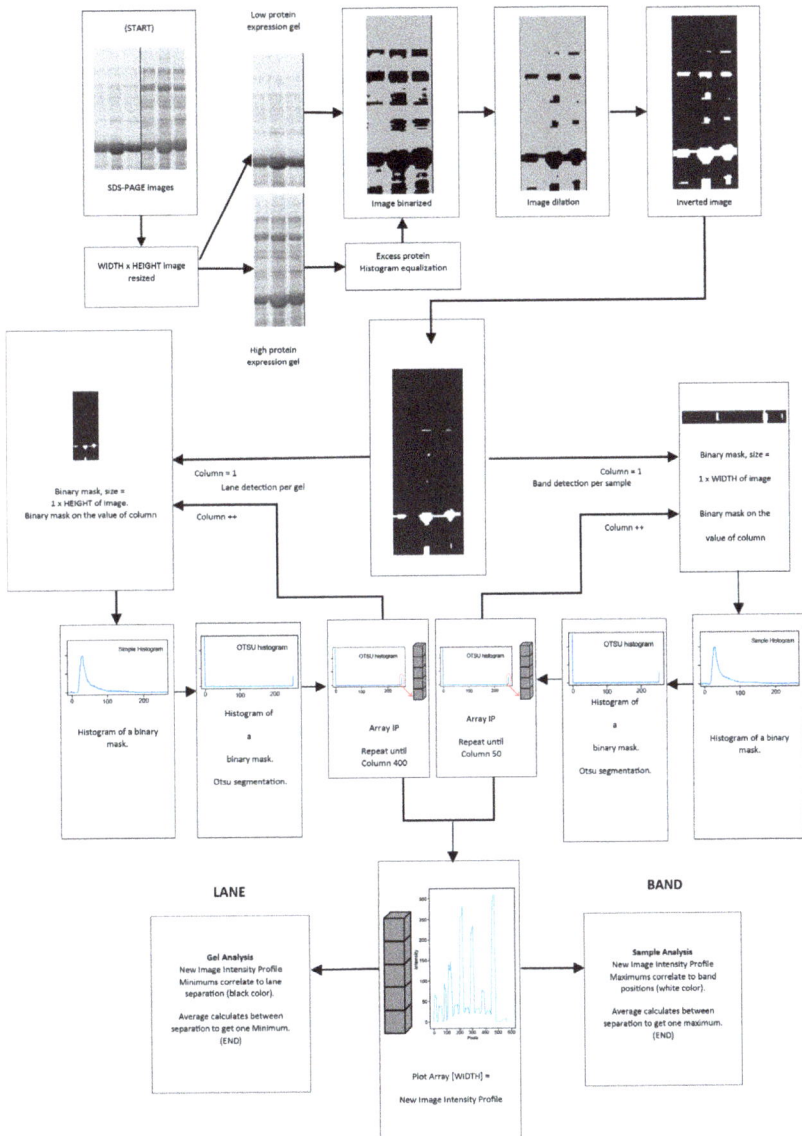

Figure 3. Preprocessing and feature extraction to analyze the complete gel image.

Application of the binary mask: The histogram of the image region delimited by the binary mask was calculated. Since the bands in this process are white, the value corresponding to the number of white pixels in the histogram of the binarized image (position 255 of the histogram) was used and stored in an array. The array data were plotted and called the "new image profile".

Interpretation of the "new image profile": For the analysis of the full SDS-PAGE image, the multiple minimum values present in the new image profile are related to the separation between every lane of the gel image. In this region, the number of white pixel values in the array obtained decreases, which helps find the number of lanes in the image. On the other hand, when the analysis was performed for each lane, the multiple maximum values of the new profile of the image were related to the position of the proteins, as they are represented in white. Places of maximum intensity represent the presence of proteins, and places of lower intensity represent the absence of proteins (between two protein separations, the average was calculated to obtain a single maximum), as shown in Figure 3.

This procedure uses several parameters. Nonetheless, more of them are set in the program and worked well for most of the experiments carried out in this research. The value of the structuring element for erosion and dilation was 25 for lanes and 3 for bands, which allowed the detection of the total number of samples and bands per protein per sample present in an SDS-PAGE gel. The threshold values for the segmentation operation were calculated using the OTSU method. The program was developed in Python with the Pytorch framework, OpenCV to evaluate the images and histograms, and numpy with matplotlib to calculate the graphs.

Figure 3 presents an overall summary of the preprocessing and feature extraction. The pseudocode is shown in Algorithm 1.

Algorithm 1. Pseudocode to find the number of lanes and bands in the polyacrylamide gel image.

Algorithm for band and lane detection
1: Resize the image to 600 × 400 px for light processing
2: **if** Excess_of_protein:
3:     Histogram equalization
4: **end if**
5: Obtain a binarized Image
6: Image dilation
7: Image invert
8: Image erosion
9: Column = 1
10: **If** Lane detection:
11:     MAXWIDE = 400 px
12: **else**:   # band detection
13:     MAXWIDE = 50 px
14: **end else**
15: **end if**
16: Apply Binary Mask on the resized image
17: Initialize Array to zero
18: **while** Column <= MAXWIDE:
19:     Get the Histogram_of_image
20:     Otsu_Segmentation_Applied_to_Binary_Mask_Size_zone
21:     Get the number of white pixels in the Histogram of the segmented region, Histogram[white_position]
    # get the quantity of white color in the histogram binarized
22:     Array [Column] = Number_White_Pixels_Histogram [255]
23:     Column++
24: **end while**
25: Plott Array
26: **if** Lane_Analysis:
27:     Multiple_Minimum_correlate_Lane_Separation(Array)
28:     Multiple_Maximus_related_Band_Separation(Array)
29: else#band_analysis
30:     Average_Multiples_Maximums_between_separations_To_Get_One_Maximum
31: **end else**
32: **end if**

3. Results and Discussion

3.1. Traditional Analysis of GPN Protein Gels at Different Concentrations

Lanes 5 to 9 of Figure 2A reveal an image of polyacrylamide gel untreated with pure recombinant GPN protein. These lanes show the protein concentration at different values (see Figure 2B). An image of polyacrylamide gel untreated with GPN protein in various concentrations is presented in Figure 4A. The thicker bands (highlighted in red) correspond to the GPN protein with higher concentration.

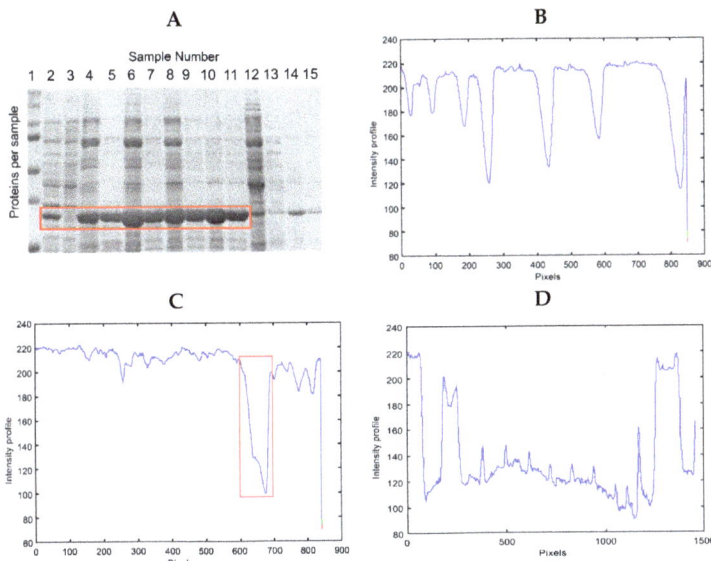

Figure 4. (**A**) SDS-PAGE gel containing GPN protein expressed at different concentrations. Lane 1, molecular weight control; lanes 2–11, GPN at the following concentrations: 2.0, 0.0, 14.5, 9.5, 27, 18, 30, 18.5, 26, and 14 µg/mL, respectively. (**B**) The intensity profile of lane 1 of (**A**) (weight control). (**C**) The intensity profile of lane 4 for (**A**). (**D**) The intensity profile of the bands for the recombinant GPN protein region, highlighted in red from the gel in (**A**).

Images of the SDS-PAGE gels, revealed with Coomassie blue, were analyzed using the intensity profile as shown in Figure 4. The gel image in Figure 4A includes samples of GPN protein at different concentrations (lanes 2 to 11 enclosed in a red box). Preliminary analyses were performed on the image to verify whether the intensity profile plots can detect the presence of the protein and its overexpression. Figure 4B shows the intensity profile of the molecular weight control or ladder (lane 1, Figure 4A). The minimum values of the graph coincide with the position of the protein bands of lane one or the control sample, which is used to determine the molecular weight of the samples analyzed in the rest of the gel. Figure 4C shows the intensity profile of lane 4, where the minimum enclosed in a red box indicates the presence of the protein with the highest expression or concentration within the lane. Finally, Figure 4D shows the intensity profile of the region containing the GPN protein at different concentrations (graph corresponding to the proteins enclosed in the red box in Figure 4A, lanes 2 to 11). The background noise generated by the proteins in the total extract and the different concentrations of GPN in the samples can be seen. The multiple maxima in the graph cannot be related to each different expression of the GPN protein.

3.2. Preprocessing and Feature Extraction Using the Proposed Algorithm

Before analyzing SDS-PAGE images with the new methodology: "Image Profile Based on Binarized Image Segmentation" (IPBBIS), the images were adjusted to the size defined by the variables MAXWIDE = 600 (width in pixels), MAXHIGH = 400 (height in pixels).

The SDS-PAGE protein gel image was converted to greyscale and then binarized. In addition, an erosion operation was performed to increase the spacing between samples and bands (see Figure 5). Next, the IPBBIS method was applied to perform feature extraction.

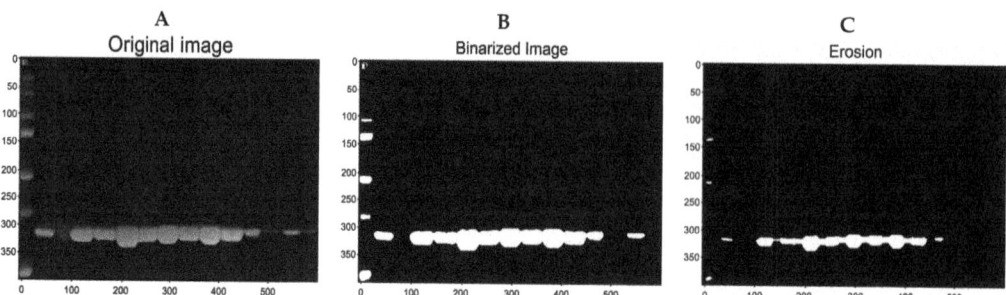

Figure 5. Preprocessing of SDS-PAGE gel images. (**A**) Grayscale image. (**B**) Binarized image. (**C**) Eroded image.

The image from the preprocessing (Figures 5C and 6) was filtered with a binary mask consisting of a matrix of 1×400 pixels (Figure 6B). The IPBBIS method was applied, and different binarization techniques such as Niblack, Sauvola, and Otsu were used. There were no significant variations in the obtained results.

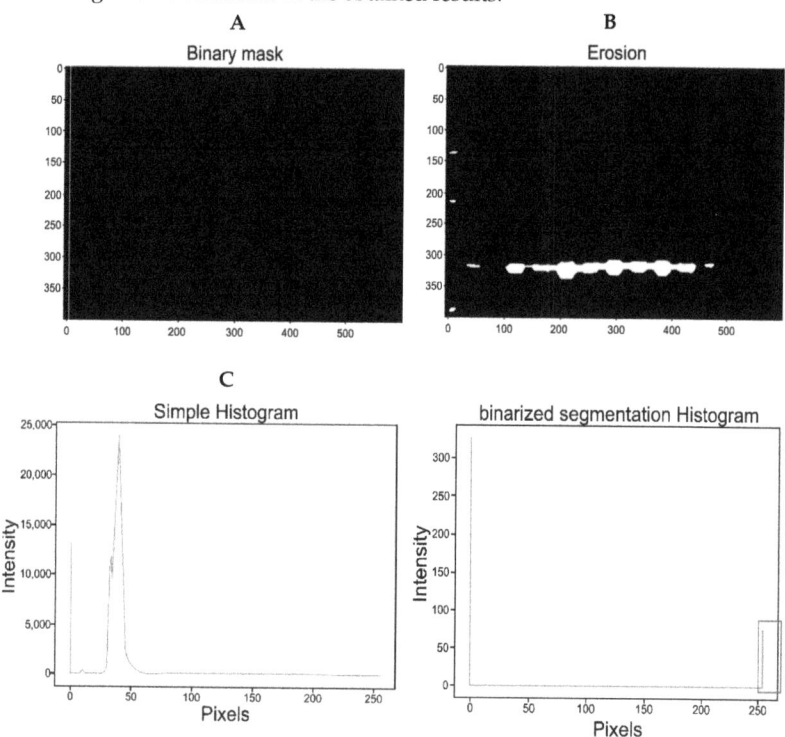

Figure 6. IPBBIS. (**A**) Binarized and eroded image. (**B**) Representation of the binary mask with a size of 1×400 pixels placed at pixel 7. (**C**) A simple histogram of the region contained in the binary mask. (**D**) Histogram of the region after applying Otsu segmentation to (**C**).

By calculating the histogram in the lane region covered by the binary mask (MAXWIDE = $1 \times$ MAXHIGH = 400, for complete gel analysis or MAXWIDE = $1 \times$ MAXHIGH = 50, for band analysis), the pattern in Figure 6C was obtained, which shows the distribution of the pixels. On the other hand, by performing the binary segmentation of the same region (inside the mask), the histogram shown in the graph in Figure 6D was generated. As the image is binarized, this pattern only shows the maximum intensity values for black (pixel 0) and white (pixel 255). The number of white pixels is stored in an array with all gel values selected by the binary mask.

The array obtained by applying the binary mask in Figure 6A generated the new image intensity profile. Figure 7 shows the maxima representing the center of each analyzed band and the multiple minima separating them.

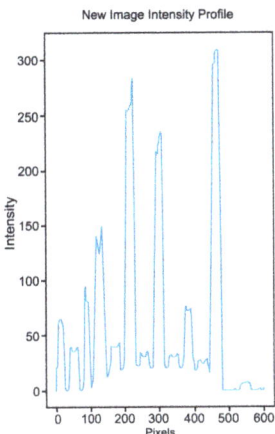

Figure 7. A new image intensity profile was obtained by plotting the array's values containing only the white pixels generated by the IPBBIS method.

3.3. Detection of Protein Overexpression in Gels Using the IPBBIS Algorithm

3.3.1. Application of the New Intensity Profile on the Complete Gel Image

The IPBBIS method was used to identify GPN protein overexpression in the gel shown in Figure 8A. In this Figure, lane 1 contains the ladder or molecular weight control. Lane 2 is the negative control, lane 3 is the positive control for GPN protein expression, lane 4 is the concentrated cell extract control, and lanes 5 to 15 represent extracts with the same amount of endogenous proteins to which different concentrations of recombinant protein have been added. The thickness of the spots indicates a higher concentration of GPN in lanes 5, 6, 11, 12, and 14, while lanes 9 and 10 show a lower concentration of the protein. These values were identified in the plot of the new image intensity profile (Figure 8B,C), where maximum peaks are observed in lanes 5, 6, 11, 12, and 14, and lower peaks correspond to the bands with lower concentrations of recombinant protein, i.e., lower overexpression (lanes 9 and 10). Image analysis was performed by equalizing the image (Figure 8B) and not equalizing it (Figure 8C).

The data of the average values resulting from the multiple maxima of the graph corresponding to the different samples in Figure 8C are shown in Table 1. It can be corroborated that the maximum intensity value was obtained in lane 5, indicating the highest overexpression of GPN protein in that sample. The lowest expressions were recorded in lanes 9 and 10, with maximum values of 23 and 29, respectively. The maxima assigned to lanes 1 to 4 were not shown as they correspond to the ladder, controls for GPN, and GPN with total protein extract expression.

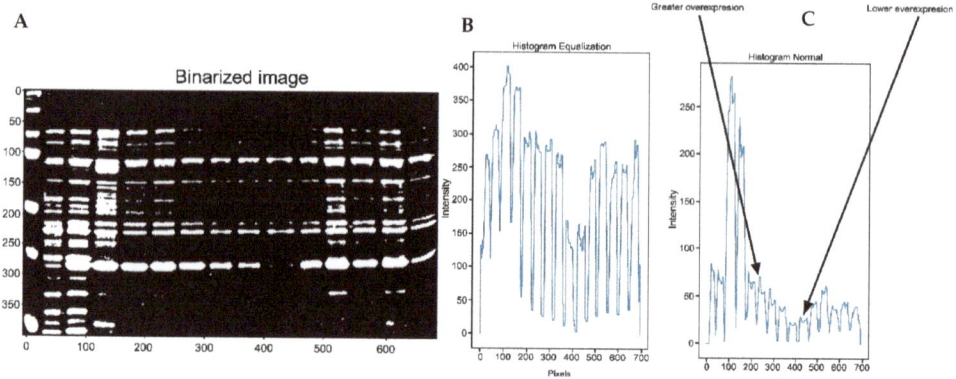

Figure 8. (**A**) A complete gel is shown. (**B**) A plot was generated using the image profile based on binarized image segmentation after image equalization. (**C**) The IPBBIS plot was obtained from the non-equalized Figure 6A.

Table 1. Peak maximum values obtained from the graph in Figure 8C.

Lane	5	6	7	8	9	10	11	12	13	14	15
Value	79	71	58	40	23	29	45	61	41	45	41

3.3.2. Application of the IPBBIS Method on a Sample with Controlled Concentrations

A controlled protein concentration was expressed to evaluate the method's effectiveness. Figure 9A shows the gel obtained, where bandwidth (or stain) increases as the protein is concentrated. Subsequently, the image was pre-treated (binarization, dilation, color inversion, and erosion; Figure 9B) for image analysis using IPBBIS. This procedure originated the new image profile graph shown in Figure 9C, where the maximum peak corresponds to the highest concentration of BSA protein (2 mg/mL). The lowest peak is related to the lowest BSA concentration (0.5 mg/mL).

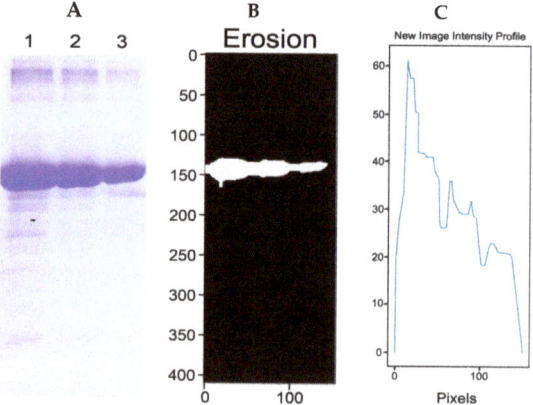

Figure 9. (**A**) SDS-PAGE gel of BSA protein with the concentrations 2 mg/mL (lane 1), 1 mg/mL (lane 2), and 0.5 mg/mL (lane 3). (**B**) Image of binary, dilated, segmented, and eroded (**A**). (**C**) Plot of the new intensity profile of the (**A**) gel using IPBBIS.

3.3.3. Effectiveness of the IPBBIS Method Using Known Concentrations

An experiment was carried out to evaluate the effectiveness of the method. GPN protein concentrations with known values (Table 2) were prepared by adding diverse cell extracts to include different proteins and increase the background noise, obtaining the samples presented in the gel image of Figure 10A by placing other concentrations in randomly chosen positions.

Table 2. GPN protein concentrations with known values.

Lane	2	3	4	5	6	7	8	9	10	11
Concentration mg/mL	2.0	0.0	14.5	9.5	27	18	30	18.5	26	14
Gel										

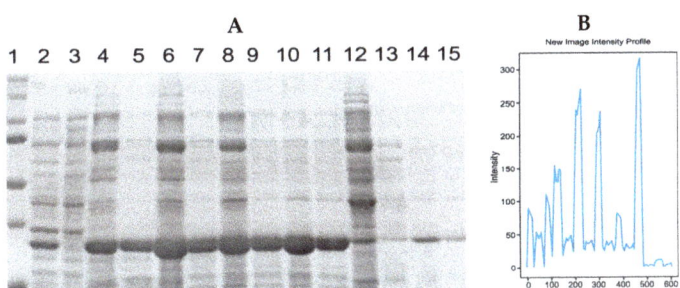

Figure 10. (**A**) Initial image of the gel with GPN samples at different concentrations before preprocessing. (**B**) A plot was generated using the IPBBIS method applied to (**A**).

Samples with different concentrations of recombinant GPN protein are distributed on the polyacrylamide gel in Figure 10A.

The results of applying the IPBBIS method to the image in Figure 10A (GPN protein at different concentrations) indicated that samples with a higher protein concentration had peaks with higher intensity values (Figure 10B). However, when comparing lane 3 (which has no GPN protein concentration, Figure 10A), the graph showed that it has a higher peak or maximum compared to peaks 5, 7, 9, and 11 (which correspond to different GPN protein concentrations), indicating a higher overexpression of the protein. This fact does not agree with the prepared concentrations, with lane 3 having a concentration of 0.0 mg/mL, lane 5 of 9.5 mg/mL, lane 7 of 18 mg/mL, lane 9 of 18.5 mg/mL, and lane 11 of 14 mg/mL (see Table 2).

This inconsistency in the results is because lane 3 had more proteins in the total extract than in samples 5, 7, 9, and 11. This behavior corroborated that the IPBBSI method can only calculate the overexpression of the proteins of interest when the background noise decreases, i.e., only when there is the same amount of proteins in the total extract can the level of overexpression of the protein of interest be detected. Proteins that are not of interest, those on the top and bottom of the recombinant protein (GPN), can be considered contaminants and should, therefore, be removed.

3.3.4. Elimination of Impurities through the Determination of the Molecular Weight of the Target Protein

A procedure is proposed to eliminate contaminants. To this end, it is necessary to find the molecular weight of the protein of interest (GPN) and separate it from the rest of the proteins present in the same sample to apply the IPBBIS method and find its level of overexpression.

Considering that the molecular weight control does not present background noise, the ladder or control in Figure 10A (lane 1) was selected, separated from the rest of the gel, changed to a horizontal orientation (see Figure 11A), and the IPBBIS method was applied, which provided the graph shown in Figure 11B. The multiple maxima represent each protein position in the ladder (Figure 11C).

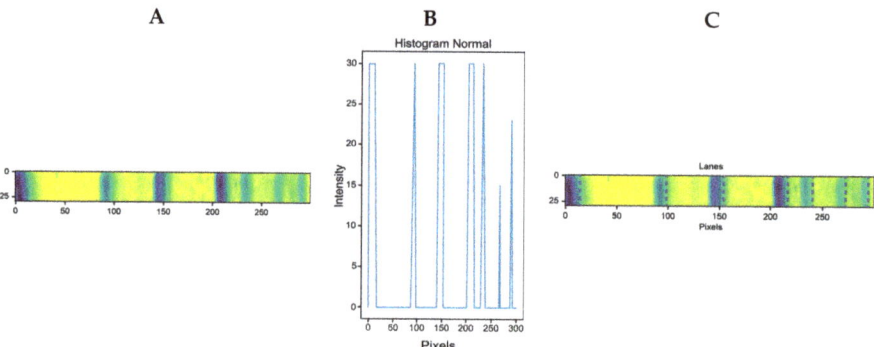

Figure 11. (**A**) Molecular weight control was obtained from lane one or the ladder in Figure 10A. (**B**) The IPBBSI plot applied to (**A**). (**C**) Bands automatically detected by IPBBIS marked with blue lines.

A relationship was established between the molecular weights of the ladder or control (provided by the manufacturer, Bio-Rad S.A. Mexico D.F.) and the multiple maxima obtained by the IPBBIS method. Due to the absence of background noise, the positions of the control proteins were automatically detected and marked with blue lines for identification, as shown in the gel image of Figure 11C. At the same time, the data were stored in an array. The stored data represent the positions of the control proteins and the values of their molecular weights. They were processed by applying numerical methods of linear interpolation, nearest interpolation, and cubic interpolation to obtain an equation. This equation was used to know and predict the molecular weight of proteins present in any position of the remaining samples of the gel image to be analyzed (see Figure 12 and Table 3). Figure 12 shows the interpolation methods used to predict the molecular weight of the detected proteins. The X-axis corresponds to the pixel positions of the bands, while the Y-axis represents the molecular weights.

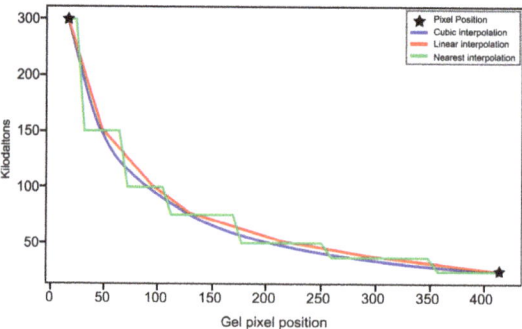

Figure 12. Results of the interpolation methods for the molecular weight of the detected proteins.

Table 3. The table shows the error obtained by applying three numerical interpolation methods to the polyacrylamide gel ladder in Figure 11A.

Interpolation Method	Calculated Weight (kDa)	Total Error %
Linear	33.4	3.35648148
Nearest	37.0	7.060185185
Cubic	31.38	9.194960019

The results obtained in Table 3 indicate that the linear interpolation method showed the lowest error, with a value of 3.35%, when correlating the actual weight of the GPN protein (34.56 kilodaltons, kDa) with that predicted by the interpolation method used.

Therefore, the formula used to identify the molecular weight of the proteins detected in the rest of the samples corresponds to the linear interpolation method and is defined by Equation (1):

$$y(x) = y_i + \frac{(y_{i+1} - y_i)(x - x_i)}{(x_{i+1} - x_i)} \quad (1)$$

where $0 \leq x_i \leq 400$ corresponds to the interval containing the number of pixels corresponding to the image height of each sample, and $0 \leq y_i \leq 250$ corresponds to the molecular weight of the proteins.

3.3.5. Choice of Threshold for the Elimination of Multiple Maximums

IPBBIS automatically detected the number of samples present in a gel image. The multiple minimum values between the various maxima indicated the number of samples in the polyacrylamide gel. Sometimes, when matching these values, marked by blue dotted vertical lines, it was impossible to detect the number of samples in the gel image (see Figure 13A). The amount of contaminating proteins caused the excess background noise in the samples. We chose a threshold containing the minimum values for this case to eliminate the background noise. In this way, we allowed the elimination of the multiple maxima detected in the graph obtained by the new image intensity profile. These multiple maxima were removed with a low-pass filter (represented by the black line in Figure 13B) applied to the latest image intensity profile, and a new graph was obtained in which the minima corresponded perfectly to the existing separations with every sample present in the gel image (Figure 13C). This method allowed us to automatically select the regions with the tiniest white pixels and link them to the regions where the samples are separated. As a result, the experiments present in the gel were automatically detected and identified with blue dotted vertical lines that perfectly matched the sample separations in the image (Figure 13D). They were identified with blue lines drawn on the gel image to verify their correspondence with the different samples. The data were stored in an array to determine the position of each of the samples in subsequent analyses.

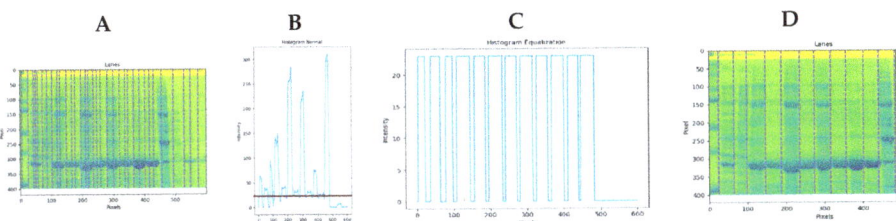

Figure 13. (**A**) Detected maxima (blue dash lines). (**B**) Threshold that allows obtaining a cut-off region that includes only the minima that represent the separation of the samples. (**C**) Graph obtained from the cut-off region. (**D**) Total of automatically detected samples (blue dash lines).

3.3.6. Analysis of the Region of Interest Using the IPBBIS Methods, Manual Area Calculation, Area Calculation by K-Means Segmentation, and Area Calculation by Otsu Segmentation

A random sample was selected on the gel to repeat the molecular weight detection process (lane 5, Figure 14A). The developed IPBBIS method was applied to detect the bands present per sample, as indicated in Figure 14B; knowing the molecular weight of the GPN protein, approximately 34 KDa, the interpolation method was applied to detect it automatically within the gel, and it was identified as indicated in Figure 14C.

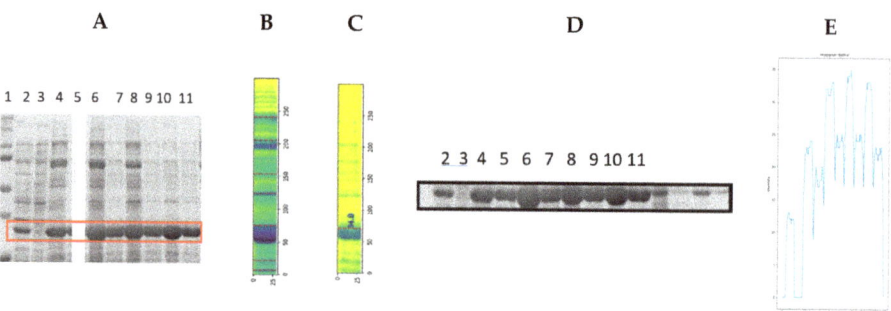

Figure 14. (**A**) Random sample selection within the gel (red box in the image). (**B**) Automatic band detection using image profiling based on binarized image segmentation. (**C**) Molecular weight detection using IPBBIS for GPN protein. (**D**) Selection of GPN protein bands from different samples for ROI. (**E**) The application of the image profile is carried out in the binarized image segmentation of (**D**).

Once the positions separating the samples in the gel were detected (Figure 14B), the molecular weight of the GPN protein and its place within the gel (Figure 14C) were identified, then the region of interest (ROI) that included the expression of the recombinant protein at different concentrations from all samples (red box in Figure 14A including lanes 2 to 11) was selected and isolated as shown in Figure 14D. Analysis was then performed to compare protein overexpression and to identify whether image profiling based on binarized image segmentation can detect proteins with higher or lower overexpression. Before this, the image color in Figure 14D was changed from RGB to HSV. The result is shown in Figure 14E, where the height of the peaks is related to the level of overexpression of the protein analyzed. Every sample with protein contained in the region of interest (ROI) in Figure 14D (lanes 2, 4, 5, 6, 6, 7, 8, 9, 10, and 11) was isolated to obtain a higher precision in the analysis. IPBBIS was applied separately to obtain one graph per sample, as shown in Figure 15. The average of the multiple maxima obtained was calculated to produce a single maximum value per sample (Figure 15A–I), which was related to the amount of protein concentrated in each of the different samples analyzed; these values were aggregated in Table 4 for future reference (row ROI-GPN Table 4). Image profiling based on binarized image segmentation was applied to each of the nine samples containing GPN protein at different concentrations in the ROI region of interest gel in Figure 14D. As shown in the Table 4, we obtained additional data by analyzing the different samples at different concentrations of recombinant GPN protein distributed on the polyacrylamide gel in Figure 14D.

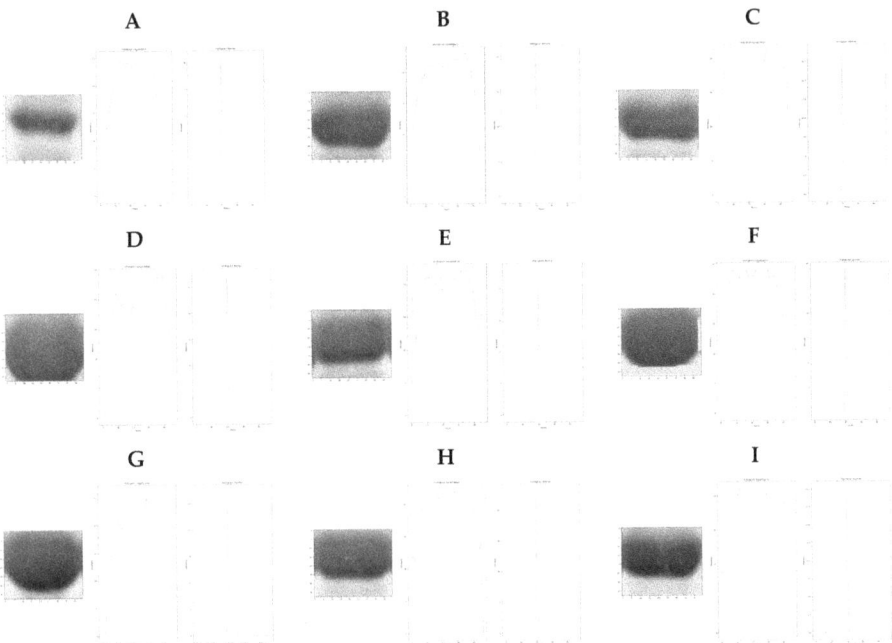

Figure 15. Image profiling of Figure 14D, based on binarized image segmentation in the region of interest. (**A**) lane 2, (**B**) lane 4, (**C**) lane 5, (**D**) lane 6, (**E**) lane 7, (**F**) lane 8, (**G**) lane 9, (**H**) lane 10, and (**I**) lane 11.

Table 4. Different samples Data of GPN concentrations.

Lane	2	3	4	5	6	7	8	9	10	11
Concentration mg/mL	2.0	0.0	14.5	9.5	27	18	30	18.5	26	14
ROI-GPN	10.38	0.0	20.55	18.67	27.34	22.87	28.82	23.08	27.26	20.11
Area Manual	513.62	0.0	830.25	680.12	1373.80	830.00	1061.50	934.50	983.50	869.25
Area K-means segmentation	495.50	0.0	969.62	933.00	1457.20	1132.20	1336.20	1134.90	1422.10	993.13
Area Otsu segmentation	535.75	0.0	1189.8	1023.4	1646.8	1318.2	1557.1	1269.1	1585.9	1174.6

The array data obtained (ROI-GPN row in Table 4) revealed that the lane 2 sample (with a maximum value of 10.38 and a concentration of 2 mg/mL) shows the lowest overexpression, followed in order of overexpression by lane 5 (18.67, 9.5 mg/mL), lane 11 (20.11, 14 mg/mL), lane 4 (20.55, 14.5 mg/mL), lane 7 (22.87, 18 mg/mL), lane 9 (23.08, 18.5 mg/mL), lane 10 (27.26, 26 mg/mL), lane 6 (27.34, 27 mg/mL), and lane 8 (28.82, 30 mg/mL). These results show that the array data obtained for the new image profile calculated by the IPBBIS method for each sample are related to the concentration of each protein and can be used to calculate the level of GPN overexpression by comparing every sample present in the gel image.

3.3.7. IPBBIS Study on the Image Dataset Using the Confusion Matrix

After verifying that the IPBBIS method can automatically identify the level of overexpression in each sample, the polyacrylamide gel image dataset was divided into homogeneous and heterogeneous gels to test their efficiency.

The so-called homogeneous gels (dataset of 44 gels with a total of 669 samples) showed similar characteristics, such as the same color and quality, and no imperfections, such as breaks or distortions due to incorrect preparation. In this research, a confusion matrix defines true positives (TP) as cases where IPBBIS correctly detected the lane or protein band. False negatives (FN) occurred when the lane existed but was not found, false positives (FP) when the lane did not exist but was detected, and true negatives (TN) when the lane did not exist and was also not detected. The confusion matrix obtained after analyzing the 669 samples is presented in Table 5. The precision obtained was 0.985052 (see Table 6, the precision of homogeneous gels).

Table 5. Confusion matrix obtained by analyzing 669 samples expressing GPN protein at different concentrations on homogeneous SDS-PAGE gels.

		Predicted	
		Positive	Negative
Real	Positive	TP = 310	FN = 8
	Negative	FP = 2	TN = 349

Table 6. The Table shows the accuracy results obtained from the confusion matrices in Tables 5 and 7.

Accuracy of Homogeneous Gels	Accuracy of Heterogeneous Gels
0.985052	0.91736

Table 7. Confusion matrix obtained by analyzing 1561 samples with GPN protein expressed at different concentrations on heterogeneous SDS-PAGE gels.

		Predicted	
		Positive	Negative
Real	Positive	TP = 671	FN = 105
	Negative	FP = 24	TN = 761

Then, the analysis was repeated using the IPBBIS method on the heterogeneous gels, a total of 90 gels with different conditions, which included distorted (smiley face effect), broken, or incorrectly stained gels. In total, 1561 samples were analyzed for GPN protein overexpression. The accuracy of this analysis was measured using the same confusion matrix with the TP, FN, TN, and FP values defined above for the homogeneous gels. The confusion matrix is shown in Table 7, and the precision obtained was 0.91736 (see Table 6, the precision of heterogeneous gels). This precision was lower than that obtained with homogeneous gels since the SDS-PAGE gels analyzed present characteristics that make them different, such as breaks, distortions due to incorrect preparation, or insufficient Coomassie blue staining.

3.3.8. Functionality of the Methods Analyzed to Find GPN Protein Overexpression: IPBBIS, Manual Area Calculation, Area Calculation by K-Means Segmentation, and Area Calculation by Otsu Segmentation

To calculate protein overexpression, the areas of each of the GPN protein bands expressed at different concentrations in Figure 14D were measured manually by outlining the contour of the band and using the K-means segmentation and Otsu segmentation techniques (see Figure 16) and then compared with the measurement performed by the IPBBIS method.

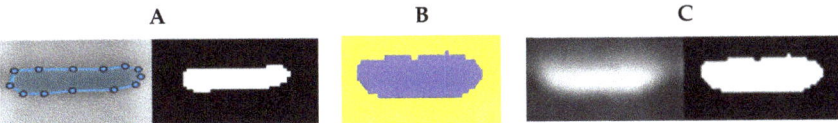

Figure 16. (**A**) Manual calculation of the band area by outlining the spot contour. (**B**) Area calculated by K-means segmentation. (**C**) Area calculated by Otsu segmentation.

For the manual measurements, K-means segmentation and Otsu segmentation, it was necessary to cut out each of the bands and separate them from the gel image as neither of these methods can analyze the whole gel. The results of the measurements of each of the bands at different concentrations are aggregated in Table 4, indicating the type of methodology used in the row.

The data in Table 4 were normalized to verify the functionality of the methodologies used to assess protein overexpression within the gel in the different samples (Figure 14D). The data in Table 4, corresponding to the intensity values in Figure 14D, were sorted according to the amount of expressed protein from lowest to highest overexpression and placed in Table 8, normalized (see Table 9), and used to measure the correct level of GPN protein overexpression for each of the methods used by the following operation:

Table 8. The data in Table 4 are ordered according to the amount of protein expressed from lowest to highest overexpression.

Concentration mg/mL	0	2	9.5	14	14.5	18	18.5	26	27	30
Lane	3	2	5	11	4	7	9	10	6	8
ROI-GPN	0	10.38	18.67	20.11	20.55	22.87	23.08	27.26	27.34	28.82
Manual Area	0	513.6	680.1	869.25	830.25	830	934.5	983.5	1373.8	1061.5
K-means segmentation Area	0	495.5	933	993.13	969.62	1132	1134.9	1422.1	1457.2	1336.2
Otsu segmentation Area	0	535.8	1023	1174.6	1189.8	1318	1269.1	1585.9	1646.8	1557.1

Table 9. Normalization of the data from Table 8.

	Normalized Data									
Concentration mg/mL	0.00	0.07	0.32	0.47	0.48	0.60	0.62	0.87	0.90	1.00
Lane	3	2	5	11	4	7	9	10	6	8
ROI-GPN	0.00	0.36	0.65	0.70	0.71	0.79	0.80	0.95	0.95	1.00
Manual Area	0.00	0.37	0.50	0.63	0.60	0.60	0.68	0.72	1.00	0.77
K-means segmentation Area	0.00	0.34	0.64	0.68	0.67	0.78	0.78	0.98	1.00	0.92
Otsu segmentation Area	0.00	0.33	0.62	0.71	0.72	0.80	0.77	0.96	1.00	0.95

Let x_n be the n-th value of the measured area, and x_m be the value of the measured area in the band with the highest protein concentration.

If $x_m > x_n \Rightarrow x_m - x_n > 0$. This expression indicates that the measurement of the overexpression level is correct since positive values are expected if the concentration is increasing. On the other hand, if $x_m < x_n \Rightarrow x_m - x_n < 0$, it indicates that the overexpression level was miscalculated. The analyzed method presents errors in its measurement because the protein concentration is increasing and not decreasing.

As seen in Table 10, the results of the above analysis, applied to each of the measurements, indicate that the proposed image profiling method based on binarized image segmentation (ROI-GPN) does not show any negative values. In contrast, the manual method shows two negative values (lanes 4 and 8 in Table 10), the K-means segmentation shows two negative values (lanes 4 and 8 in Table 10), and the Otsu segmentation shows two negative values (lanes 8 and 9 in Table 10).

Table 10. Data from Table 9 normalized and compared to the predecessor to identify if there are negative variations corresponding to mismeasurement in protein overexpression.

Concentration mg/mL	0.00	0.07	0.32	0.47	0.48	0.60	0.62	0.87	0.90	1.00
Lane	3	2	5	11	4	7	9	10	6	8
ROI-GPN	0.00	0.36	0.29	0.05	0.02	0.08	0.01	0.15	0.00	0.05
Comparison of Manual Area	0.00	0.37	0.12	0.14	−0.03	0.00	0.08	0.04	0.28	−0.23
Comparison of K-means Segmentation Area	0.00	0.34	0.30	0.04	−0.02	0.11	0.00	0.20	0.02	−0.08
Comparison of Otsu Segmentation Area	0.00	0.33	0.30	0.09	0.01	0.08	−0.03	0.19	0.04	−0.05

These results demonstrate that the developed IPBBIS profiling allows for discovering overexpression and correctly identifying the level of overexpression related to GPN protein concentration. On the other hand, the manual methods, K-means segmentation and Otsu segmentation, presented errors in the measurements.

In addition, the IPBBIS method can analyze the ROI region without cutting out each of the samples present in the gel. In contrast, manual techniques, like K-means segmentation and Otsu segmentation, require the samples to be separated, as they cannot analyze the whole gel.

The ROI-GPN has values of the array data with the number of white pixels analyzed by binary mask. The other methods have values in pixel areas, and normalization was performed to realize a comparison. The normalization was made by taking the maximum value measured by each method when applying four methods to identify the overexpression manually (since there is no automatic one) and dividing it by each of the values of its respective method. Thus, the maximum value for all samples is unity. When sorted by the designed concentration value (lowest to highest), the normalized values increase and do not decrease, as did all the previous methods except for the IPBBIS method developed in this work. The increase only occurs if the methods correctly calculate the size of the bands (by area or by intensity).

These results indicate that traditional methods can identify the position of proteins. However, they cannot identify a particular protein band nor determine the concentration or molecular weight, and the rest of the new intensity profile plot cannot be related to any of the proteins present in the same sample. The IPBBIS method identified the most minor and most overexpressed GPN protein and even detected the order of overexpression.

These results indicate that the IPBBIS method can be used to identify GPN protein overexpression related to IDC and ILC Her2+ breast cancer and can also be applied to identify overexpression of other proteins of biological interest and to detect the progression of cancer stages in different samples from the same patient.

In summary, the IPBBIS method applies a binary mask pixel by pixel, choosing the white intensity value and storing it in an array. The array contains multiple maxima and multiple minima. The intensity value of the minima is related to the separation of the number of samples when analyzing a full gel. As the number of targets decreases, this indicates the separation between proteins. When analyzing per sample for proteins in SDS-PAGE gels, the new image profile values of the multiple minima quantify the level of overexpression of proteins present per sample.

Current methods for searching for proteins in SDS-PAGE gels perform image profiling, processing techniques, threshold, and brightness changes but require the analyst to select the region of interest. The IPBBIS method automatically identifies the number of samples in the gel and the amount of proteins in a sample. It also detects the level of overexpression based on molecular weight.

4. Conclusions

A new methodology called IPBBIS was developed to identify the number of samples present in an SDS-PAGE gel and the molecular weight of the recombinant protein of interest, with a margin of error of 3.35%. An accuracy of 0.985052 was obtained when the gels analyzed were homogeneous, i.e., free of errors such as smiley face distortion, breaks, or poor staining. For gels with such errors, the accuracy was 0.91736.

The IPBBIS method enables the identification of the target protein in the gel by its molecular weight, allowing confirmation of overexpression levels. In contrast to manual area calculation, K-means segmentation, and Otsu segmentation, the IPBBIS approach demonstrated the capability to detect overexpression across the entire gel, eliminating the need to isolate specific areas as other methods require.

Thus, image profiling based on binarized image segmentation can be an auxiliary tool to detect protein overexpression at a lower cost than other molecular techniques, helping to ascertain whether cancer treatment is working.

Future Work

It is hoped that the IPBBIS method will be applied to identify any overexpression of proteins present in polyacrylamide gels.

In future work, we would like to apply this methodology to detect separations in close objects, such as a cell cluster or tissue images, and identify cellular overexpression.

Since the IPBBIS method allows the calculation of the gaps between protein bands in a polyacrylamide gel, its application is sought in imaging samples with cells corresponding to different stages of cancer. Since the cells increase in number during each phase, the spaces between them decrease.

Author Contributions: Conceptualization, J.J.-L. and L.A.-R.; methodology and software and polyacrylamide gel preparation, M.G.-V. and J.J.-L.; validation and formal analysis, R.D.-H.; investigation, A.S.-S.; writing—original, J.J.-L. and A.S.-S.; draft preparation A.S.-S.; writing—review and editing, L.A.-R., R.D.-H. and A.S.-S.; supervision, L.A.-R. All authors have read and agreed to the published version of the manuscript.

Funding: This research received no external funding.

Data Availability Statement: Data are contained within the article.

Conflicts of Interest: The authors declare no conflicts of interest.

References

1. Kaabouch, N.; Schultz, R.R.; Milavetz, B. An analysis system for DNA Gel Electrophoresis images based on automatic thresholding and enhancement. In Proceedings of the 2007 IEEE International Conference on Electro/Information Technology, Chicago, IL, USA, 17–20 May 2007; pp. 1–6.
2. Ferrari, M.; Cremonesi, L.; Carrera, P.; Bonini, P. Diagnosis of genetic disease by DNA technology. *Pure Appl. Chem.* **1991**, *63*, 1089–1096. [CrossRef]
3. Goez, M.M.; Torres-Madroñero, M.C.; Röthlisberger, S.; Delgado-Trejo, E. Preprocessing of 2-Dimensional Gel Electrophoresis Images Applied to Proteomic Analysis: A Review. *Genom. Proteom. Bioinform.* **2018**, *16*, 63–72. [CrossRef] [PubMed]
4. Intarapanich, A.; Kaewkamnerd, S.; Shaw, P.J.; Ukosakit, K.; Tragoonrung, S.; Tongsima, S. Automatic DNA diagnosis for 1D Gel Electrophoresis Images using Bio-image Processing Technique. *BMC Genom.* **2015**, *16*, S15. [CrossRef] [PubMed]
5. Jian-Derr, L.; Chung-Hsien, H.; Neng-Wei, W.; Chen-Song, L. Automatic DNA sequencing for electrophoresis gels using image processing algorithms. *J. Biomed. Sci. Eng.* **2011**, *4*, 523–528.
6. Taher, R.S.; Jamil, N.; Nordin, S.; Bahari, U.M. A new false peak elimination method for poor DNA gel images analysis. In Proceedings of the 2014 14th International Conference on Intelligent Systems Design and Applications, Okinawa, Japan, 28–30 November 2014; pp. 180–186.
7. Koprowski, R.; Wróbel, Z.; Korzynska, A.; Chwialkowska, K.; Kwasniewski, M. Automatic analysis of 2D polyacrylamide gels in the diagnosis of DNA polymorphisms. *Biomed. Eng.* **2013**, *12*, 68. [CrossRef] [PubMed]
8. Cai, F.; Liu, S.; Dijke, P.T.; Verbeek, F.J. Image analysis and pattern extraction of proteins classes from one-dimensional gels electrophoresis. *Int. J. Biosci. Biochem. Bioinform.* **2017**, *7*, 201–212. [CrossRef]
9. Ahmed, N.E. EgyGene GelAnalyzer4: A powerful image analysis software for one-dimensional gel electrophoresis. *J. Genet. Eng. Biotechnol.* **2021**, *19*, 18. [CrossRef]

10. Alnamoly, M.H.; Alzohairy, A.M.; Mahmoud, I.; El-Henawy, I.M. EGBIOIMAGE: A software tool for gel images analysis and hierarchical clustering. *IEEE Access* **2019**, *8*, 10768–10781. [CrossRef]
11. Juárez, J.; Guevara-Villa, M.; Sánchez-Sánchez, A.; Díaz-Hernández, R.; Altamirano-Robles, L. Tridimensional structure prediction and purification of human protein GPN2 to high concentrations by nickel affinity chromatography in presence of amino acids for improving impurities elimination. In *Transactions on Computational Science & Computational Intelligence*; Springer Nature: Cham, Switzerland, 2021.
12. Abadi, M.F. Processing of DNA and Protein Electrophoresis Gels by Image Processing. *Sci. J.* **2015**, *36*, 3486–3494.
13. Abeykoon, A.; Dhanapala, M.; Yapa, R.; Sooriyapathirana, S. An automated system for analyzing agarose and polyacrylamide gel images. *Ceylon J. Sci.* **2015**, *44*, 45–54. [CrossRef]
14. Bajla, I.; Holländer, I.; Fluch, S.; Burg, K.; Kollár, M. An alternative method for electrophoresis gel image analysis in the GelMaster software. *Comput. Methods Programs Biomed.* **2005**, *77*, 209–231. [CrossRef] [PubMed]
15. Brauner, J.M.; Groemer, T.W.; Stroebel, A.; Grosse-Holz, S.; Oberstein, T.; Wiltfeang, J.; Maler, J.M. Spot quantification in two-dimensional gel electrophoresis image analysis: Comparison of different approaches and presentation of a novel compound fitting algorithm. *Bioinformatics* **2014**, *15*, 181. [CrossRef] [PubMed]
16. Efrat, A.; Hoffmann, F.; Kriegel, K.; Schultz, C.; Wenk, C. Geometric algorithms for the analysis of 2D-Electrophoresis gels. *J. Comput. Biol.* **2002**, *9*, 299–315. [CrossRef]
17. Faisal, M.; Vasiljevic, T.; Donkor, O.N. A review on methodologies for extraction, identification and quantification of allergenic proteins in prawns. *Food Res. Int.* **2019**, *121*, 307–318. [CrossRef]
18. Fernández-Lozano, C.; Seoane, J.A.; Gestal, M.; Gaunt, T.R.; Dorado, J.; Pazos, A.; Campbell, C. Texture analysis in gel electrophoresis images using an integrative kernel-based approach. *Sci. Rep.* **2016**, *6*, 19256. [CrossRef] [PubMed]
19. Kaur, N.; Sharma, P.; Jaimni, S.; Kehinde, B.A.; Kaur, S. Recent developments in purification techniques and industrial applications for whey valorization: A review. *Chem. Eng. Commun.* **2019**, *207*, 123–138. [CrossRef]
20. Labyed, N.; Kaabouch, N.; Schultz, R.R.; Singh, B.B. Automatic segmentation and band detection of protein images based on the standard deviation profile and its derivative. In Proceedings of the 2007 IEEE International Conference on Electro/Information Technology, Chicago, IL, USA, 17–20 May 2007; pp. 577–582.
21. Ramaswamy, G.; Wu, B.; MacEvilly, U. Knowledge management of 1D SDS PAGE Gel protein image information. *J. Digit. Inf. Manag.* **2010**, *8*, 223–232.
22. Rezaei, M.; Amiri, M.; Mohajery, P. A new algorithm for lane detection and tracking on pulsed field gel electrophoresis images. *Chemom. Intell. Lab. Syst.* **2016**, *157*, 1–6. [CrossRef]
23. Viswanathan, S.; Ünlü, M.; Minden, J. Two-dimensional difference gel electrophoresis. *Nat. Protoc.* **2006**, *1*, 1351–1358. [CrossRef]
24. Heras, J.; Domínguez, C.; Mata, E.; Pascual, V.; Lozano, C.; Torres, C.; Zarazaga, M. GelJ—A tool for analyzing DNA fingerprint gel images. *BMC Bioinform.* **2015**, *16*, 270. [CrossRef]
25. Alawdi, R.M.; Amer RB, M.; Alzohairy, A.M.; Khedr, W.M. The Computational Techniques Developed to Analyze DNA Gel Images. *Int. J. Adv. Eng. Res. Sci.* **2016**, *3*, 139–149.
26. Heras, J.; Domínguez, C.; Mata, E.; Pascual, V.; Lozano, C.; Torres, C.; Zarazaga, M. A survey of tools for analysing DNA fingerprints. *Brief. Bioinform.* **2015**, *17*, 903–911. [CrossRef] [PubMed]
27. Pavel, A.B.; Vasile, C.I. PyElph-a software tool for gel images analysis and phylogenetics. *BMC Bioinform.* **2012**, *13*, 9. [CrossRef] [PubMed]
28. Khakabimamaghani, S.; Najafi, A.; Ranjbar, R.; Raam, M. GelClust: A software tool for gel electrophoresis images analysis and dendrogram generation. *Comput. Methods Programs Biomed.* **2013**, *111*, 512–518. [CrossRef] [PubMed]
29. Alnamoly, M.H.; Alzohairy, A.M.; El-Henawy, I.M. A survey on gel image analysis software tools. *J. Intell. Syst. Internet Things* **2020**, *1*, 40–47.
30. Juárez-Lucero, J.; Guevara-Villa, M.G.; Sánchez-Sánchez, A.; Díaz-Hernández, R.; Altamirano-Robles, L. Development of a Methodology to Adapt an Equilibrium Buffer/Wash Applied to the Purification of hGPN2 Protein Expressed in Escherichia coli Using an IMAC Immobilized Metal Affinity Chromatography System. *Separations* **2022**, *9*, 164. [CrossRef]
31. Lara-Chacón, B.; Guerrero-Rodríguez, S.L.; Ramírez-Hernández, K.J.; Robledo-Rivera, A.Y.; Velazquez MA, V.; Sánchez-Olea, R.; Calera, M.R. Gpn3 is essential for cell proliferation of breast cancer cells independent of their malignancy degree. *Technol. Cancer Res. Treat.* **2019**, *18*. [CrossRef]
32. Juárez, J.; Guevara-Villa MD, R.; Sánchez, A.; Díaz, R.; Altamirano, L. Image Segmentation Applied to Line Separation and Determination of GPN2 Protein Overexpression for Its Detection in Polyacrylamide Gels. In *Progress in Artificial Intelligence and Pattern Recognition*; Lecture Notes in Computer Science; Springer: Cham, Switzerland, 2021; pp. 303–315.

Disclaimer/Publisher's Note: The statements, opinions and data contained in all publications are solely those of the individual author(s) and contributor(s) and not of MDPI and/or the editor(s). MDPI and/or the editor(s) disclaim responsibility for any injury to people or property resulting from any ideas, methods, instructions or products referred to in the content.

Article

Point-Sim: A Lightweight Network for 3D Point Cloud Classification

Jiachen Guo and Wenjie Luo *

School of Cyber Security and Computer, Hebei University, Baoding 071000, China; guozai97@163.com
* Correspondence: lwj12111@hbu.edu.cn

Abstract: Analyzing point clouds with neural networks is a current research hotspot. In order to analyze the 3D geometric features of point clouds, most neural networks improve the network performance by adding local geometric operators and trainable parameters. However, deep learning usually requires a large amount of computational resources for training and inference, which poses challenges to hardware devices and energy consumption. Therefore, some researches have started to try to use a nonparametric approach to extract features. Point-NN combines nonparametric modules to build a nonparametric network for 3D point cloud analysis, and the nonparametric components include operations such as trigonometric embedding, farthest point sampling (FPS), k-nearest neighbor (k-NN), and pooling. However, Point-NN has some blindness in feature embedding using the trigonometric function during feature extraction. To eliminate this blindness as much as possible, we utilize a nonparametric energy function-based attention mechanism (ResSimAM). The embedded features are enhanced by calculating the energy of the features by the energy function, and then the ResSimAM is used to enhance the weights of the embedded features by the energy to enhance the features without adding any parameters to the original network; Point-NN needs to compute the similarity between each feature at the naive feature similarity matching stage; however, the magnitude difference of the features in vector space during the feature extraction stage may affect the final matching result. We use the Squash operation to squeeze the features. This nonlinear operation can make the features squeeze to a certain range without changing the original direction in the vector space, thus eliminating the effect of feature magnitude, and we can ultimately better complete the naive feature matching in the vector space. We inserted these modules into the network and build a nonparametric network, Point-Sim, which performs well in 3D classification tasks. Based on this, we extend the lightweight neural network Point-SimP by adding some trainable parameters for the point cloud classification task, which requires only 0.8 M parameters for high performance analysis. Experimental results demonstrate the effectiveness of our proposed algorithm in the point cloud shape classification task. The corresponding results on ModelNet40 and ScanObjectNN are 83.9% and 66.3% for 0 M parameters—without any training—and 93.3% and 86.6% for 0.8 M parameters. The Point-SimP reaches a test speed of 962 samples per second on the ModelNet40 dataset. The experimental results show that our proposed method effectively improves the performance on point cloud classification networks.

Keywords: deep learning; point cloud; attention mechanism; pattern recognition

Citation: Guo, J.; Luo, W. Point-Sim: A Lightweight Network for 3D Point Cloud Classification. *Algorithms* **2024**, *17*, 158. https://doi.org/10.3390/a17040158

Academic Editors: Chih-Lung Lin, Bor-Jiunn Hwang, Shaou-Gang Miaou and Yuan-Kai Wang

Received: 1 March 2024
Revised: 9 April 2024
Accepted: 10 April 2024
Published: 15 April 2024

Copyright: © 2024 by the authors. Licensee MDPI, Basel, Switzerland. This article is an open access article distributed under the terms and conditions of the Creative Commons Attribution (CC BY) license (https://creativecommons.org/licenses/by/4.0/).

1. Introduction

In recent years, significant advancements have been witnessed in the field of 3D computer vision, which has become a subject of extensive research. Various formats, including meshes, volumetric meshes, depth images, and point clouds, can be utilized to represent 3D data [1]. Point clouds offer an unorganized sparse depiction of a 3D point set while preserving the original geometric information of an object in 3D space. Their representation is characterized by its simplicity, flexibility, and retention of most information without the need for discretization. The rapid development of 3D sensor technology, including various 3D scanners and LiDARs, has facilitated the acquisition of point cloud data [2]. Owing to its abundant geometric, shape, and scale information, 3D point clouds are crucial for scene understanding and find application in diverse fields such as autonomous driving, robotics, 3D reconstruction, and remote sensing, such as through RN4 and RN5.

However, the disorder and irregularity inherent in 3D point cloud data present challenges for deep learning-based point cloud feature extraction methods, which play a vital role in various point cloud processing tasks. Numerous approaches have been proposed to transform point clouds into regular structures, such as projecting into multiview images [3,4] and voxelization [5,6]. Although these methods have shown superior results in point cloud classification and segmentation tasks compared to traditional manual feature extraction techniques, they compromise the intrinsic geometric relationships of 3D data during processing. Moreover, the computational complexity of voxelization, being proportional to the cube of the volume, limits its application in more complex scenes.

To address these challenges, researchers have started considering the direct processing of raw point cloud data to reduce computational complexity and to fully leverage the characteristics of 3D point cloud data. PointNet [7] directly processes raw data by extracting point cloud features through MLP (MultiLayer Perceptron) and max pooling, thereby ensuring permutation invariance of the point cloud. Although the processing method is simple, it yields significant results and has become an important theoretical and ideological foundation in 3D point cloud processing. PointNet++ [8] extends PointNet by considering both global and local features. It obtains key point sets through farthest point sampling (FPS) and constructs a local graph using k-nearest neighbors (k-NN). Subsequently, MLP and max pooling are employed to aggregate the local features.

Since PointNet++, the main trend in deep learning-based point cloud processing methods has been to add advanced local operators and extend the trainable parameters, and while the performance gain rises by the amount of parameters added, so does the cost of computing resources, and deep learning training is often time-consuming. Many previous works have approached deep learning from a lightweight perspective in order to efficiently address the training and inference time issues of deep learning. For example, MobileNet [9] uses depthwise separable convolution to build a lightweight network, which improves the overall network accuracy and speed; UL-DLA [10] proposes an ultralightweight deep learning architecture. It forms a Hybrid Feature Space (HFS), which is used for tumor detection using a Support Vector Machine (SVM), thereby culminating in high prediction accuracy and optimum false negatives. Point-NN [11] proposes a new approach to nonparametric point cloud analysis that employs simple trigonometric functions to reveal local spatial patterns and a nonparametric encoder for networks to extract the training set features, which are cached as a point feature repository. Finally, the point cloud classification is accomplished using naive feature matching. However, its simple use of trigonometric functions in the process of feature embedding is blind and may lead to the neglect of key features. And because of its feature magnitude change in vector space during feature extraction, this will affect the stability of the model and have an impact in the final naive feature matching stage.

Inspired by the above work, we propose a nonparametric network model for point cloud classification task, which is composed of nonparametric modules, and uses the nonparametric attention block ResSimAM(Residual Simple Attention Module) to derive the attention weights, as well as the features during the feature extraction process, in order to enhance the weights of features with higher energy. In the feature extraction stage, a nonlinear feature transformation is achieved by using the Squash operation to squeeze the input features to a certain range without changing the direction in the vector space. The Squash operation helps to preserve the directional information of the feature vectors while eliminating the effect of magnitude, thereby allowing the network to better learn the structure and patterns in the data and better preserving the relationships between the feature vectors, which helps reduce numerical instability due to vector length variations for subsequent naive similarity matching.

The key contributions of our contributions can be summarized as follows:

1. Aiming at the problem that there is some blindness in Point-NN when using trigonometric functions to encode features for mapping features into high dimensional space, we calculate the energy of each feature by utilizing an energy function and then add weights for each feature according to its energy, which improves the model's ability to extract features without adding any trainable parameters to the original model.
2. In order to alleviate the influence of feature magnitude in the final naive feature matching, we use the Squash operation in the stage of feature extraction so that the features are squeezed to a certain range without changing the direction in the vector space, thereby eliminating the instability brought by the feature magnitude. This enables the network to better learn the structure and patterns in the data and improve the model classification ability.
3. We extend a lightweight parametric model by adding a small number of MLP layers to the nonparametric model feature extraction stage and applying the MLP to the final global features to obtain the final classification results, and we validate the performance of the model in the absence of other state-of-the-art operators.

The remainder of the paper is structured as follows. Section 2 gives related work. Section 3 describes the nonparameter network Point-Sim and the lightweight network Point-SimP methods in detail. We evaluate our methods in Section 4. Section 5 concludes the paper.

2. Related Work

To effectively handle 3D data, scholars have conducted diverse and significant endeavors aimed at addressing the challenges posed by the inherent sparsity and irregularity of point clouds. Such endeavors can be categorized into multiview-based, voxel-based, and point-based methodologies. Initially, we review the learning methodologies grounded in multiview representation and voxelization. Subsequently, we scrutinize point-based learning methodologies, which encompass graph-based and attention-based strategies.

2.1. Multiview-Based Methods

The Multiview Convolutional Neural Network (MVCNN) [4] projects point clouds or 3D shapes onto 2D images, thereby subsequently employing Convolutional Neural Networks (CNNs) for processing the projected 2D images. This methodology integrates feature information from multiple viewpoints into a compact 3D shape descriptor via convolutional and pooling layers. These aggregated features are then fed into a fully connected layer for classification. Zhou [12] proposed the Multiview PointNet (MVPointNet), where the views are acquired through a Transformation Network (T-Net) [7] to generate transformation matrices that determine multiple views captured at identical rotational angles, thereby ensuring the network's robustness against geometric transformations.

Despite the efficacy of projecting point clouds into multiple views for point cloud segmentation and classification tasks compared to conventional manual feature extraction methods, notable limitations persist. Firstly, predetermined viewpoints are required when projecting 3D point clouds into multiple 2D views. View variations result in differential contributions to the final shape descriptor; similar views yield akin contributions, whereas significantly distinct views offer advantages for shape recognition. Secondly, 2D projections are confined to modeling the surface attributes of objects, which are unable to capture the 3D internal structure adequately. This partial representation disrupts intrinsic geometric relationships within 3D data, thus failing to exploit contextual information comprehensively within 3D space and incurring information loss, which is particularly unsuitable for large-scale scenes. Furthermore, feature extraction via multiview approaches often necessitates pretraining and fine-tuning, thereby consequently escalating workload demands.

2.2. Voxel-Based Methods

The VoxNet framework, as introduced by [5], initially employs an occupancy grid algorithm to represent the original point cloud as multiple 3D grids. Each grid corresponds to a voxel, and subsequent 3D convolutions are applied for feature extraction. Le proposed a hybrid network named PointGrid in the work of [13], which integrates both point and grid representations for efficient point cloud processing. This approach involves sampling a constant number of points in each embedded volumetric grid cell, thus allowing the network to utilize 3D convolutions to extract geometric details.

In contrast, voxel-based methods follow a two-step process. Firstly, the original point cloud undergoes voxelization, thereby converting the unordered point cloud into an ordered structure. Subsequently, 3D convolution is applied for further processing. This approach is more direct and simpler, thus drawing inspiration directly from 2D convolutional neural networks. However, it comes with a significant computational cost, and due to the uniformity of each voxel postvoxelization, there is a loss of information regarding fine structures.

2.3. Point-Based Methods

2.3.1. Graph-Based Methods

Point cloud data, which are characterized by an irregular and a disordered distributions of points, inherently lack explicit interconnections among individual points. Nevertheless, these non-Euclidean geometric relationships can be effectively modeled through graph structures. PointNet [7] stands out as the pioneering network specifically designed for the direct processing of point clouds. Despite the groundbreaking achievements of the PointNet network in tasks such as point cloud classification or segmentation, it remains afflicted by the limitation of its inadequate capture of local neighborhood information. To address this limitation and extract more nuanced local features, Qi [8] extended PointNet by introducing the PointNet++ network framework. The fundamental concept involves the construction of a local hierarchical module within the network. Each layer within this module comprises a sampling layer, a grouping layer, and a feature extraction layer. By selecting the local neighborhood center of mass through the *FPS* layer—forming a local neighborhood subset via the k-NN layer—and deriving local neighborhood feature vectors through the PointNet layer, the framework adeptly captures local features across a multi-level hierarchical structure. Nonetheless, PointNet++ faces challenges due to its isolation of individual point sample features within the local neighborhood and the adoption of a greedy max pooling strategy for feature aggregation, thereby risking information loss and presenting certain constraints. In response to these issues, Wang [14] proposed Softpoolnet, which introduces the concept of soft pooling by substituting max pooling with a soft pooling mode. Unlike the exclusive retention of maximal features in max operation, soft pooling retains more features by preserving the first N maximal features during pooling. Meanwhile, Zhao [15] introduced 3D point cloud capsule networks, and they created an autoencoder tailored for processing sparse 3D point clouds while preserving spatial alignment and consolidating the outcomes of multiple maximal pool feature mappings into an informative latent representation through unsupervised dynamic routing.

Nevertheless, it fails to adequately handle neighborhood points information, thus resulting in inadequate interactions between points. In order to enhance direct information exchange and foster better communication, DGCNN [16] employs the k-NN algorithm to construct local graphs, thus grouping points in semantic space and facilitating global feature extraction through continuous feature updates of edges and points. Notably, this approach enables the capture of geometric features of the local neighborhood while maintaining permutation invariance. Furthermore, DeepGCN [17] leverages deep Convolutional Neural Network (CNN) principles emphasizing deep residual connections, extended convolution, and dense connections, thereby enabling reliable training in deep models. GACNet [18], on the other hand, enhances segmentation results in edge areas by constructing a graph structure for each point based on its neighboring points and integrating an attention mechanism to compute edge weights between the central point and its neighbors.

Wang [19] proposed a method for training deformed convolution kernels in local feature extraction, wherein an anchor point is initially selected, followed by the selection of neighboring points through k-NN. Subsequently, a set of displacement vectors is constructed to represent features in this region, thereby facilitating continuous updates of these displacement vectors to extract local point cloud features. Finally, multiple sets of learned displacement vectors are weighted and summed to construct the convolutional kernel for feature extraction, which is then applied to perform feature extraction on the image. Notably, Point-NN [11] introduced a nonparametric network for 3D point cloud analysis comprising purely nonparametric components such as FPS, k-NN, trigonometric functions, and pooling operations. Remarkably, it demonstrates exceptional performance across various 3D tasks without any parameters or training, even outperforming existing fully trained models.

2.3.2. Attention-Based Methods

SENet [20] introduces an efficient and lightweight gating mechanism that explicitly constructs correlations between channels. This consideration stems from the acknowledgment that pixels carry varying degrees of importance across different channels. These importance weights are then leveraged to amplify useful features while suppressing less relevant ones. CBAM [21] derives attention mappings separately along two dimensions, channel and spatial, within the feature mapping. Subsequently, these attention mappings are applied to the input for adaptive feature refinement of the feature map. With the demonstrated success of self-attention and transformer mechanisms in natural language understanding [22], there has been a proliferation of efforts in computer vision to substitute convolutional layers with self-attention layers. However, despite its accomplishments, self-attention incurs computational costs that scale quadratically with the size of the input image. PAT [23] employs a self-attention-like mechanism to capture correlation information between points and extract the most salient global features via Gumbel downsampling. Transformer [24] devises a point transformer layer and builds a residual point transformer block around it, thus enabling information exchange between local feature vectors and the generation of new feature vectors for all points. PCT [25] encodes input coordinates into the feature space to generate features and conducts feature learning through the offset attention mechanism. PoinTr [26] processes the point cloud into a series of point proxies, which represent features of local areas within the point cloud. These proxy points encapsulate neighborhood information, which is then inputted into a transformer for further processing. Subsequently, an encoder–decoder architecture is employed to accomplish the point cloud completion task.

3. Methods

In this section, we will present the details of the nonparametric network Point-Sim and the lightweight neural network Point-SimP. We will show the overall structure of the proposed method, which consists of multiple reference-free components and incorporates the operations of the nonparametric attention mechanism and the feature Squash in the process of feature extraction.

3.1. Overall Structure

The nonparametric modeling of point cloud classification method known as Point-Sim is shown in Figure 1. In the classification model, nonparametric feature embedding is first performed using trigonometric functions(the Trigo block). Subsequently, in the hierarchical feature extraction stage, the centroids are selected using FPS, and from these centroids, the point clouds are grouped using k-NN. We apply trigonometric functions to map the local geometric coordinate. In order to better match the feature naive similarity, the geometric and local features are added and fed into the Squash block, the features are squeezed to make them smoother, and then the smoothed features are fed into the ResSimAM block so that the model can pay better attention to the features with higher energy; this improves the classification ability of the encoder, and then finally the global features are obtained by using the pooling operation.

The nonparametric point cloud classification model has been extended by integrating neural network layers at various stages within Point-Sim. The constructed Point-SimP network, outlined in Figure 2, introduces a lightweight framework. To enhance the model, the raw embedding layer within the nonparametric network was substituted with an MLP. Furthermore, MLP layers were incorporated post the Feature Expansion and Geometry Extraction phases during feature extraction and applied to the ultimate global feature to obtain the classification outcomes.

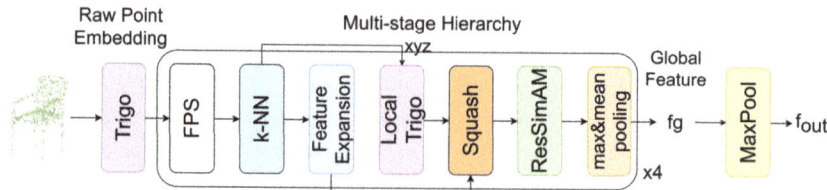

Figure 1. Overall structure of Point-Sim. Different colors represent different module types in the network.

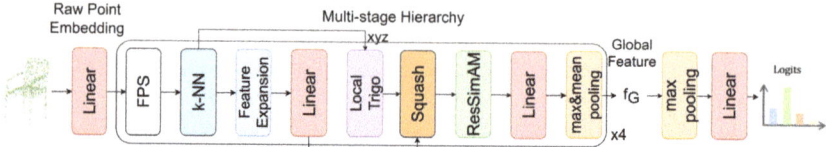

Figure 2. Overall structure of Point-SimP. Different colors represent different module types in the network.

3.2. Basic Components

Our approach begins from the local structure, thereby extracting features layer by layer. We select a certain number of key points within the point clouds, utilizing k-NN to select the nearest neighboring points to generate local regions, and update the features of this local region. By repeating multiple stages, we gradually expand the sensory field and obtain the global geometric information of the point clouds. In each stage, we represent the input point clouds of the previous stage as $\{p_i, f_i\}_{i=1}^M$, where $p_i \in \mathbb{R}^{1\times 3}$ represents the coordinates of point i, and $f_i \in \mathbb{R}^{1\times C}$ represents the features of point i. To begin, the point set is downsampled using FPS to choose a subset of points from the original set. In this case, we select $\frac{M}{2}$ local centroids from the M points, where M is an even number.

$$\{p_c, f_c\}_{c=1}^{\frac{M}{2}} = FPS(\{p_i, f_i\}_{i=1}^M) \tag{1}$$

Afterward, by employing the k-NN algorithm, groups of localized 3D regions are established by selecting the k-nearest neighbors from the original M points for each centroid c (Figure 3).

$$\mathcal{N}_c = k - NN(p_c, \{p_i\}_{i=1}^M) \tag{2}$$

where $\mathcal{N}_c \in \mathbb{R}^{k\times 1}$ represents the k-nearest neighbors.

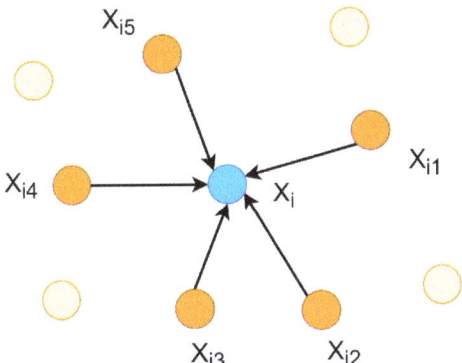

Figure 3. K-nearest neighbors of point X_i. Where X_i represents the center point of the local region, X_{i1}, X_{i2}, ..., X_{i5} represent the nearest neighbors of X_i, and the rest of the points are not included in the local region.

After obtaining the local information, we perform feature expansion (Figure 4) to obtain the features $f_l \in \mathbb{R}^{C \times K}$ of the local points. These are obtained by repeating the centroid point k times and concatenating it with the local features.

$$f_l = Concat\left(Repeat(f_c), \{f_n\}_{n=1}^{k}\right) \quad (3)$$

where $f_c \in \mathbb{R}^{C \times 1}$ represents the features of the center point, $f_n \in \mathbb{R}^{C \times 1}$ denotes the features of the remaining local points, and $C = 2 \times D$.

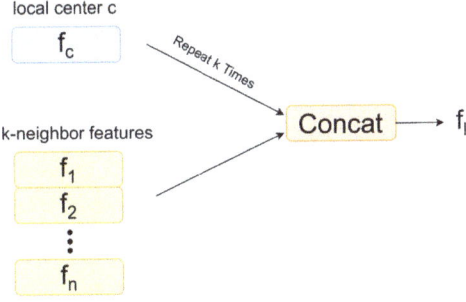

Figure 4. Feature expansion for a local group.

Furthermore, the operator $\Phi(\cdot)$ is utilized to extract the geometry features $\mathcal{N}_c$ of each local neighborhood, which comprises trigonometric functions, Squash, and ResSimAM.

$$\Phi(\cdot) = ResSimam(Sqush(\text{Trigonometric}(\cdot) + f_l)) \quad (4)$$

Local features f_l are processed using $\Phi(\cdot)$, thus resulting in the enhanced local features $f_j \in \mathbb{R}^{C \times K}$.

$$f_j = \Phi(f_l) \quad (5)$$

MaxPooling and MeanPooling are performed to aggregate the data, thus producing $f_g \in \mathbb{R}^{C \times 1}$, which signifies the global features of the chosen key points.

$$f_g = MaxP\left(\{f_j\}_{j \in \mathcal{N}_c}\right) + MeanP\left(\{f_j\}_{j \in \mathcal{N}_c}\right) \quad (6)$$

Following this, after the above feature extraction stage, max pool aggregation is used to obtain the final high-dimensional global feature $f_{out} \in \mathbb{R}^{1 \times C_G}$:

$$f_{out} = MaxP(f_g) \tag{7}$$

Finally, the resulting feature f_{out} is cached in the memory bank F_{mem}, and we construct a corresponding label memory bank T_{mem} as follows:

$$F_{mem} = \text{Concat}(\{f_{out}\}_{n=1}^{N}) \tag{8}$$

$$T_{mem} = \text{Concat}(\{table_i\}_{n=1}^{N}) \tag{9}$$

where $table_i$ is the ground truth as one-hot encoding, and n represents the serial number of each point cloud object in training set n from 1 to N.

3.3. Trigonometric Functions Embedding

Referring to positional encoding in the transformer [22], for a point in the input point cloud, we use trigonometric functions to embed it into a C-dimensional vector:

$$\text{Trigonometric}(p_i) = \text{Concat}(f_i^x, f_i^y, f_i^z) \in \mathbb{R}^{1 \times C_i} \tag{10}$$

where $f_i^x, f_i^y, f_i^z \in \mathbb{R}^{1 \times \frac{C_i}{3}}$ denote the embeddings of three axes, and C_i represents the initialized feature dimension. Taking f_i^x as an example, for channel index $m \in [0, \frac{C_i}{6}]$, we have the following:

$$\begin{aligned} f_i^x[2m] &= \sin\left(\alpha x_i / \beta^{\frac{6m}{C_i}}\right), \\ f_i^x[2m+1] &= \cos\left(\alpha x_i / \beta^{\frac{6m}{C_i}}\right) \end{aligned} \tag{11}$$

where α and β respectively control the magnitude and wavelength. Due to the inherent properties of trigonometric functions, the transformed vectors can effectively encode the relative positional information between different points and capture fine-grained structural changes in the three-dimensional shape.

3.4. Nonparametric Attention Module (Squash and ResSimAM)

SimAM [27] devises an energy function to discern the importance of neurons based on neuroscience principles, with most operations selected according to this energy function to avoid excessive structural adjustments. SimAM has been verified to have good performance in 2D parametric models. Due to its nonparametric character, we are considering incorporating this attention mechanism into our 3D point cloud network.

To successfully implement attention, we need to estimate the importance of individual features. In visual neuroscience, neurons that exhibit unique firing patterns from surrounding neurons are often considered to have the highest information content. Additionally, an active neuron may also inhibit the activity of surrounding neurons, which is a phenomenon known as spatial suppression [28]. In other words, neurons that exhibit significant spatial suppression effects during visual processing should be assigned higher priority. As with SimAM, we use the following equation to obtain the minimum energy for each position:

$$e_t^* = \frac{4(\hat{\sigma}^2 + \lambda)}{(t - \hat{\mu})^2 + 2\hat{\sigma}^2 + 2\lambda} \tag{12}$$

where $\hat{\mu} = \frac{1}{M}\sum_{i=1}^{M} x_i$, $\hat{\sigma}^2 = \frac{1}{M}\sum_{i=1}^{M}(x_i - \hat{\mu})^2$, and M denote the feature dimensions.

The above equation indicates that the lower the energy e_t^*, the greater the difference between the neuron and its surrounding neurons, which is also more important in visual processing. The importance of neurons is represented by $1/e_t^*$. To enhance the features, we construct a residual network. Firstly, we apply the Squash operation to smooth the features, and then we add the ResSimAM attention operation to the squashed features:

$$X = Squash(f_i + f_c)$$
$$\tilde{X} = sigmoid\left(\frac{1}{E}\right) \odot X + X \quad (13)$$

where E groups all e_t^* across all dimensions, and a sigmoid is added to restrict too large values in E.

Algorithm 1 denotes the pseudocode for the implementation of ResSimAM using PyTorch, where $X = Squash(f)$ as $X = \frac{\|f\|^2}{1+\|f\|^2}\frac{f}{\|f\|}$, and $\|f\|$ denotes the module of f.

Algorithm 1: A PyTorch-like implementation of our ResSimAM

    Input: f_i, f_c, λ
    Output: X
1 **def forward (f_i, f_c, λ):**
2   X = Squash(f$_i$ + f$_c$);
3   n = X.shape[2] − 1;
4   d = (X − X.mean(dim = [2])).pow(2);
5   v = d.sum(dim = [2])/n;
6   E_inv = d /(4 ∗ (v + lambda)) + 0.5;
7   **return X* sigmoid(E_inv) + X;**

The Squash operation enables a nonlinear feature transformation by squeezing the input features to a certain range without changing the direction in the vector space. This squeezing helps to preserve the directional information of the feature vectors while eliminating the effect of magnitude, thereby allowing the network to better learn the structure and patterns in the data and be able to better preserve the relationships between the feature vectors. The feature squeezing operation makes each feature vector have a unit length, which helps with better similarity computation between the vectors, and by normalizing the vectors to a unit length, the magnitude difference between the vectors can be reduced, which helps in reducing the numerical instability due to the change in the length of the vectors for the subsequent similarity matching of the features and improves the generalization ability of the network.

It is worth mentioning that Algorithm 1 does not introduce any additional parameter and can therefore work well in a nonparametric network. The energy function involved in the algorithm only requires computing the mean and variance of features, which are then brought into the energy function for calculation. This allows for the computation of weights to be completed in linear time.

3.5. Naive Feature Similarity Matching

In the naive feature similarity matching stage (Figure 5), for a test point cloud, we similarly utilize a nonparametric encoder to extract its global feature $f_{out}^t \in \mathbb{R}^{1 \times C_G}$.

Firstly, we calculate the cosine similarity between the test feature f_{out}^t and F_{mem}:

$$S_{cos} = \frac{f_{out}^t F_{mem}}{\|f_{out}^t\|\|F_{mem}\|} \in \mathbb{R}^{1 \times N} \quad (14)$$

The above equation represents the semantic relevance between the test point cloud and N training samples. By weighting with S_{cos}, we integrate the one-hot labels from the label memory T_{mem} as:

$$logits = \varphi(S_{cos}T_{mem}) \in \mathbb{R}^{1 \times K} \quad (15)$$

where $\varphi(x) = \exp(-\gamma(1-x))$ serves as an activation function from Tip-adapter [29].

In S_{cos}, the higher the score of a similar feature memory pair, the greater its contribution to the final classification logits and vice versa. Through this similarity-based label integration, the point memory bank can adaptively differentiate different point cloud instances without any training.

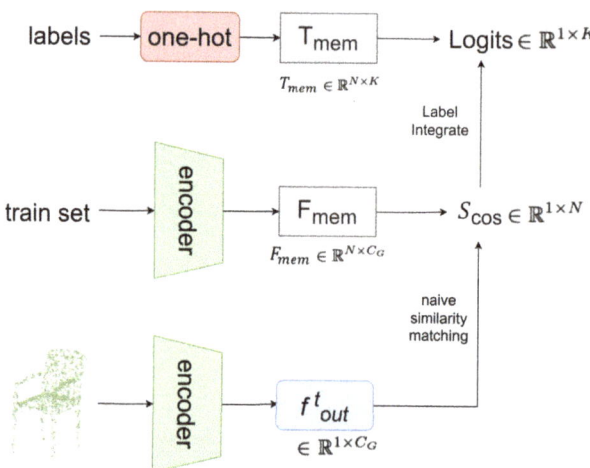

Figure 5. Naive feature similarity matching.

4. Experiments

To validate the effectiveness, we evaluated the efficacy and versatility of the proposed methods for the shape classification task on the ModelNet40 dataset and ScanObjectNN dataset.

4.1. Shape Classification Task on ModelNet40 Dataset

Dataset: We evaluated our method on the ModelNet40 dataset for the classification task. This dataset comprises a total of 12,311 CAD mesh models, with 9843 models assigned for training and 2468 models for testing. The dataset covers 40 different classes.

In order to optimize memory usage and improve computational speed, we followed the experimental configuration of PointNet [7]. We uniformly selected 1024 points from the mesh surface using only the 3D coordinates as input data. We used the overall accuracy (OA) and the number of parameters for evaluation.

For the parametric network, we applied data augmentation; the data were augmented by adding jitter, point random dropout, and random scale scaling to each coordinate point of the object, where the mean value of jitter is 0, and its standard deviation is 0.1. The random scale scaling was between 0.66 and 1.5, and the probability of each point dropping out ranged from 0 to 0.875. The data were augmented with a weight decay of 0.0001 using an initial learning rate of 0.003 for the Adam optimizer with an initial learning rate of 0.001, and a weight decay of 0.0001 was used. In addition, training was performed using crossentropy loss. The batch size set for training was 32, and the maximum epoch was set to 300.

Experimental Results: The classification results on ModelNet40 are shown in Table 1. We compared our results with some recent methods on a RTX 3090 GPU. This comparison signifies that our proposed model generally outperformed several other models. We compared our results with respect to overall accuracy (OA), the number of parameters (Params), training time, and test speed (samples/second) with some recent methods. The proposed nonparametric method achieved an OA of 83.9% with 0 M parameters and without any training time, and the proposed parametric method achieved an OA of 93.3% with 0.8 M parameters, while our light parametric model test speed reached 926 samples per second. And because of the Squash module, our model was able to converge in a relatively short time of 3.1 h. Based on these comparisons with our method and related works, we have reached the conclusion that the network has advantages in terms of training speed and accuracy, as well as device requirements.

Table 1. Classification results on ModelNet40.

Method	Overall Accuracy (%)	Parameters	Train Time	Test Speed
PointNet	89.2	3.5 M	-	-
PointNet++	90.7	1.7 M	3.4 h	521
GBNet	93.8	8.4 M	-	189
DGCNN	92.9	1.8 M	2.4 h	617
PointMLP	94.1	12.6 M	14.4 h	189
Point-NN	81.80	0 M	0	275
Point-Sim	83.9	0 M	0	231
Point-SimP	93.3	0.8 M	3.1 h	962

Our results are visualized on the ModelNet40 dataset, and the results are shown in Figure 6. For the nonparameterized model Point-Sim, the model OA was improved compared to Point-NN with similar inference speed. For the parameterized model Point-SimP, it was able to greatly improve the inference speed while maintaining the accuracy and had an advantage in the network training time.

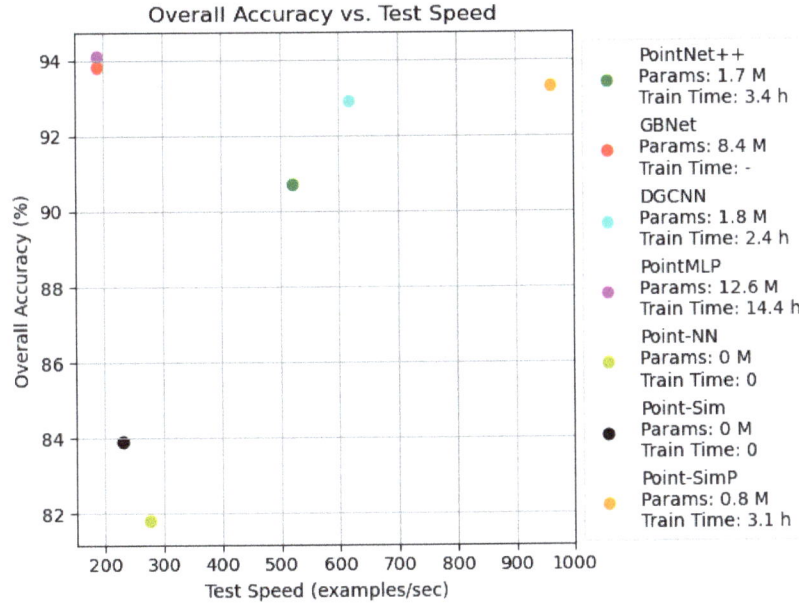

Figure 6. Visualization results on the Modenet40 dataset.

We generated a 40 × 40 confusion matrix for our classification results, and the results are shown in Figure 7, in which there are 40 categories, with the horizontal axis representing the predicted labels and the vertical axis representing the ground truth labels (including airplane, bathtub, bed, bench, etc.). By visualizing the confusion matrix, we can see that most of the categories were classified well; for example, all the classifications on label 1 (airplane) and label 19 (keyboard) are correct, but the accuracies on label 16 (flower pot) and label 32 (stairs) still need to be improved. Figure 8 shows some representative results on ModelNet40.

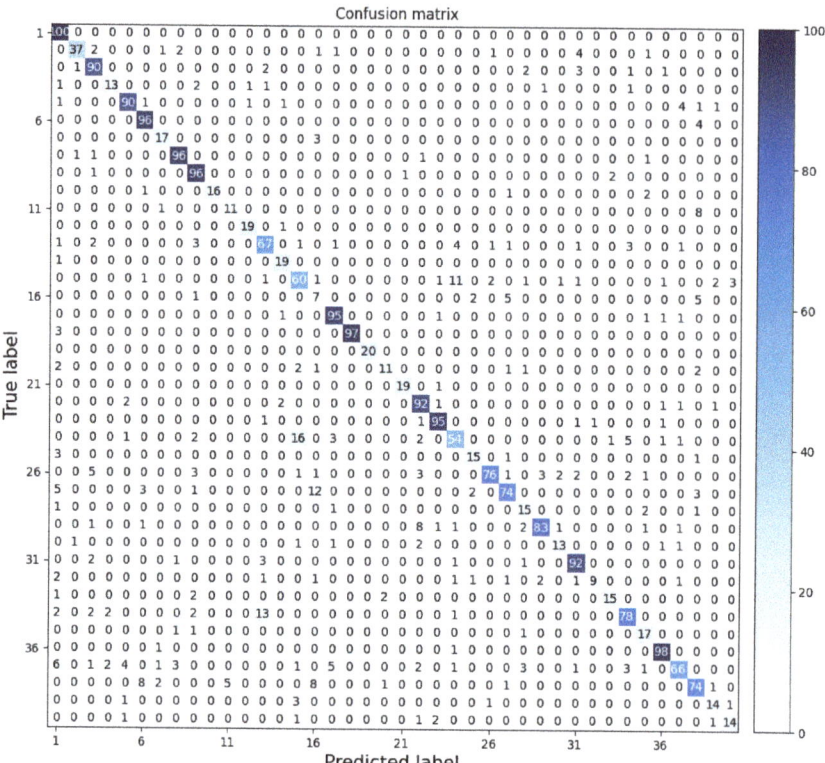

Figure 7. Confusion matrix on Point-Sim result. The class containing the most test objects in the test dataset has 100 objects.

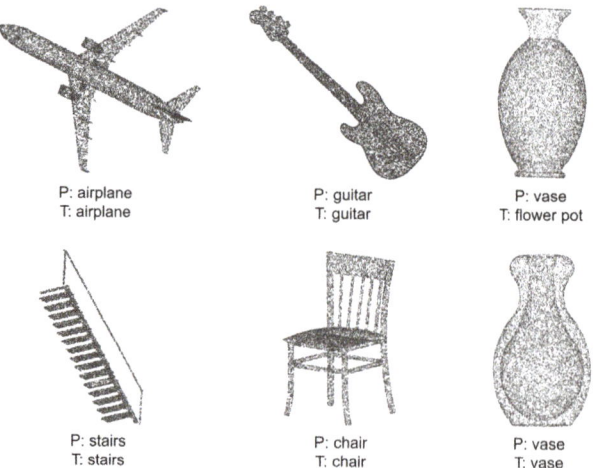

Figure 8. Some representative classification results. P represents predicted label, and T represents the ground truth.

4.2. Shape Classification Task on ScanObjectNN Dataset

Dataset: Although ModelNet40 is a widely adopted benchmark for point cloud analysis, its synthetic nature and the fast-paced advancements in this field may not fully address the requirements of current research. Thus, we have also undertaken experiments utilizing the ScanObjectNN [30] benchmark.

The ScanObjectNN dataset consists of 15,000 objects, with 2902 unique instances found in the real world. These objects belong to 15 different classes. However, analyzing this dataset using point cloud analysis methods can be challenging due to factors such as background interference, noise, and occlusion. Our loss function, optimizer, learning rate evolution scheduler, and data augmentation scheme maintained the same settings as the ModelNet40 classification task. We used the overall accuracy (OA) and the number of parameters for evaluation.

Experimental Results: The classification results obtained from ScanObjectNN are shown in Table 2. We assessed the accuracy of all methods by reporting the performance on the official split of PB-T50-RS. The model achieved an OA of 66.3% with 0 M parameters and 86.6% with 0.8 M parameters, thereby demonstrating the versatility of our proposed method and the robustness of our model under background interference, noise, and occlusion.

Table 2. Classification results on ModelNet40.

Method	Overall Accuracy (%)	Parameters
PointNet	68.2	3.5 M
PointNet++	77.9	1.7 M
GBNet	80.5	8.4 M
DGCNN	78.1	1.8 M
PointMLP	85.2	12.6 M
Point-NN	64.9	0 M
Point-Sim	66.3	0 M
Point-SimP	86.6	0.8 M

4.3. Ablation Study

To showcase the efficacy of our approach, we conducted an ablation study on the classification task in ModelNet40. Furthermore, we performed separate ablation experiments on the ResSimAM and the Squash to assess the impact of removing each component.

In our settings (Table 3), W/O R means no ResSimAM interaction, and W/O S means no Squash. The corresponding results are shown in Table 4.

Table 3. Settings of ResSimAM and Squash. where ✓ means that the module is included, and - means that the module is not included.

Method	Res-SimAM	Squash
W/O R&S	-	-
W/O S	✓	-
W/O R	-	✓
Point-Sim	✓	✓

Table 4. ResSimAM and Squash ablation results.

Method	Overall Accuracy (%)
W/O R&S	81.8
W/O S	82.4
W/O R	83.2
Point-Sim	83.9

We utilized ResSimAM, which resulted in an improvement of the overall accuracy by 0.6%. We employed Squash to squeeze the features, thus leading to a 1.4% improvement in the overall accuracy. And we employed both operations—leading to a 2.1% improvement—and obtained a state-of-the-art result of 83.9% in no-parametric point cloud classification. It has been proven that using ResSimAM can better focus on higher energy features during the feature extraction stage, which can enhance features useful for subsequent processing, while the Squash module enables the input features to be squeezed to a certain range without changing the direction in the vector space, which realizes a nonlinear feature transformation and reduces the numerical instability due to the change of vector length. With ResSimAM, we can indeed better capture features with higher energy for feature enhancement, but it is possible that features with higher energy are not the most appropriate choice in the subsequent processing, so this approach brings some enhancement to the model's classification ability but with some limitations. For the Squash operation, although squeezing the features facilitates the network to capture the relationship between the features and better perform the naive similarity matching, squeezing the features also brings some loss of feature information. These aspects still need to be improved.

5. Conclusions

This study introduces an innovative approach aimed at improving the efficiency of existing point cloud classification methods. The methods for deep learning-based point cloud processing have become increasingly intricate and often requiring long training times and high costs. We propose a new network model: a nonparametic point cloud classification network. We utilized trigonometric functions for embedding and apply Squash to smooth the features for subsequent processing. Then, we enhanced the features using the nonparametic attention mechanism ResSimAM, thereby leading to significant improvements in the purely nonparametric network for 3D point cloud analysis. Based on this, we also extended a lightweight parametric network, which allows for efficient inference with a small number of parameters. For the nonparametric model, our model achieved 83.9% accuracy on the ModelNet40 dataset without any training, which greatly saves time in training the model for the point cloud classification task. For the lightweight parametric model, we achieved 93.3% accuracy using only 0.8 M parameters, the training time was only 3.1 h, and the inference speed reached 962 samples per second, which will greatly reduce the pressure on hardware devices and keep the inference speed relatively high. Various tasks like autonomous vehicles, virtual reality, and aerospace fields demand real-time data handling, and our lightweight models could work efficiently in these tasks.

Although our method has achieved promising results, there is still room for improvement. For nonparametric network models, the feature extraction ability of our network on diverse datasets still needs to be tested and improved. For the lightweight parametric model, although the Squash operation was used to accelerate the convergence of the network, it brings some impact on the feature extraction ability of the network. In future research, we will focus on enhancing the generality and robustness of the proposed network. Future work needs to consider the computational efficiency of the network and the feature extraction capability of the model, as well as propose more effective and concise lightweight methods. This can be achieved by designing new nonparametric modules and combining them with a small number of neural networks, as well as adopting more efficient computational methods. In future work, we will explore nonparametric models with a wider range of application scenarios.

Author Contributions: Conceptualization, J.G. and W.L.; methodology, J.G. and W.L.; validation, W.L.; investigation, J.G.; writing—original draft preparation, J.G.; writing—review and editing, W.L.; visualization J.G. All authors have read and agreed to the published version of the manuscript.

Funding: This research was funded by the Natural Science Foundation of Hebei Province (F2019201451).

Data Availability Statement: The data presented in this study are available in this article.

Conflicts of Interest: The authors declare no conflicts of interest.

Symbol

The list of abbreviations and symbols is shown below.

Symbols	Definition
$FPS()$	farthest point sampling
$k-NN()$	k-nearest neighbor
$Concat()$	concatnate the feature
$MaxP()$	max pooling
$MeanP()$	mean pooling
$sigmoid()$	sigmoid activation
F_{mem}	feature memory
T_{mem}	label memory
Acronyms	**Full Form**
FPS	farthest point sampling
k-NN	k-nearest neighbor
MLP	multilayer perceptron
CNN	convolutional neural networks
OA	overall accuracy

References

1. Guo, Y.; Wang, H.; Hu, Q.; Liu, H.; Liu, L.; Bennamoun, M. Deep Learning for 3D Point Clouds: A Survey. *IEEE Trans. Pattern Anal. Mach. Intell.* **2021**, *43*, 4338–4364. [CrossRef] [PubMed]
2. Liang, Z.; Guo, Y.; Feng, Y.; Chen, W.; Qiao, L.; Zhou, L.; Zhang, J.; Liu, H. Stereo Matching Using Multi-Level Cost Volume and Multi-Scale Feature Constancy. *IEEE Trans. Pattern Anal. Mach. Intell.* **2021**, *43*, 300–315. [CrossRef] [PubMed]
3. Chen, X.; Ma, H.; Wan, J.; Li, B.; Xia, T. Multi-view 3d object detection network for autonomous driving. In Proceedings of the IEEE Conference on Computer Vision and Pattern Recognition, Honolulu, HI, USA, 21–26 July 2017; pp. 1907–1915.
4. Su, H.; Maji, S.; Kalogerakis, E.; Learned-Miller, E. Multi-view Convolutional Neural Networks for 3D Shape Recognition. In Proceedings of the 2015 IEEE International Conference on Computer Vision (ICCV), Santiago, Chile, 7–13 December 2015; pp. 945–953. [CrossRef]
5. Maturana, D.; Scherer, S. VoxNet: A 3D Convolutional Neural Network for real-time object recognition. In Proceedings of the 2015 IEEE/RSJ International Conference on Intelligent Robots and Systems (IROS), Hamburg, Germany, 28 September–3 October 2015; pp. 922–928. [CrossRef]
6. Wu, Z.; Song, S.; Khosla, A.; Yu, F.; Zhang, L.; Tang, X.; Xiao, J. 3d shapenets: A deep representation for volumetric shapes. In Proceedings of the IEEE Conference on Computer Vision and Pattern Recognition, Boston, MA, USA, 7–12 June 2015; pp. 1912–1920.
7. Charles, R.Q.; Su, H.; Kaichun, M.; Guibas, L.J. PointNet: Deep Learning on Point Sets for 3D Classification and Segmentation. In Proceedings of the 2017 IEEE Conference on Computer Vision and Pattern Recognition (CVPR), Honolulu, HI, USA, 21–26 July 2017; pp. 77–85. [CrossRef]
8. Qi, C.R.; Yi, L.; Su, H.; Guibas, L.J. Pointnet++: Deep hierarchical feature learning on point sets in a metric space. *Adv. Neural Inf. Process. Syst.* **2017**, *30*, 5105–5114.
9. Howard, A.G.; Zhu, M.; Chen, B.; Kalenichenko, D.; Wang, W.; Weyand, T.; Andreetto, M.; Adam, H. Mobilenets: Efficient convolutional neural networks for mobile vision applications. *arXiv* **2017**, arXiv:1704.04861.
10. Qureshi, S.A.; Raza, S.E.A.; Hussain, L.; Malibari, A.A.; Nour, M.K.; Rehman, A.U.; Al-Wesabi, F.N.; Hilal, A.M. Intelligent Ultra-Light Deep Learning Model for Multi-Class Brain Tumor Detection. *Appl. Sci.* **2022**, *12*, 3715. [CrossRef]
11. Zhang, R.; Wang, L.; Wang, Y.; Gao, P.; Li, H.; Shi, J. Parameter is not all you need: Starting from non-parametric networks for 3d point cloud analysis. *arXiv* **2023**, arXiv:2303.08134.
12. Zhou, W.; Jiang, X.; Liu, Y.H. MVPointNet: Multi-view network for 3D object based on point cloud. *IEEE Sens. J.* **2019**, *19*, 12145–12152. [CrossRef]
13. Le, T.; Duan, Y. Pointgrid: A deep network for 3d shape understanding. In Proceedings of the IEEE Conference on Computer Vision and Pattern Recognition, Salt Lake City, UT, USA, 18–23 June 2018; pp. 9204–9214.
14. Wang, Y.; Tan, D.J.; Navab, N.; Tombari, F. Softpoolnet: Shape descriptor for point cloud completion and classification. In Proceedings of the Computer Vision–ECCV 2020: 16th European Conference, Glasgow, UK, 23–28 August 2020; Proceedings, Part III 16; Springer: Berlin/Heidelberg, Germany; pp. 70–85.
15. Zhao, Y.; Birdal, T.; Deng, H.; Tombari, F. 3D point capsule networks. In Proceedings of the IEEE/CVF Conference on Computer Vision and Pattern Recognition, Long Beach, CA, USA, 15–20 June 2019; pp. 1009–1018.
16. Wang, Y.; Sun, Y.; Liu, Z.; Sarma, S.E.; Bronstein, M.M.; Solomon, J.M. Dynamic graph cnn for learning on point clouds. *ACM Trans. Graph. (Tog)* **2019**, *38*, 1–12. [CrossRef]
17. Li, G.; Muller, M.; Thabet, A.; Ghanem, B. Deepgcns: Can gcns go as deep as cnns? In Proceedings of the IEEE/CVF International Conference on Computer Vision, Seoul, Republic of Korea, 27 October–2 November 2019; pp. 9267–9276.

18. Wang, L.; Huang, Y.; Hou, Y.; Zhang, S.; Shan, J. Graph attention convolution for point cloud semantic segmentation. In Proceedings of the IEEE/CVF Conference on Computer Vision and Pattern Recognition, Long Beach, CA, USA, 15–20 June 2019; pp. 10296–10305.
19. Wang, Y.; Tan, D.J.; Navab, N.; Tombari, F. Learning local displacements for point cloud completion. In Proceedings of the IEEE/CVF Conference on Computer Vision and Pattern Recognition, New Orleans, LA, USA, 18–24 June 2022; pp. 1568–1577.
20. Hu, J.; Shen, L.; Sun, G. Squeeze-and-excitation networks. In Proceedings of the IEEE Conference on Computer Vision and Pattern Recognition, Salt Lake City, UT, USA, 18–23 June 2018; pp. 7132–7141.
21. Woo, S.; Park, J.; Lee, J.Y.; Kweon, I.S. Cbam: Convolutional block attention module. In Proceedings of the European Conference on Computer Cision (ECCV), Munich, Germany, 8–14 September 2018; pp. 3–19.
22. Vaswani, A.; Shazeer, N.; Parmar, N.; Uszkoreit, J.; Jones, L.; Gomez, A.N.; Kaiser, L.; Polosukhin, I. Attention is all you need. *Adv. Neural Inf. Process. Syst.* **2017**, *30*, 6000–6010.
23. Yang, J.; Zhang, Q.; Ni, B.; Li, L.; Liu, J.; Zhou, M.; Tian, Q. Modeling point clouds with self-attention and gumbel subset sampling. In Proceedings of the IEEE/CVF Conference on Computer Vision and Pattern Recognition, Long Beach, CA, USA, 15–20 June 2019; pp. 3323–3332.
24. Zhao, H.; Jiang, L.; Jia, J.; Torr, P.H.; Koltun, V. Point transformer. In Proceedings of the IEEE/CVF International Conference on Computer Vision, Montreal, BC, Canada, 11–17 October 2021; pp. 16259–16268.
25. Guo, M.H.; Cai, J.X.; Liu, Z.N.; Mu, T.J.; Martin, R.R.; Hu, S.M. Pct: Point cloud transformer. *Comput. Vis. Media* **2021**, *7*, 187–199. [CrossRef]
26. Yu, X.; Rao, Y.; Wang, Z.; Liu, Z.; Lu, J.; Zhou, J. Pointr: Diverse point cloud completion with geometry-aware transformers. In Proceedings of the IEEE/CVF International Conference on Computer Vision, Montreal, BC, Canada, 11–17 October 2021; pp. 12498–12507.
27. Yang, L.; Zhang, R.; Li, L.; Xie, X. Simam: A simple, parameter-free attention module for convolutional neural networks. In Proceedings of the International Conference on Machine Learning, Virtual Event, 18–24 July 2021; pp. 11863–11874.
28. Webb, B.S.; Dhruv, N.T.; Solomon, S.G.; Tailby, C.; Lennie, P. Early and late mechanisms of surround suppression in striate cortex of macaque. *J. Neurosci.* **2005**, *25*, 11666–11675. [CrossRef] [PubMed]
29. Zhang, R.; Fang, R.; Zhang, W.; Gao, P.; Li, K.; Dai, J.; Qiao, Y.; Li, H. Tip-adapter: Training-free clip-adapter for better vision-language modeling. *arXiv* **2021**, arXiv:2111.03930.
30. Uy, M.A.; Pham, Q.H.; Hua, B.S.; Nguyen, T.; Yeung, S.K. Revisiting point cloud classification: A new benchmark dataset and classification model on real-world data. In Proceedings of the IEEE/CVF International Conference on Computer Vision, Seoul, Republic of Korea, 27 October–2 November 2019; pp. 1588–1597.

Disclaimer/Publisher's Note: The statements, opinions and data contained in all publications are solely those of the individual author(s) and contributor(s) and not of MDPI and/or the editor(s). MDPI and/or the editor(s) disclaim responsibility for any injury to people or property resulting from any ideas, methods, instructions or products referred to in the content.

MDPI
Grosspeteranlage 5
4052 Basel
Switzerland
Tel.: +41 61 683 77 34
www.mdpi.com

Algorithms Editorial Office
E-mail: algorithms@mdpi.com
www.mdpi.com/journal/algorithms

Disclaimer/Publisher's Note: The statements, opinions and data contained in all publications are solely those of the individual author(s) and contributor(s) and not of MDPI and/or the editor(s). MDPI and/or the editor(s) disclaim responsibility for any injury to people or property resulting from any ideas, methods, instructions or products referred to in the content.